PROGRESS IN COLLOID & POLYMER SCIENCE

Editors: H.-G. Kilian (Ulm) and G. Lagaly (Kiel)

Volume 90 (1992)

Physics of Polymer Networks

Guest Editors:
S. Wartewig and G. Helmis (Merseburg)

ISBN 3-7985-0914-X (FRG)
ISBN 0-387-91411-0 (USA)
ISSN 0340-255 X

This work is subject to copyright. All rights are reserved, whether the whole or part of the material is concerned, specifically these rights of translation, reprinting, reuse of illustrations, recitation, broadcasting, reproduction on microfilms or in other ways, and storage in data banks. Duplication of this publication or parts thereof is only permitted under the provisions of the German Copyright Law of September 9, 1965, in its version of June 24, 1985, and a copyright fee must always be paid. Violations fall under the prosecution act of the German Copyright Law.

© 1992 by Dr. Dietrich Steinkopff Verlag GmbH & Co. KG, Darmstadt.
Chemistry editor: Dr. Maria Magdalene Nabbe; English editor: James Willis; Production: Holger Frey.

Printed in Germany.

The use of registered names, trademarks, etc. in this publication does not imply, even in the absence of specific statement, that such names are exempt from the relevant protective laws and regulations and therefore free for general use.

Type-Setting: Graphische Texterfassung, Hans Vilhard, D-6126 Brombachtal
Printing: betz-druck gmbh, D-6100 Darmstadt 12

Preface

The 29th Europhysics Conference on Macromolecular Physics "Physics of Polymer Networks" was organized by the Department of Physics of the Technical University Merseburg, Germany under the auspices of the Macromolecular Division of the European Physicsl Society and took place at Alexisbad/Harz (Saxony-Anhalt) 9—14 September 1991. The scientific programme which was devoted to the topics: theory of networks (equilibrium and dynamical properties), formation of networks, sol-gel transition, experimental investigations on structure and properties of polymer networks attracted more then 100 scientifists from 14 European countries and the USA. Fortunately, one fifth of all participants came from the East European countries.

This progress volume presents the major part of the lectures and posters from the conference. We hope that the objective of the conference, which was to reflect the present level of the research in the field of polymer networks and to provoke discussions, is being expressed. Furthermore, we expect that physicists, chemists and materials engineers employed in the network research can obtain valuable stimuli for their future work.

S. Wartewig
G. Helmis

Contents

Preface V

Vilgis TA: Theory of phantom networks — topology, structure, elasticity: New and open problems 1
Duering ER, Kremer K, Grest GS: Structural properties of randomly crosslinked polymer networks 13
Heinrich G: The dynamics of tire tread compounds and their relationship to wet skid behavior 16
Kraus V, Kilian H-G, v. Soden W: Relaxation in permanent networks 27
Lairez D, Adam M, Raspaud E, Emery JR, Durand D: Do local motions influence rheological properties near the gelation threshold? 37
Sommer J-U: On the dynamics of moderately and lightly crosslinked polymer networks 43
Heinrich G, Beckert W: A new approach to polymer networks including finite chain extensibility, topological constraints, and constraints of overall orientation 47
Schulz M: Formation of networks — a lattice model for kinetic growth processes 52
Babayevsky PG: Gelation and 1,1-transition in three-dimensional condensation and chain polymerization ... 57
Doublier JL, Coté I, Llamas G, Charlet G: Effect of thermal history on amylose gelation.................. 61
Garnier C, Axelos MAV, Thibault JF: Rheological, potentiometric and ^{23}Na NMR studies on pectin-calcium systems.. 66
Strehmel B, Anwand D, Timpe H-J: The formation of semiinterpenetrating polymer networks by photoinduced polymerization 70
Wetzel H, Häusler K-G, Fedtke M: Observation of the curing process of epoxy resins by inverse gas chromatography... 78
Strehmel B, Younes M, Strehmel V, Wartewig S: Fluorescence probe studies during the curing of epoxy systems 83
Mel'nichenko Yu, Klepko V: Conditions of formation and equilibrium swelling of polymer networks formed by protein macromolecules 88
Brereton MG: Cross-link fluctuations: NMR properties and rubber elasticity 90
Baumann K, Gronski W: Segmental orientation in filled networks 97
Weber H-W, Kimmich R, Köpf M, Ramik T, Oeser R: Field-cycling NMR relaxation spectroscopy of molten linear and cross-linked polymers. Observation of a $T_1 \propto \nu^{0,25}$ law for semi-global chain fluctuations 104
Chapellier B, Deloche B, Oeser R: Segmental orientation of "long" and "short" chains in strained bimodal PDMS networks: A ^{2}H-NMR study 111
Zielinski F, Buzier M, Lartigue C, Bastide J, Boué F: Small chains in a deformed network. A probe of heterogeneous deformation? 115
Oeser R: Aggregation of free chains within a deformed network: A SANS study........................ 131
Klüppel M: Trapped entanglements in polymer networks and their influence on the stress-strain behavior up to large extensions 137
Apekis L, Pissis P, Christodoulides C, Spathis G, Niaounakis M, Kontou E, Schlosser E, Schönhals A, Goering H: Physical and chemical network effects in polyurethane elastomers 144
Rogovina L, Vasiliev V, Slonimsky G: Influence of the thermodynamical quality of the solvent on the properties of polydimethylsiloxane networks in swollen and dry states 151
Halperin A, Zhulina EB: Triblock copolymers, mesogels and deformation behavior in poor solvents 156
Eicke H-F, Hofmeier U, Quellet C, Zölzer U: Microemulsion mediated polymer networks 165
Solovjev ME, Raukhvarger AB, Ivashkovskaya TK, Irzhak VI: Theory of the mechanical and swelling properties of elastomers with chemical and physical networks 174
Borisov OV, Birshtein TM, Zhulina EB: The effect of free branches on the collapse of polyelectrolyte networks... 177
Walasek J, Grela S: Local order and statistics of a polymer chain in an external field 182
Tamulis A, Bazhan L: Charge photogeneration in carbazole-containing compounds and valency bands of oligomers 186
Rozenberg BA, Irzhak VI: The peculiarities and nature of large-scale motion of highly crosslinked polymers. 194
Scherzer T, Strehmel V, Tänzer W, Wartewig S: FTIR spectroscopy studies on epoxy networks 202

Strehmel V, Zimmermann E, Häusler K-G, Fedtke M: Influence of imidazole on the structure of epoxy amine networks 206

Kulik SG, Babayevsky PG, Borovko VV: Epoxy polymer networks: Relaxation processes and crack resistance 209

Pekcan Ö: Fluorescence study of interpenetrating network morphology of polymer films 214

Smirnov LP, Volkova NN: Kinetic regularities of polymer network thermal degradation 222

Schulze U, Janke A, Pompe G, Meyer E, Rätzsch M: Interpenetrating polymer networks based on EVA copolymer and PMMA 227

Rizos AK, Fytas G, Wang CH, Meyer GC: Fast segmental dynamics in poly(methyl methacrylate)-polyurethane interpenetrating networks 232

Sandakov GI, Smirnov LP, Sosikov AI, Summanen KT, Volkova NN: Thermally and mechanically activated degradation of polyesterurethane networks. Analysis of molecular weight distribution functions 235

Janik H, Foks J: The solidification of bulk and solution cast segmented polyurethanes 241

Author Index 247

Subject Index 248

Theory of phantom networks — topology, structure, elasticity: New and open problems

T. A. Vilgis

Max-Planck-Institut für Polymerforschung, Mainz, FRG

Abstract: The effects of structural elements on the elastic behavior of networks is discussed. Heterogeneities of fractal nature embedded in networks will effect the elastic behavior and the neutron scattering as long as the fractal regions are not saturated. This saturation depends on the topology (connectivity) via the spectral dimension of the fractal. This has the effect that, if fractals are embedded in the network, their internal crosslinks do not count to the modulus if the fractals are saturated. The phase behavior of crosslinked blends and semi-interpenetrating networks is also discussed. For example, a critical crosslink density of thenetwork is found for which free chains solved in the network (semi IPN) are expelled from the network.

Key words: Networks; fractals; heterogeneities; elasticity; IPN; semi IPN

Introduction

It was recently realised that the theory of rubber elasticity and gels has to go in new directions, where novel ingredients play an important role. After the great success of phantom-type theories [1—3], i.e., non-interacting chains, the discussion in the past decade has been dominated by the role of entanglements and their effects of elasticity from different points of view [2, 4—6]. The long dispute in the literature can now probably be solved by computer simulation where networks with and without entanglements can be generated very simply [7].

We do not go into these details here, but we pick up possible new directions in the theory of networks. The basic effects of such new directions can be studied first along the lines of phantom-type models. Phantom-type models consider networks composed of phantom chains, i.e., chains which do not have excluded volume interactions. Therefore, the problem of entanglements is irrelevant. The only fact which matters is the topology of crosslinking.

It has been reported recently that the topology of crosslinkage deeply influences the behaviour of networks. Model systems have been considered, such as fractal networks, e.g., a network of the Sierpinski connectivity [8]. It has been suggested that such fractal "left overs" from a vulcanization or gelation process creates heterogeneities within the rubber sample which might affect the macroscopic behavior of rubbers. Clearly, this will only be the case if no entanglements are present between two crosslinks. Trapped entanglements will introduce a new relevant length scale, i.e., the distance between them, and the topological distance between crosslinks will no longer matter.

A simple mean field theory of heterogeneous rubbers has been considered by the present author [9] and independently by Heinrich and Schimmel [10]. It was found that such inhomogeneities do not affect the deformation behaviour, but strongly influence the neutron scattering [9]. Moreover, inhomogeneities reduce the effective modulus [10]. We will not discuss this in detail here, but leave it for a separate paper [11].

So far, we have spoken about the real topological structure. Now, we want to introduce a "thermodynamic topology". By this term we mean that thermodynamic interactions, as present in blends, strongly influence the phase behavior and elasticity of crosslinked blends. It has been suggested [12] that crosslinkage of partially miscible polymers prevent complete phase separation. Crosslinked blends cannot phase separate macroscopically, but form a microdomain structure since they undergo a microphase separation. The statistical mechanics of such materials is completely unknown and only first attempts have been made to solve this question [13].

The paper is organised as follows. First, the classical theories will be reviewed on their structure and topology relationship. We then turn to fractal networks and fractal containing networks. Finally, we discuss the case of crosslinked blends and semi IPNs.

Classical models of regularly crosslinked polymers

The classical models can be divided into two limiting cases. The first is the Kuhn model which is essentially a single-chain approximation where it turns out that the free energy is a sum over all contributions from each individual chain. Without going into details, we quote the result for the free energy of deformation

$$\beta F = \frac{1}{2} N \sum_{i=1}^{3} \lambda_i^2 , \qquad (2.1)$$

where N is the number of subchains between two crosslinks and λ_i is the deformation ratio in the i-th cartesian direction. β is the inverse temperature. The above model is quoted as an affine model since the crosslink distance deforms in the same way as the macroscopic dimensions. It is important to realise that Eq. (2.1) does not depend on any structural parameter such as functionality of the crosslinks, etc.

To find a dependence on structural parameters a more sophisticated theory has to be employed. The first step in this direction has been made by James and Guth [2] for four functional crosslinks. Thier basic assumption was that several crosslinks of the rubber are fixed on the walls of the sample, forming the surface, whereas the inner crosslinks are free to move.

The result of their calculation is given by

$$\beta F = \frac{1}{4} N \sum_{i=1}^{3} \lambda_i^2 , \qquad (2.2)$$

which is one-half of the free energy of the Kuhn model given by Eq. (2.1). This reduction of the free energy is due to additional motion of the crosslinks.

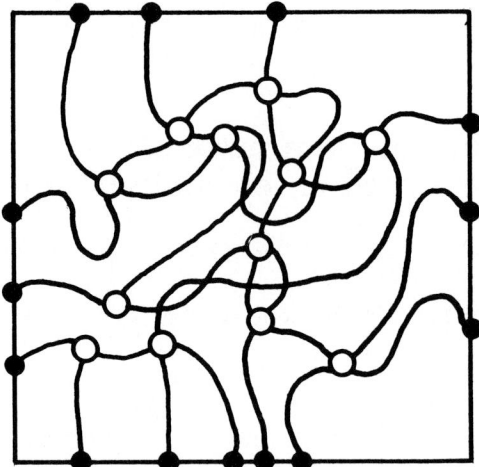

Fig. 1. The basic assumptions in the James and Guth model. The surface crosslinks are fixed, whereas internal crosslinks are free to move

For arbitrary functionalities f this result has been generalized by Graessley [14] for an exact calculation for networks of tree-like structure.

The macroscopic free energy is given by

$$\beta F = \frac{1}{2} N (1 - 2/f) \sum_{i=1}^{3} \lambda_i^2 , \qquad (2.3)$$

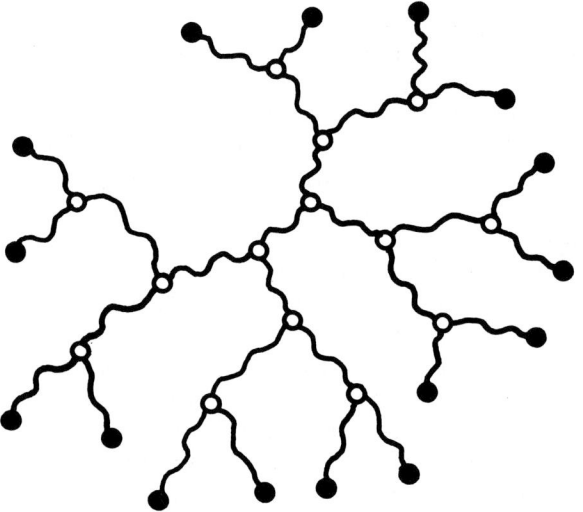

Fig. 2. Tree-like structure of a network. The internal crosslinks are free to move. the surface crosslinks are kept fixed. Here, $f = 3$

which has become the basic formula for phantom networks. A nice independent proof of Eq. (2.3) has been given recently by Higgs and Ball [15], who have used an electrical analogy. It has to be noted that Eq. (2.3) is only exact for tree-like structures, corresponding to mean field solutions where no fluctuations of the structure, i.e., loops are present. This can be seen as an elementary example (see [15]). On a cubic lattice ($f = 6$), one finds $\beta F = 1/3N \sum \lambda_i^2$ ($\neq 1 - 2/f$), whereas on a tetrahedral lattice one finds $\beta F = 1/2N \sum \lambda_i^2$ ($= 1 - 2/f$). Equation (2.3) has to be considered as a mean field solution which is exact for tree-like structures. We will come back to this point later in the case of rod networks.

Nevertheless, in an exact solution on a Bethe lattice a dependence on a structural and topological parameter (functionality f) has been found. Note that a general answer to a structure-modulus relation is unknown.

Flory [16] has generalized Eq. (2.3) to

$$\beta F = \frac{1}{2} \xi \sum_{i=1}^{3} \lambda_i^2 , \qquad (2.4)$$

where ξ is the cycle rank, i.e., the number of independent loops in the network which determines the elastically active network strands and the elastically active crosslinks in the following graph theoretical manner:

$$\xi = N + 1 - M \simeq N - M , \qquad (2.5)$$

where M is the number of crosslinks. For perfect networks, one has

$$fM = 2N , \qquad (2.6)$$

and Eq. (2.4) reduces to Eq. (2.3) immediately. Hence, Eq. (2.3) is exact for perfect phantom networks. The reduction of the modulus by $(2/f)k_B TN$ is due to fluctuations of the crosslink points. This can be made explicit in several ways [15—17].

The Kuhn model and the James and Guth model are considered as two limiting cases, since in the first model one has purely affine deformation, whereas in the latter the deformatin is now affine since crosslinks are fluctuating in a range given by the distance between two crosslinks. Attempts have been made to interpolate between the two cases (see [2] for a review), but such concepts become unreliable if entangled networks are considered [4, 5].

Networks of non classical connectivity

The physics of fractals has been explored deeply during the last decase. Fractals are objects with non-classical connectivity, i.e., they represent spaces with a non-integer dimension (fractal dimension). Typical examples are the infinite percolation cluster [18], an example for a random fractal or the Sierpinski gasket as an example for a regular fractal. The fractals are normally treated as lattice fractals, i.e., that the bonds between the junction points are rigid. It is now tempting to replace these rigid bonds by flexible Gaussian chains. This will lead to polymeric fractals. This idea is due to Cates [19]. The properties of such polymeric fractals are very different from those of rigid lattice fractals, especially regarding the question of elasticity. In lattice fractals each path through the fractal is rigid and the elasticity is enthalpic, whereas in polymeric fractals one has entropic elasticity since each path through the fractal is Brownian. Examples for polymeric networks are shown in Fig. 3.

The Sierpinski network is a typical example of a network with non-classical connectivity. It has functionality 4 but there exist holes on all scales, i.e., larger fluctuations in structure. The other example — the hierarchical fractal network — is based on the idea of modeling the backbone of the percolation cluster [20], and it has been demonstrated that it has the property of multifractality of the voltage distribution if a voltage is applied at both ends of the network.

We hve to first define the size and the topological behavior of such (non-entangled) fractal networks before we discuss typical network properties of them. The basic quantities we need for later discussions are the relations between size and topology, i.e., connectivity of the network and the swelling or collapse behavior of such networks in their own melts or in fractal solvent. We first need a relation between the connectivity (spectral dimension) and the ideal Gaussian fractal dimension of the polymeric fractal. It can be shown that the Gaussian fractal dimension is given by

$$d_f = \frac{2 d_s}{2 - d_s} , \qquad (3.1)$$

where d_s is the spectral dimension of the lattice fractal which is, of course, the same as that of the polymeric fractal since the topology is preserved.

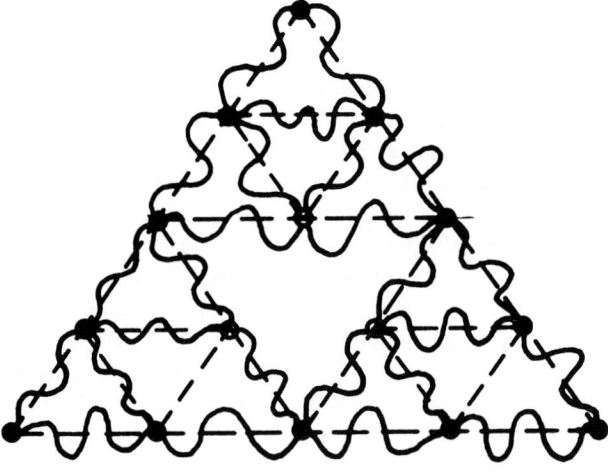

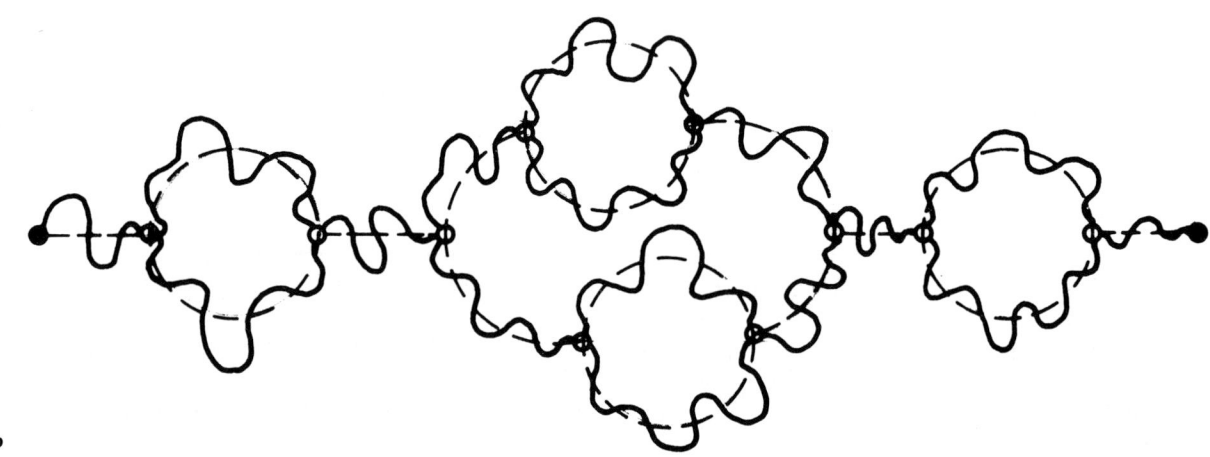

Fig. 3. a) Sierpinski network. The connectivity is given by that of the Sierpinski gasket, but in reality the structure is highly folded. b) A hierachical fractal network which has been used as a model of the 2-d percolation cluster. The next generation can be achieved if each chain becomes replaced by the sequence of a chain, two chains forming a loop and another chain

(3.1) suggests that the Gaussian dimension of the network is given by

$$R_0^{d_f} \sim m, \tag{3.2}$$

where m is the total mass in the fractal.

If this fractal is put in solvent, we expect it to swell and a new fractal dimension is obtained. This phenomenon is well known in linear chains, which are a special case of a polymeric fractal with $d_s = 1$ (see Eq. (3.1)). Here, we are interested in more general situations, i.e., we can put the fractal into ordinary solvent or in more complex solvent such as linear chains or another fractal with Gaussian dimension δ_f [21]. Excluded volume becomes screened on the scale of the size of the different fractals and we obtain a generalized Flory free energy [21]

$$F = k_B T \frac{R^2}{R_0^2} + \frac{v}{N^{\delta_f/d_f}} \frac{N^2}{R^d}. \tag{3.3}$$

The first term is the usual elastic free enrgy ($\sim R^2$ since the fractal is Brownian) and the second term is the screened excluded volume. Note that if $\delta_f = 0$, we obtain the ordinary Flory free energy of swelling in point like (low molecular weight solvent) and if $\delta_f = d_f$ we recover the classical expression for a melt of fractals, i.e., for $R_0^2 \sim \sqrt{m}$ the Daoud-Family expression for swelling of lattice animals ($d_f = 4$) is recovered [22].

Minimization of the free energy predicts a fractal dimension

$$D_f = \frac{d+2}{2 - \frac{1}{d_f}(\delta_f - 2)}, \tag{3.4}$$

where d is Euclidian space dimension. The reader may be convinced that Eq. (3.4) contains all well known cases [21]. Most remarkable is the result that if any fractal with d_f is mixed in linear chains $\delta_f = 2$ a universal size exponent $D_f = (d + 2)/2$ is predicted. Another case which is important to notice is the case if $D_f > d$ which is unphysical. For such cases the fractals become saturated. For melts, this condition is given by $d_f > d$, i.e., if that Gaussian fractal dimension is larger than the space dimension. In terms of the topological connectivity this means

$$d_s > \frac{2d}{2 + d} \:. \qquad (3.5)$$

Therefore, in $d = 3$ melts with $d_s > 6/5$ are always saturated and the fractal nature does not matter since $D_f = 3$, i.e., $R \sim m^{1/3}$.

In conclusion, one can say that a melt of percolation clusters consists always of saturated clusters, as well as a melt of Sierpinski networks where $d_s = 4/3$ and $d_s = 2 \dfrac{\log(d + 1)}{\log(d + 3)}$, respectively. We will see that this point becomes very important when we discuss elasticity in the following sections.

The results of the scaling theory can be summarized in Fig. 4, where D_f (Eq. (3.4)) and the upper critical dimensions $d_{uc} = 2d_f - \delta_f$ are shwon. The axis of $\delta_f = 0$ corresponds to the swelling of a frctal d_f in low molecular weight solvent, i.e., $d_f = 2$ are linear chains $d_f = 4$ are animals. Since $D_f = 3$ is the maximum value the curves are cut. If the upper critical dimension d_{uc} is less than 3, the fractal takes its unperturbed Gaussian dimension.

We want to mention an experiment which can be discussed in the framework above. Antonietti et al. [23] synthesized microgels which are probably of fractal nature, and their connectivities are close to percolation cluster. In good solvent they obey $R^2 \sim m$ as isolated branched molecules (or percolation networks) with $d_f = 4/3$ or $d_s = 4$.

Deuterated microgels in protonated microgels behave as $R^3 \sim m$, i.e., they are saturated, whereas microgels solved in linear chains behave as $R^{2.6} \sim m$, although Eq. (3.4) suggests $R^{2.5} \sim m$ which is in very good agreement (see Fig. 5) with neutron scattering data by Antonietti et al.

This fact now has implications on networks containing fractal heterogeneities. Such fractal heterogeneities can be "left overs" from the vulcanization or gelation process. Let us first look at a model example proposed by Boué and Bastide [8, 24, 25] which we reproduce in Fig. 6.

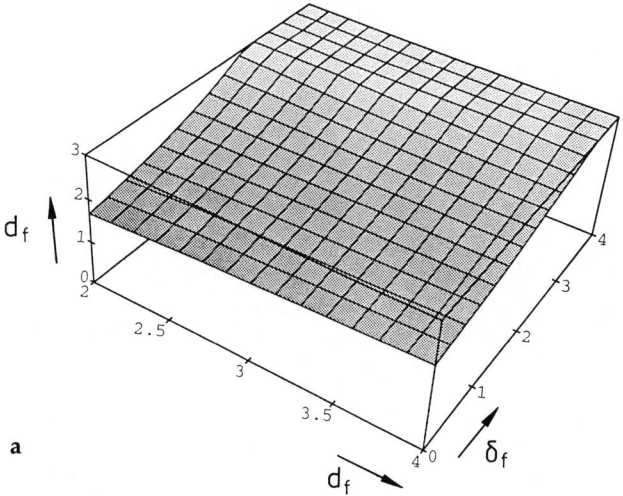

a

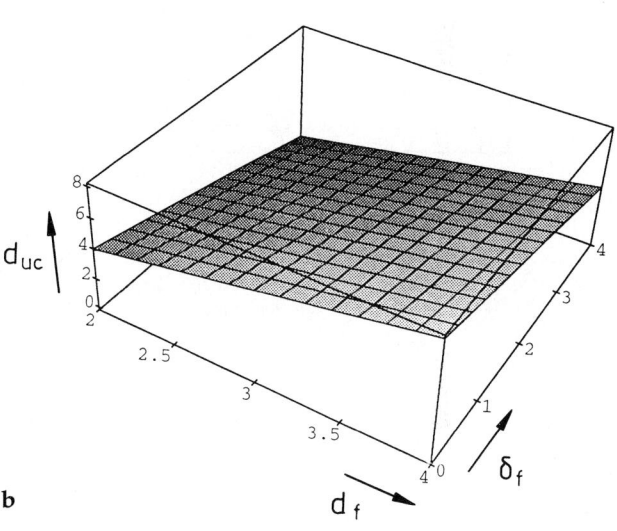

b

Fig. 4. $D_f(d_f, \delta_f)$ [4a] and $d_{uc}(d_f, \delta_f)$ [4b] are shown in three-dimensional plots

The structure is drawn open, but it really represents a melt of Sierpinski networks which are crosslinked at the edges. Since the spectral dimension of the d-dimensional Sierpinski gasket is always larger than $2d/(2 + d)$, i.e.,

$$d_s = 2 \frac{\log(d + 1)}{\log(d + 3)} \:, \qquad (3.6)$$

the polymeric gaskets are always saturated, i.e., their fractal nature does not matter and they form

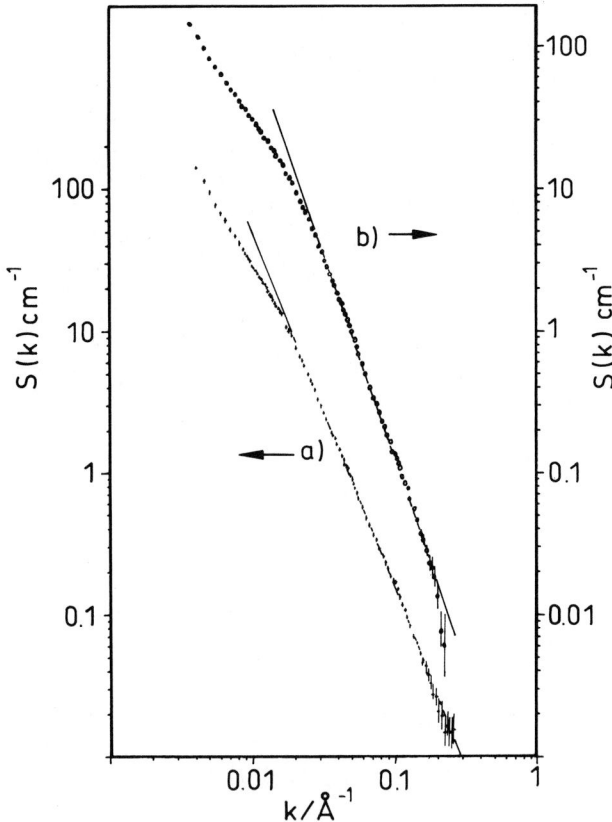

Fig. 5. Experimental results of [23], structure factor vs. wave vector k. Curve a: Microgels in linear chains, slope = 2.6 (D_f = 2.6). Curve b: Melt of microgels, slope = 3.0 (D_f = 3.0)

Fig. 6. Heterogeneous network formed by joined Sierpinski networks (reproduced from [8]). This structure is drawn open, but it would be a melt of Sierpinski networks

balls with $R^d \sim m_g$, where m_g is the mass of monomers in a single gasket. Therefore, one may conclude that the elastic properties of a network given in Fig. 5 are determined by the number of crosslinks which join the gasket rather than by the total number of crosslinks, including those within the individual gaskets. We expect this to be valid at least at small deformations.

More generally, this applies for more complicated situations. Imagine again a network with some fractal "left overs", i.e., parts of percolation clusters or vulcanization clusters at some scale d_ξ which are crosslinked to other chains and other clusters. Therefore, the size and physical behavior of such systems is given by Eq. (3.4) where the clusters with $d_f (d_s \simeq 4/3)$ are surrounded by linear chains with δ_f = 2. In d = 3, we have D_f = 5/2, i.e., the clusters are swollen and their internal crosslinks would contribute to elasticity.

Such considerations are only simple models and may not be applicable for realistic systems, but such models give some insight into the behavior of phantom networks containing fractal heterogeneities. Note further that such conclusions become invalid if entanglements are present. Entanglements introduce a new length scale, the entanglement distance which determines the physical behavior of such networks [5]. The effect of heterogeneities becomes blurred, due to entanglement sliding.

Let us now discuss the elastic properties of pure fractal polymer networks. We will see that there are many open problems as far as this point is concerned. Consider first a Sierpinski network as shown in Fig. 3a. The figure represents a gasket in its third generation. The first generation is just the upper triangle. By doubling itself, we arrive at the second generation. Doubling this gasket again, we find the third generation, and so on. The toplogy of the Sierpinski network is given by the number of chains and the number of crosslinks as a function of its generation p. The number of chains is given by

$$N_p = 3^p \ . \tag{3.7}$$

The number of crosslinks is more complicated. In the first generation these are three crosslinks, i.e., the edges. In p = 2 there are $3 \cdot 3 - 3$ crosslinks, in p = 3 there are $3(3 \cdot 3 - 3) - 3$, etc. For arbitrary p, we have

$$M_p = 3^p - \sum_{n=1}^{p-1} 3^n = \frac{3^p + 3}{2} \tag{3.8}$$

crosslinks. Therefore, the cycle rank of the structure in its p-th generation is given by (see Eq. (2.5))

$$\xi_p = N_p + 1 - M_p = 1 + \sum_{n=1}^{p-1} 3^n = \sum_{n=0}^{p-1} 3^n . \quad (3.9)$$

The sum is a simple geometric series and is given by

$$\xi_p = \frac{1}{2}(3^p - 1) . \quad (3.10)$$

The structure has ξ_p independent loops and follows a recursion law

$$\xi_{p+1} = 3\xi_p + 1 , \quad (3.11)$$

which can be deduced from simple geometry of the gasket. Note that (3.10) is a solution of (3.11). According to Flory, the free energy is then

$$F_p \sim \frac{1}{2} \xi_p k_B T \sum_{i=1}^{3} \lambda_i^2 , \quad (3.12)$$

and the modulus is given by

$$G_p = \frac{\xi_p}{\Omega_p} k_B T , \quad (3.13)$$

where Ω_p is the volume in the p-th generation. If we assume a closed packed structure, we assume $\Omega_p \sim N_p v_0$, where v_0 is the volume of the chain we would predict for the modulus

$$G_p = \left(\frac{1}{2} - \frac{1}{2} 3^{-p}\right) k_B T \frac{1}{v_0} \quad (3.14)$$

and in the limit of $p \to \infty$

$$G_\infty = \frac{1}{2} k_B T \frac{1}{v_0} . \quad (3.15)$$

This is just the modulus of the James and Guth-network for functionality $f = 4$, where we have one chain in the volume $1/v_0$. The Kuhn modulus would be $k_B T/v_0$. Note that this predicts the classical modulus of the James and Guth model. Note further that if we calculate the ratio of the number of chains and the number of crosslinks, we find

$$\frac{N_p}{M_p} = 2 \cdot \frac{1}{1 + \frac{3}{3^p}} \xrightarrow{p \to \infty} 2 \equiv \frac{f}{2} , \quad (3.16)$$

since the functionality $f = 4$ for the Sierpinski network. Therefore, this method predicts a modulus which is in accordance with the James and Guth model $G \sim \frac{kT}{v_0}\left(1 - \frac{2}{f}\right)$.

This observation is in contradiction with calculations by Bastide and Boué [24] who suggest an ultraweak modulus for the Sierpinski micro network. Their argument can be summarized as follows. The starting point is a Sierpinski network, where the three summits are fixed. Thus, the Kuhn modulus is given by $G_p = N_K k_B T/\Omega_p$, where N_K is the number of chains attached to the fixed crosslinks, i.e., the edges. This is just $N_K = 3$, i.e., only the outer chains count. (This has been concluded from application of the star-triangle equivalence [24]). Theredore, the volume Ω_p is increasing as $3^p v_0$ and the modulus is given by

$$G_p = 3(k_B T/v_0)3^{-p} \xrightarrow{p \to \infty} 0 . \quad (3.17)$$

Therefore, the gasket network would represent an "ultraweak solid".

This is because in Bastide's and Boué's calculation only three points are fixed and they are the edges. This number remains constant for each generation, whereas the number of network elements (chains) is increasing exponentially. Bastide's and Boué's calculation is equivalent to Graessley's theory [14] where the free energy of a tree micro network in p-th generation is given by

$$F_p = \frac{1}{2} \nu k_B T \frac{N_{kp}}{N_p} \sum_{i=1}^{3} \lambda_i^2 , \quad (3.18)$$

where N_{kp} is the number of fixed crosslinks and N_p the number of chains in total. For tree-like structures this term goes as $(f-2)/(f-1)$, which is different from the macroscopic network. ν is the number of strands per unit volume, $\nu \simeq N_p/V$. The calculation of the macroscopic modulus now requires a similar procedure as given in [14] and leads directly to the cycle rank. A detailed statistical mechanical theory on the modulus of a fractal micro network of the Sierpinski type is also an open problem and is left to a separate publication.

Remarks on rigid rod networks

Another example of the effect of the influence of structure is the rigid rod network. Here, the topology of crosslinking can be the same as in flexible chain networks, but the flexible chains have been replaced by rigid rods. There are basically two different possibilities. The first one is if the rods are connected rigidly or flexibility as shown in Fig. 7.

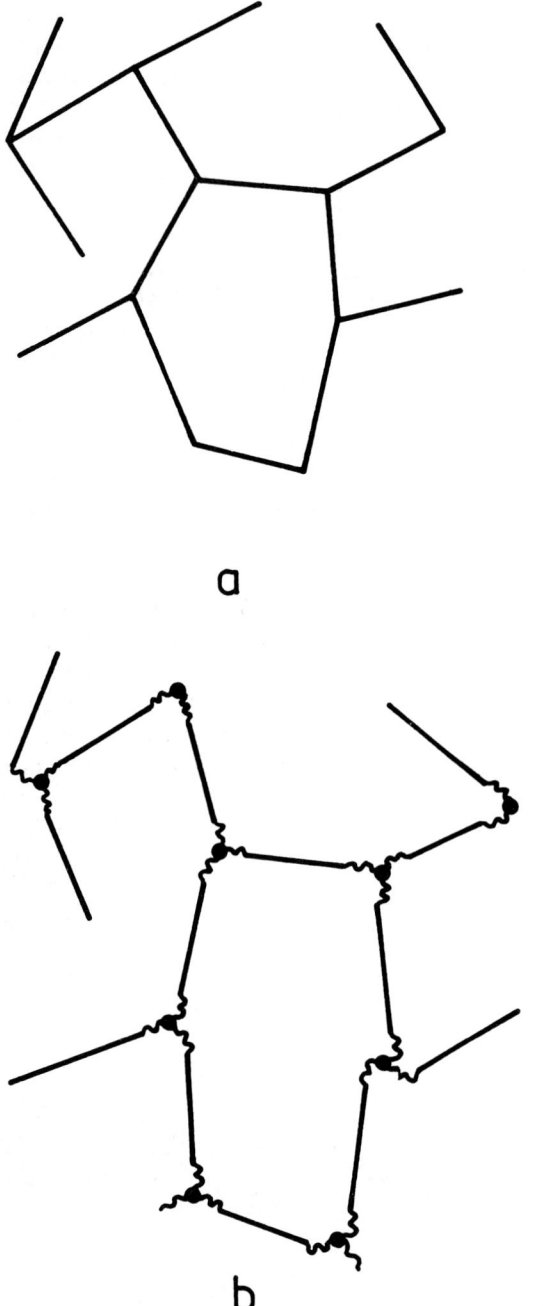

Fig. 7. Two rigid rod networks: a) enthalpic; b) entropic

In Fig. 7a the rods are rigidly hinged at the junction points, whereas in Fig. 7b the rods are freely joined. The first one is an example of an enthalpic network [26] and the latter one is purely entropic [27]. The rigid enthalpic network has been discussed in [26] on the basis of scaling arguments. One essential difference of both types of networks is that the scaling of the modulus with the volume fraction of the rods is significantly different [26], i.e.,

$$G \sim \phi^2 \quad \text{frozen crosslinks}$$
$$G \sim \phi^{3/2} \quad \text{flexible crosslinks} \quad (4.1)$$

The flexibly hinged network gave rise to a field theory formulation for networks [27] which enables a general description for networks with arbitrary elements. We do not go into mathematical details and refer the reader to [27].

The first question for the flexible rod network is: Does it deform anyhow? The answer is given by counting the number of the degrees of freedom [28]. If the degrees of freedom are greater than zero the network can be deformed. To count the degrees of freedom consider N rods in d dimensions. These free rods have $N(2d - 1)$ degrees of freedom. Now, there are M crosslinks which remove Mf degrees of freedom, where f is the functionality, but give Md translational degrees of freedom. Thus, one has the condition

$$N(2d - 1) - Mfd + Md \geqslant 0 \ . \quad (4.2)$$

For a perfect network Eq. (2.6) can be employed and we find a limiting functionality

$$f_{\max} = 2d \ . \quad (4.3)$$

Thus, if $f < f_{\max}$ the networks can be deformed entropically, whereas for $f > f_{\max}$ the network becomes enthalpic, i.e., the crosslinks localize. Another argument has been put forward in [26], i.e., M free crosslinks have Md degrees of freedom, but there are $Mf/2$ rigid rod constraints in the sample (for a perfectly linked network), so that $f_{\max} = 2d$ as before.

The field theory can be set up for flexible rigid rod networks and its Hamiltonian is given by

$$H = \mu \int d^3r \int d^3r' \phi(r) g(r - r') \phi(r')$$
$$+ v \int d^3r (\phi^*(r))^f - \int d^3r \phi(r) \phi^*(r) \ , \quad (4.4)$$

and the partition function is given by

$$Z = \oint \oint \frac{d\mu N!}{\mu^{N+1}} \frac{dvM!}{v^{M+1}} \int \delta\phi \int \delta\phi^* e^{-H} \ . \quad (4.5)$$

μ and v are fugacities which encounter for the addition of rods and crosslinks. The propagator $g(r)$ has, for rods, the form

$$g(r) = \frac{1}{4\pi} \delta(|r| - L), \quad (4.6)$$

where L is the rod length. Equations (4.5) and (4.4) describe a non Hermitian field theory for *endlinking* of rods, and all correlation functions are non-zero if Eq. (2.6) is satisfied.

The partition function can be evaluated in tree approximation (saddle point) and we obtain for small deformation the free energy

$$F = \frac{1}{2} NkT \left(1 - \frac{2}{f}\right) \sum \lambda_i^2. \quad (4.7)$$

That is the James and Guth result. This is not surprising since there is enough entropy in the crosslinks to account for Eq. (4.7). The typical form of the modulus is an artefact of the tree approximation. Note that Eq. (4.7) is exact for tree-like networks. If loops are present the degrees of freedom become reduced and Eq. (4.7) becomes invalid. A tree can always be deformed even at $f \to \infty$ (see 4.7), whereas a loopy structure cannot. Thus, fluctuations in the structure (loops) play an important role. The modulus for $f < f_{max}$ beyond the tree approximation is an open problem.

Thermodynamic topology — crosslinked blends and semi IPNs

Crosslinked blends are technically very important and have a wide range of applications. The physics of such materials, i.e., their statistical mechanics is still unknown and is an open problem. In this paper, we want to introduce a new concept to this problem. Let us first describe the system. Imagine a partially miscible blend of two polymers, say A and B. These polymers interact with an excluded volume interaction

$$\int_0^{L_\sigma} ds \int_0^{L_\tau} ds' V_{\sigma\tau}(R_\sigma(s) - R_\tau(s')), \quad (5.1)$$

where $\sigma, \tau = A, B$ and L_σ is the length of the polymer in species σ. Usually, $V_{\sigma\tau}(r)$ is taken to be short ranged, i.e., $V_{\sigma\tau}(r) = V_{\sigma\tau}\delta(r)$. The phase behavior is ruled by de Gennes famous RPA equation:

$$\frac{1}{S_A^0(k)} + \frac{1}{S_B^0(k)} - 2\chi_F > 0, \quad (5.2)$$

where the Flory χ-parameter $2\chi_F = 2V_{AB} - (V_{AA} + V_{BB})$ and $S_\sigma^0(k)$ are the bare structure factors of the species σ. A similar equation can be derived for block polymer melts [30, 31]

$$\frac{S_A^0(k) + S_B^0(k) + S_{AB}^0(k)}{S_A^0(k)S_B^0(k) - (S_{AB}^0(k))^2} - 2\chi_F > 0 \quad (5.3)$$

for stable melts. S_A^0, S_B^0 are the structure factors of the A and B block, whereas S_{AB}^0 is the structure factor of the AB correlations. If χ_F becomes larger a phase separation takes place. For an ordinary blend this is a macroscopic phase separation, whereas for block copolymer melts this happens at finite k-values which indicates a micro-phase separation.

Under the term crosslinked blend we understand a partially miscible blend, which is homogeneous at large temperatures and is then crosslinked in the one-phase region. Therefore, we get different types of crosslinks $M_{\sigma\tau}$ which will behave differently. If now the χ_F-parameter is increased the uncrosslinked system phase separates, but the crosslinked system cannot, since $M_{AB} \neq 0$. Therefore, we expect a microphase separation where the result is A-rich regions and B-rich regions, separated by surfaces formed by the AB crosslinks. Note that AA and BB crosslinks can move cooperatively into the A-rich or B-rich phases. But, we expect a renormalization of the χ_F parameter due to the presence of crosslinks. The question is, how can such a problem be quantified with the simplest version? Here, the theory is only outlined and we have the technical details for a separate discussion [32].

What we first need is an appropriate Hamiltonian of the one-component network. To derive it one can follow the method of Goldbart and Goldenfeld [33]. The partition function for N chains with M crosslinks can be written as

$$Z = \int \prod_{a=1}^N \mathscr{D}R_a(s) e^{-\beta H} \prod_{e=1}^M \delta(R_{a_e}(s_e) - R_{\beta_e}(s'_e)), \quad (5.4)$$

where the Hamiltonian of the uncrosslinked chains is given by

$$\beta H = \sum_{a=1}^N \int_0^L \left(\frac{\partial R_a}{\partial s}\right)^2 + \sum_{\alpha\beta} \int_0^L ds \int_0^L ds' V(R_\alpha(s) - R_\beta(s')), \quad (5.5)$$

where $V(r - r')$ is the excluded volume interaction. The δ-function in Eq. (5.4) represents the crosslink constraint. Now, one has to perform an average over all crosslink constraints and since the crosslinks are quenched degrees of freedom, the replica method has to be employed [5]. For a randomly crosslinked system (radiation crosslinked) a uniform distribution can be assumed, and one has

$$\langle Z^n \rangle = \int \prod_{a=1}^{n} \prod_{a=1}^{N} \mathscr{D}R_a^a(s) e^{-\beta H_{\text{eff}}} , \quad (5.6)$$

with

$$\beta H_{\text{eff}} = \sum_{a=1}^{n} \sum_{a=1}^{N} \int_0^L ds \left(\frac{\partial R_a^a}{\partial s} \right)$$
$$+ \sum_{a=1}^{n} \sum_{a,\beta}^{N} \int_0^L ds \int_0^L ds' V(R_a^a(s) - R_\beta^a(s'))$$
$$- \mu \sum_{a\beta} \int_0^L ds \int_0^L ds' \prod_{a=1}^{n} \delta(R_a^a(s) - R_\beta^a(s')) , \quad (5.7)$$

where n is the number of replicas and μ is proportional to the mean number of crosslinks. The replicas are highly coupled due to the crosslink term. Equation (5.7) can now be readily generalized to the effective replicated Hamiltonian of a crosslinked blend by addition of the other interactions. The crosslink term gives rise to a definition of new collective variables $\Omega_{\hat{k}}$ which depends on a super wave vector $\hat{k} = (k^1, k^2, ..., k^n)$

$$\Omega_{\hat{k}} = \sum_{a=1}^{N} \int_0^L ds \, e^{i\hat{k} \cdot \hat{R}_a(s)} , \quad (5.8)$$

where $\hat{R}(s)$ is a super vector $\hat{R}_a(s) = (R_a^1(s), R_a^2(s), ..., R_a^n(s))$ and the scalar product $\hat{k} \cdot \hat{R}$ is defined by

$$\hat{k} \cdot \hat{R}_a = \sum_{a=1}^{n} k^a \cdot R_a^a . \quad (5.9)$$

Together with the ordinary density variables

$$\rho_k^a = \sum_{a=1}^{N} \int_0^L ds \, e^{ikR_a^a(s)} \equiv \Omega_{(0,0,...k^a,0...0)} , \quad (5.10)$$

the Hamiltonian can be rewritten

$$\beta H = \sum_{a=1}^{n} \sum_{a=1}^{N} \int_0^L \left(\frac{\partial R_a^a}{\partial s} \right)^2$$
$$+ (V - \mu) \sum_{k_a} \Omega_{(0,...k_a...0)} \Omega_{(0,...-k_a...0)}$$
$$- \mu \sum_{\hat{k}}{}' \Omega_{\hat{k}} \Omega_{-\hat{k}} . \quad (5.11)$$

This is an appropriate Hamiltonian which can be generalized to crosslinked blends. Note that in the one replica sector the potential becomes renormalized by the crosslink term λ. The one replica sector, i.e., these replicas which are not coupled now describe the thermodynamics of the system, whereas the coupled replicas describe the elastic properties and the rigidity of the system. The prime on the sum over $\hat{k}$ means that the one replica sector is excluded.

With this information the RPA calculation can be performed [32], which is quite technical, and so only the main results are quoted. The Hamiltonian for the incompressible-crosslinked blend is given by

$$H = \sum_{\hat{k}} \{F(k^a) - (2\chi_F - 2\xi_x) \mid \Omega_{(0,...k_a...0)} \mid^2$$
$$+ 2\zeta_x \mid \Omega_{\hat{k}} \mid^2 + \text{higher orders}\} , \quad (5.12)$$

where $F(k^a)$ is a complicated function of the wavevector k^a which contains information of the bare and the cross form factor of the network (similar to block copolymers, see Eq. (5.3)) and the structure factor of the chain origins. The definition of the second ζ_x parameter is given by

$$2\zeta_x \sim (M_{AB} - (M_{AA} + M_{BB}))\mu_0 . \quad (5.13)$$

There will be a microphase separation at a finite wavevector k^* given by

$$F(k^*) - 2(\chi_F - \zeta_x) = 0 , \quad (5.14)$$

which depends on the number of crosslinks M_{AA}, M_{BB}, and M_{AB}, the chain length L and the number of chains $N = N_A + N_B$. $1/|k^*|$ determines the size of the microdomain structure (see Fig. 8).

A few special cases can be deduced.

a) $M_{AB} = 0$

$$\frac{1}{S(k)} = F_0(k) - 2\chi_F - \mu_0(M_{AA} + M_{BB}) . \quad (5.15)$$

If the last term is large the system will always phase separate into two unentangled networks.

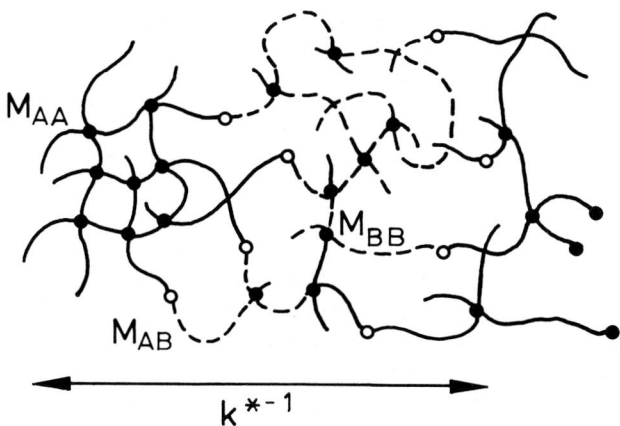

Fig. 8. Microphase separation in crosslinked blends (unentangled). The M_{AB} crosslinks from the interface between two phases A-rich and B-rich. The M_{tt} crosslink can separate cooperatively into their regions

b) $M_{AB} \gg M_{AA} + M_{BB}$

$$\frac{1}{S(k)} = F_1(k) - 2\chi_F + \mu_0 M_{AB} > 0 \quad (5.16)$$

is always positive. The system is always stable at $k = 0$.

c) $M_{AB} = 0$, $M_{AA} > 0$ $M_{BB} = 0 = $ semi IPN

The structure factor is given by

$$\frac{1}{S(k)} = \frac{1}{G_{AA}^0(k) S_{AA}^0(k)} + \frac{1}{S_{BB}^0(k)} - 2\chi_F - \mu_0 M_{AA}, \quad (5.17)$$

where $G_{AA}^0(k)$ is the structure factor of the A-chain ends. The system will phase separate at $k = 0$ at

$$\frac{1}{\phi N_A} - \frac{1}{(1-\phi)N_B} - 2\chi_F - \mu_0 M_{AA} = 0. \quad (5.18)$$

Then, there will be a critical crosslink density of the network where the free chains will phase separate even at $\chi_F = 0$. Since $1/\phi N_A$ is the total amount of network monomers it can be neglected $\phi N_A \gg 1$, and we find

$$\mu_0 M_{AA}^{crit} = \frac{1}{(1-\phi)N_B} - 2\chi_F. \quad (5.19)$$

For a larger number of crosslinks the free chains will separate from the network, whereas for $M_{AA} < M_{AA}^{crit}$ the free chains will be solved in the network.

Conclusion

In this paper we have discussed a few new ideas and relevant questions for phantom networks. The basic problem is to relate the modulus of a given network to its topology which can be self-similar (fractal). Such considerations are relevant for heterogeneous networks, where the inhomogeneities are of fractal topology. We sometimes found that the fractal nature is important, and sometimes not, depending on the spectral dimension. Thermodynamic topology is also relevant in connection with the discussion about IPNs and semi IPNs. Throughout the paper entanglements have been neglected. These are relevant, especially for IPNs and a detailed theory is in progress to include these influence on the phase behavior.

Acknowlegement

The author thanks François Boué for sending him ref. [24] prior to publication and also for numerous discussions and onging collaboration.

References

1. Treloar LRG (1975) "The physics of rubber elasticity", Oxford Clarendon Press
2. Mark JE, Erman B (1988) "Rubberlike Elasticity — a molecular primer", John Wiley, New York
3. Vilgis TA (1989) In: Comprehensive polymer science, vol 6, Eastmond et al. (eds) Pergamon Press, Oxford
4. Heinrich G, Straube E, Helmis G (1988) Adv Pol Sci 85:33
5. Edwards SF, Vilgis TA (1988) Rep Prog Phys 51:243
6. Edwards SF, Vilgis TA (1987) Polymer 27:483
7. Duering E, Kremer K, Grest GS (this volume)
8. Boué F, Bastide J, Buzier M, Collette C, Lapp A, Herz J (1987) Prog Colloid Polym Sci 75:152
9. Vilgis TA (1992) Macromolecules 25:399
10. Schimmel KH, Heinrich G (1991) Coll Pol Sci 269:1003
11. Heinrich G, Vilgis TA, to be published
12. Brereton MG, Vilgis TA (1990) Macromolecules 23:2044 and refs. therein
13. Binder K, Frisch HL (1984) J Chem Phys 81:2126
14. Graessley WW (1975) Macromolecules 8:186
15. Higgs PG, Ball R (1988) J Phys France 49:1785
16. Flory PJ (1976) Proc R Soc Lond A 351:351
17. Deam R, Edwards SF (1976) Phil Trans R Soc A 280:317
18. Stauffer D (this volume)
19. Cates ME (1985) J Phys France 46:1059
20. Coniglio A (1986) Physica A 140:51

21. Vilgis TA (1988) Physica A 153:341
22. Daoud M, Family F (1981) J Phys France 45:175
23. Antonietti M, Ehlich D, Fölsch KJ, Sillescu H, Schmidt M, Lindner P (1989) Macromolecules 2:2802
24. Bastide J, Boué F, preprint
25. Bastide J, Boué F (1986) Physica 140A:251
26. Jones JL, Marquez CM (1990) J Phys Rance 51:1113
27. Boué F, Edwards SF, Vilgis TA (1988) J Phys France 49:1635
28. Vilgis TA, Boué F, Edwards SF (1989) In: Molecular basis of polymer networks, Springer Proc in Physics 42:170
29. de Gennes PG (1979) "Scaling concepts in polymer physics", Cornell University Press, Ithaca
30. Leibler L (1980) Macromolecules 13:1602
31. Vilgis TA, Borsali R (1990) Macromolecules 23:3172
32. Vilgis TA, to be published
33. Goldbart P, Goldenfeld N (1987) Phys Rev Lett 58:2676

Author's address:

T. A. Vilgis
Max-Planck-Institut
für Polymerforschung
Ackermannweg 10
D-W-6500 Mainz, FRG

Structural properties of randomly crosslinked polymer networks

E. R. Duering, K. Kremer[1]), and G. S. Grest[2])

Max-Planck-Institute für Polymerenforschung, Mainz, FRG
[1]) Institut für Festkörperforschung, Forschungszentrum Jülich, FRG
[2]) Corporate Research Science Laboratories, Exxon Research and Engineering Company, Annandale, USA

Abstract: The present work discusses structural properties of randomly crosslinked polymer melts. Using a method based on the burning method for percolation investigations, rubber networks can be characterized by their connectivity. We identify the elastically active crosslinks as well as the active polymer strands. This also reveals the effective length of the elastic strands, which turns out to be larger than the bare average strand length between two consecutive crosslinks within the percolating gel.

Key words: Random networks; polymers; melts; percolation; elasticity

Introduction

The problem of rubber elasticity is one of the outstanding theoretical and experimental problems in macromolecular physics. The elastic properties of these systems are governed by the specific structures of the crosslinked system. As a first basic concept the notion of elastically active and inactive strands were very useful for the description of these materials [1]. It is however important to note that this classification does not take into account entanglements between strands.

The structure of rubber networks obtained by random crosslinking of equilibrated polymer melts with multifunctional junctions has been theoretically studied by many authors; here, we follow Pearson and Graessley [2]; they used mean field like combinatorial arguments. However, by using computer simulations, we are not restricted to mean field assumptions. In particular, a recent extensive study of statistical properties of random crosslinked melts [3] has led to interesting results about the distribution of dangling chains, strand lengths, and links. There, deviations between expected and measured values of the above quantities were qualitatively discussed, based on short intrachain loops [4]. Here, we give a more quantitative analysis based on the actual structure of the network. Computer simulations of end-linked elastomers with trifunctional and tetrafunctional networks [5] confirm the relevance of the ineffective loops for the calculation of the number of fundamental circuits. Still, there is no exact evaluation of the mean active strand and dangling chain lengths, and other important characteristic network quantities.

It is the purpose of this work to present first results of the study of structural properties of polymer networks using a new structure-searching method derived from the burning method for percolation. In the following, we first briefly discuss the simulation and the structure searching method. Then, we present the main results obtained so far and summarize the main conclusions.

Structure searching method

Randomly crosslinked equilibrated polymer melts [6] with M polymers of N monomers and an average of p links per chain, giving $2p$ linked monomers, were studied. All monomers interact through a shifted repulsive Lenard-Jones potential, which represents the excluded volume interaction of the monomers. For nearest neighbor along the chain an additional bond potential was added. For details of the molecular dynamics equilibration method see [3, 6]. The equilibrated melts were then crosslinked. Each introduced crosslink is taken as an additional chemical bond. For the present investigation, however, we do not need any additional relaxation, since the structural properties cannot relax. During the crosslinking process, a point in space is first chosen at random. Then the closest of the MN monomers is taken and crosslinked to a randomly chosen neighboring monomer, provided, this monomer is situated within a distance r_x (of the order of the persistence length) and not a nearest or next nearest neighbor along the chain. This process was repeated pM times, independent of previous selected pairs. As no additional condition was imposed on the link distribution, free chains or even small free clusters of chains may still be present in the system. In general, one melt configuration was used and crosslinked approximately 100 times to improve the statistics. It is also possible that a monomer belongs to more than one crosslink.

The structure searching method may be summarized as follows [7][1]):

a) We search for the largest cluster, which forms the network.

b) We search, by basically applying the burning method [8], for all monomers which would divide the largest cluster into pieces, if taken away, since the common assumption is that the omitted monomers do not affect the elastic part of the network.

c) All these monomers are eliminated and the largest piece is chosen as the remaining studied network. (Near the vulcanization threshold there might be a few large clusters [7]).

d) Only those links which have at least a remaining functionality of three are active links [9]. The curvilinear distance between them is the length of the active strands. By this method, we identify all elastically active strands and elastically active crosslinks.

Results and discussion

According to [2], for the special case of tetrafunctional cross-links, the number of active strands is

$$v = pM(3p_1p_2 + 2p_2^2) , \quad (1)$$

---
[1]) In this paper, we will give a very detailed account on the numerical method, which goes beyond the scope of the present work. There, a variety of different lengths of the primary chains is considered.

and the number of active links is

$$\mu = pM(2p_1p_2 + p_2^2) , \quad (2)$$

where p_1 and p_2 are the gel fraction-dependent [2] probabilities that a randomly chosen non-crosslinked unit is connected to the network by 1 and 2 paths, respectively. They are only functions of the total number of crosslinks and the gel fraction. The values of p_1 and p_2 used in these equations are calculated according to [2], and the only imput from the simulations was the gel fraction.

In Fig. 1, we compare the estimated values of v, μ, and $\xi = v - \mu$ (which is called the number of fundamental circuits) with the simulation results. For the complete equations and their derivations, see [2].

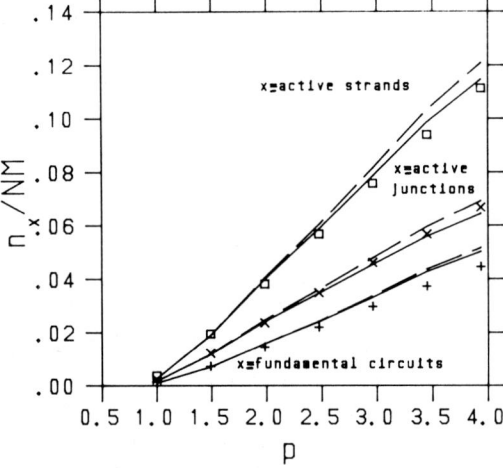

Fig. 1. Number of active strands ($n_x = v$), active cross links ($n_x = \mu$), and fundamental circuits ($n_x = \xi$) versus the mean number of crosslinks per chain for $M = 400$ and $N = 50$. The dashed line shows the estimates under the tetrafunctional crosslink approximation, see Eqs. (1), (2) and the following discussion. The solid line shows the correction introduced by the multifunctional junctions (see [2])

Another important structural feature of a network is the mean length of the active strand

$$L_{as} = p_2NM/v . \quad (3)$$

This parameter should be closely related to the maximum extensibility and tensile strength of the network. In Fig. 2 simulation results are compared with these estimates. The deviation is caused by ineffective links and similar structures [7].

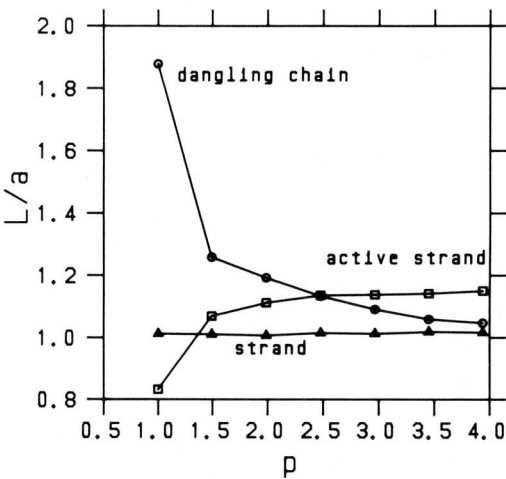

Fig. 2. The ratio of the active strand length, taken from the simulations, and estimated using Eq. (3) (□) for $M = 400$ and $N = 50$. The equivalent ratio for the dangling chains length (○) and strand length (△) are also included

The complementary part of the network, still present in the largest cluster, are the inactive strands and dangling chains.

The average number of monomers in these ends can be estimated as

$$L_{ec} = p_1 NM/\nu_{ec}, \quad (4)$$

where ν_{ec} is the number of dangling chains. In Fig. 2 simulation results are compared with the estimated values. The ratio of the mean strand length (which is the mean distance between closest crosslinks) estimated and obtained from simulations is also shown. For the estimates, we have used the real link distribution under the assumption that the links are randomly distributed along the chain. Good agreement between the mean field estimates and simulation results is observed for this case.

In this investigation, we followed the Scanlan-Case [2, 9] definition to characterize the active crosslinks and strands. This is more directly related to the network structure than the Flory definition [10]. Flory, however, considers only the number of elementary circuits to be of importance. Since it was shown by Pearson and Graessley [2] that $\xi = \nu - \mu$ is independent of the precise definition of ν and μ, respectively, our results can also be used to analyze Flory's theory.

In summary, the approximations of [2] give reasonable estimates of the number of active strands and active links. However, the difference which determines the rubber plateau modulus in the affine approximation clearly requires a better theory. The estimated mean active strand length was within around 20% in agreement with the simulation results. The main differences were observed for the characterization of dangling chains. The deviation increases near the gelation-percolation threshold.

Acknowledgement

This work was supported by the Deutsche Forschungsgemeinschaft, Sonderforschungsbereich 262 and by a CPU time grant of the German Supercomputer Center HLRZ, Jülich.

References

1. Langley NR (1968) Macromolecules 4:348—352
2. Pearson DD, Graessley WW (1978) Macromolecules 11:528—533
3. Grest GS, Kremer K (1990) Macromolecules 23:4994—5000
4. Tonelli AE, Helfand E (1974) Macromolecules 7:59—63; Helfand E, Tonelli AE (1974) Macromolecules 7:832—834
5. Leung YK, Eichinger BE (1984) J Chem Phys 80:3877—3884; Leung YK, Eichinger BE (1984) J Chem Phys 80:3885—3891
6. Kremer K, Grest GS, Carmesin I (1988) Phys Rev Lett 61:566—569; Kremer K, Grest GS (1990) J Chem Phys 92:5057—5086
7. Duering E, Kremer K, Grest GS (in preparation)
8. Herrmann HJ, Hong DC, Stanley HE (1984) J Phys A17:L261—L266
9. Scanlan J (1960) J polym Sci 43:501—508; Case LC (1960) J Polym Sci 45:397—404
10. Flory PJ (1976) Proc Roy Soc London, Ser A351:351

Authors' address:

Kurt Kremer
Institut für Festkörperforschung
Forschungszentrum Jülich
D-5170 Jülich, FRG

The dynamics of tire tread compounds and their relationship to wet skid behavior

G. Heinrich

Continental AG, Materials Research, Hannover, FRG

Abstract: The relationship between glass-transition temperature T_g, plateau modulus G_N^0, microstructure of different polymers (high cis-1,4-BR, NR, butyl rubber and various styrene-butadiene copolymers) and the wet skid resistance (WSR) of the corresponding filled treat compounds were investigated. Additionally, the relation between WSR and rolling resistance of vulcanizates was investigated. Both T_g and G_N^0 give only a poor correlation to the values of WSR. A good correlation is reached relating the WSR of elastomers to their total viscoelastic behavior in the glass-rubber transition zone. Realistic damping curves of the polymers in this region can be obtained within a bead-spring model of a tree-like entanglement network with a certain number of subchains. It is concluded that the high-frequency dependence of the loss curve is related to the local chain dynamics, whereas the cooperative large scale Rouse-like motion of many entangled strands dominates in the low-frequency region of the loss curve. In the surrounding of the loss maximum act both intra- and intermolecular correlations.

Key words: Carbon black field networks; viscoelasticity; network dynamics; glass-rubber transition; wet skid resistance

Introduction

The major application of carbon black-reinforced elastomers is in the manufacture of automotive tires. Associated with each tire component (tread, sidewall, etc.) is a rubber mix formulated specifically to achieve the optimum balance of vulcanizate properties required for that element in a particular type of tire. Relatively, the tread contains the largest amount of crosslinked and filled polymer material within the whole tire. The tread is a wear-resistant component in contact with the road. It has to be designed for treadwear resistance, traction, low noise and heat built-up.

The basic component of all rubber mixes is a single polymer or a blend of polymers. Typical recipes involve natural, butadiene, styrene-butadiene and butyl rubbers.

The use of carbon black fillers, together with accelerated sulfur vulcanization, has remained the fundamental technique for achieving the incredible range of mechanical properties required for the tread compounds in tires. Carbon black leads to the phenomenon of reinforcement of elastomers which has been defined as increased modulus, tear strength, tensile strength, cracking resistance, fatigue resistance and abrasion resistance.

It should be borne in mind that compounding of rubber mixes is still more of an art than a science. One reason is that consideration must be given to the cost of raw materials and the ability to process the resultant compound using current production equipment. But the main reason is that the relationship between the static and dynamic mechanical properties of the tread and the molecular structure of the polymer network are poorly understood at present. Nevertheless, this relationship has a substantial background of polymer physics and — chemistry; some important aspects will be discussed in this paper.

Dynamic mechanical properties and wet skid behavior of polymer networks

The two functional properties of tire tread rubbers (Table 1) of prime importance today are low rolling resistance (RR) and high wet traction during braking. It is known that RR and skid resistance are mutually opposed (Fig. 1).

Generally, the emphasis in tire development is placed dominantly on the improvement of skid properties with the least possible trade-off in abrasion and RR. For future trends, the aim of minimizing RR without sacrificing skid properties is likely to gain importance [1].

Rolling resistance is predominantly related to loss tangent ($\tan\delta$) of the bulk polymer at comparatively low frequencies in the plateau region of the storage shear modulus ($G'(\omega)$) — curve (Fig. 2). This radian low-frequency region is in the the order of the angular velocity of the rolling tire. According to the time-temperature equivalence the measurement of the coefficient of rolling friction corresponds to a measurement of $\tan\delta$ well above the glass-transition temperature [2].

The wet skid resistance (WSR) is linked with dynamic losses in the glass-rubber transition zone of the bulk polymer (typically in the range of 1 kHz — 1 MHz) [3, 4]. WSR is quantified by the laboratory wet friction coefficient using a British Pendulum Skid Tester (BPST). The basic principle of this instrument will be explained in the experimental section.

The transition zone is located between the glassy zone (characterized by T_g) and the plateau region of the storage modulus (characterized by the plateau modulus G_N^0). The T_g influences the location of the skid process within the transition zone. Therefore, this quantity is often used as a physical criterion in materials development: the higher the T_g, the higher the WSR. This reasoning can explain the WSR of many rubbers (Fig. 3). However, the most notable exception is butyl rubber (IIR), which has a low T_g ($\approx -65\,°C$), but is very good in WSR. Both IIR and NR ($T_g \approx -64\,°C$) have comparable glass-transition temperatures. However, IIR has much better WSR. Further, Fig. 3 indicates that solution styrene-butadiene rubbers (S-SBR's) show a broad scattering without any strong correlation between WSR and T_g. Clearly, there are other factors influencing the wet skid behavior of elastomers beside the glass-transition temperature.

Recently, Rahalkar [4] and Gargani et al. [5] proposed the hypothesis that WSR is influenced by the polymer motion in the glass-rubber transition

Table 1. Rubbers used

No.	Rubber	T_g (DSC) [°C]	G_N^0 [MPa]	Wet skid resistance (rating)	Rolling resistance (rating)	$\Delta T^{3)}$ [°C]	$\overline{\tan\delta}$
1	NR (cis-1,4-polyisoprene)	−64	0.71 [7]	88	88	−10 ... −29	0.38
2	Buna CB 10 (cis-1,4-polybutadiene)	−104	0.76 [7]	61	91	−28 ... −56	0.16
3	IIR (polyisobutylene)	−65	0.25 [7]	132	123	−10 ... −29	0.75
4	E-SBR 1500	−49	0.9[1)]	100	100	−6 ... −23	0.45
5	E-SBR 1516	−31	0.88[2)]	119	108	2 ... −10	0.72
6	S-SBR A (experimental SBR)	−62	0.93[2)]	114	104	−10 ... −29	0.54
7	S-SBR B (experimental SBR)	−23	0.74[2)]	139	111	5 ... −7	0.8
8	S-SBR C (experimental SBR)	−34	0.83[2)]	139	111	0 ... −14	0.83
9	S-SBR D (experimental SBR)	−35	0.84[2)]	126	109	0 ... −14	0.85

[1)] Experimental value. — [2)] (Weight average) mixing rule. — [3)] Skid region in the temperature scale.

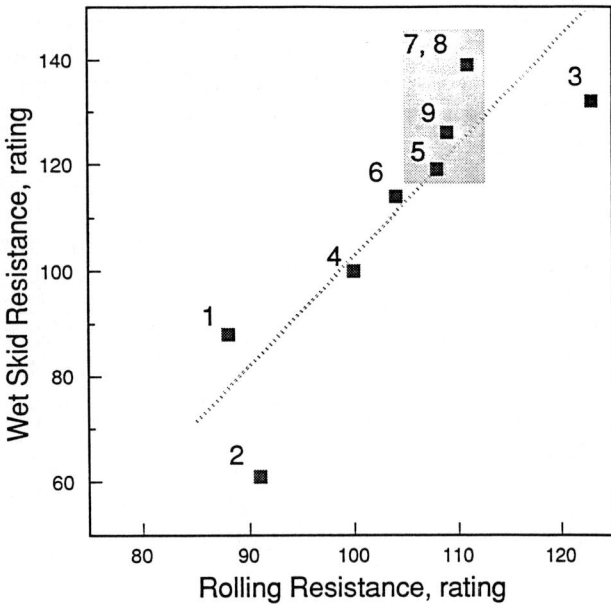

Fig. 1. Wet skid resistance as a function of rolling resistance

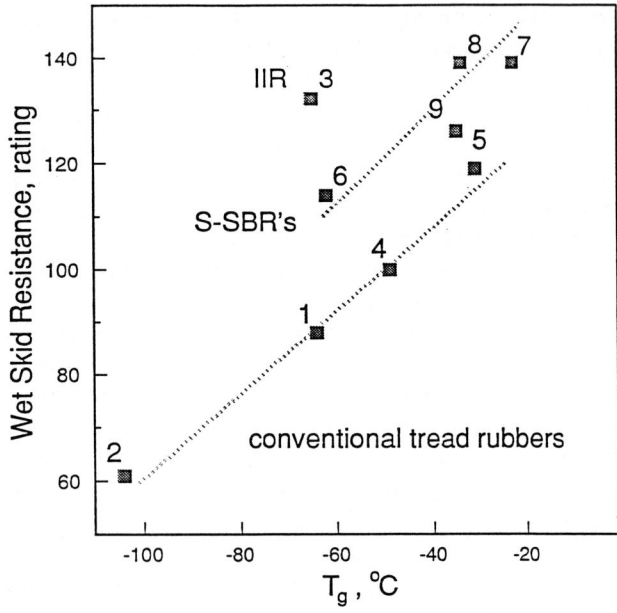

Fig. 3. Wet skid resistance as a function of the glass-transition temperature T_g

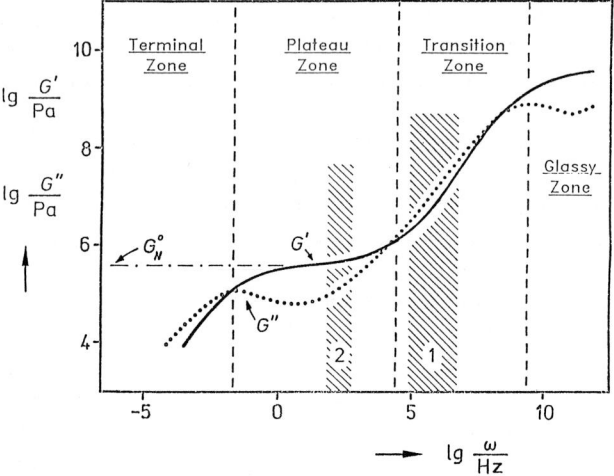

Fig. 2. Frequency dependence of storage (G') and loss (G'') shear moduli of a viscoelastic amorphous polymer. 1: region which correlates with the wet skid behavior of a tread compound; 2: region which correlates with the rolling resistance of a tire

region. Obviously, the chain mobility and not T_g is the key factor in influencing the wet skid behavior. The mobility of the chain in the transition region depends on the entanglement molecular weight M_e and the monomeric friction coefficient ζ_0. The predominant motion in the transition is the motion of the polymer chains between the entanglements. The entanglement density is directly related to the plateau modulus G_N^0 of the bulk polymer. This quantity is mainly controlled by chain tortuosity and is directly related to parameters of the microstructure of the polymer (statistical segment length, cross-sectional area of the polymer chain) [6, 7]. For that reason, we investigated relationships between WSR, G_N^0 and the chain dynamics of the corresponding network polymers.

We note that a number of scaling theories consider the problem of plateau modulus and entanglement density in bulk polymers [16]. A rather general framework was introduced by Graessley and Edwards [10], and a corresponding, more rigorous microscopic basis was discussed in [11, 17]. The models are based on the idea that the plateau modulus should be determined solely from the density of chain contour length and the Kuhn statistical segment length of a chain. Three scaling ideas about entanglement have been shown to be special cases of the GE scaling framework with particular values of the scaling exponents arising from different conjectures about the nature of entanglement [16].

Experimental

Rubbers

A number of different polymers was investigated: different types of emulsion and commercial as well as experimental solution styrene-butadiene copolymers, high cis-1,4-BR, NR and butyl rubber (Table 1). The chemical compositions listed in Table 2 were determined by infrared spectroscopic methods. All polymers were characterized by their glass-transition temperatures T_g, and their values of the plateau moduli G_N^0 of the corresponding uncrosslinked bulk systems. The T_g was measured using a differential scanning calorimeter (DSC) and operating with a heating rate of 10 K/min. The G_N^0 moduli of BR, NR and IIR were taken from literature [6, 7]. The plateau modulus of the bulk copolymer SBR 1500 (23% styrene content) was taken from the viscoelastic plateau of the storage modulus $G'(\omega)$ which has been measured by dynamic oscillations using parallel plates in a rheometrics mechanical spectrometer (Fig. 4). The quantity G_N^0 is equal to the storage modulus G' at the frequency where $\tan\delta$ is at its minimum in the plateau zone, i.e., $G_N^0 = [G']_{\tan\delta \to \text{minimum}}$ [7]. We obtained $G_N^0 \approx 0.90$ MPa. This value matches closely an independent calculation using a weight averaging mixing rule and the corresponding values of the homopolymers: G_N^0 (polystyrene) = 0.20 MPa, G_N^0 (cis, trans-polybutadiene) = 1.2 MPa [6]. The agreement between measured and calculated values of G_N^0 (SBR 1500) justified the application of the mixing rule to the estimation of the plateau moduli of the remaining copolymers. The relative amount of the components is given in Table 2.

Mixing, compounding and vulcanization

The test samples were prepared in the following way: The ingredients of the compounds were mixed on a two-roll mill, molded, and cured at 150°C for a specified time ensuring optimum cure. The cure curves were measured on a Monsanto Rheometer. The cure time was the time required to attain 95% of the maximum torque. Table 3 shows the recipes and cure conditions for the rubbers. All samples contain 50 phr (parts per hundred rubber) carbon black N 339 (ASTM classification) and are crosslinked by sulfur.

We skid and dynamic mechanical analysis of vulcanizates

The frictional forces generated when viscoelastic bodies such as rubber slide on road surfaces are made up of a component arising from adhesion between the rubber and the road surface, and a hysteresis loss component accompanying deformation in the rubber. In the case of wet surfaces, since the water film prevents direct contact between rubber and road surface, adhesion friction is greatly decreased [18, 19].

The wet friction coefficient was determined using a British Pendulum Skid Tester (BPST). The basic principle of this instrument is that the energy loss by the pendulum, as a rubber sample slides over the surface (asphalt in our case) during the test cycle, is equal to the work done in overcoming the friction between the sliding rubber sample and the surface [19]. Clearly, the skid tester yields only relative data which depend on the kind of surface employed. The coefficient of friction calculation from measurements by this instrument is referred to as BPST-number. It provides a close correlation with the wet skid coefficient of friction of actual tire tests [3]. The rating number has been introduced to relate the BPST-number to that of SBR 1500 vulcanizates (BPST (SBR 1500) = 100). This meant, vulcanizates with WSR > 100 were characterized by improved wet skid properties compared to SBR 1500.

Table 2. Chemical composition of the styrene-butadiene rubbers

No.	Type	Rubber	Chemical composition		
			styrene weight %	butadiene	
				1.2 weight %	1.4 weight %
4	E-SBR	SBR 1500	23.5	13.0	63.5
5	E-SBR	SBR 1516	39.5	10.0	50.5
6	S-SBR	S-SBR A	23.0	22.5	54.5
7	S-SBR	S-SBR B	21.5	51.5	27.0
8	S-SBR	S-SBR C	27.5	31.0	41.5
9	S-SBR	S-SBR D	23.5	34.0	42.5

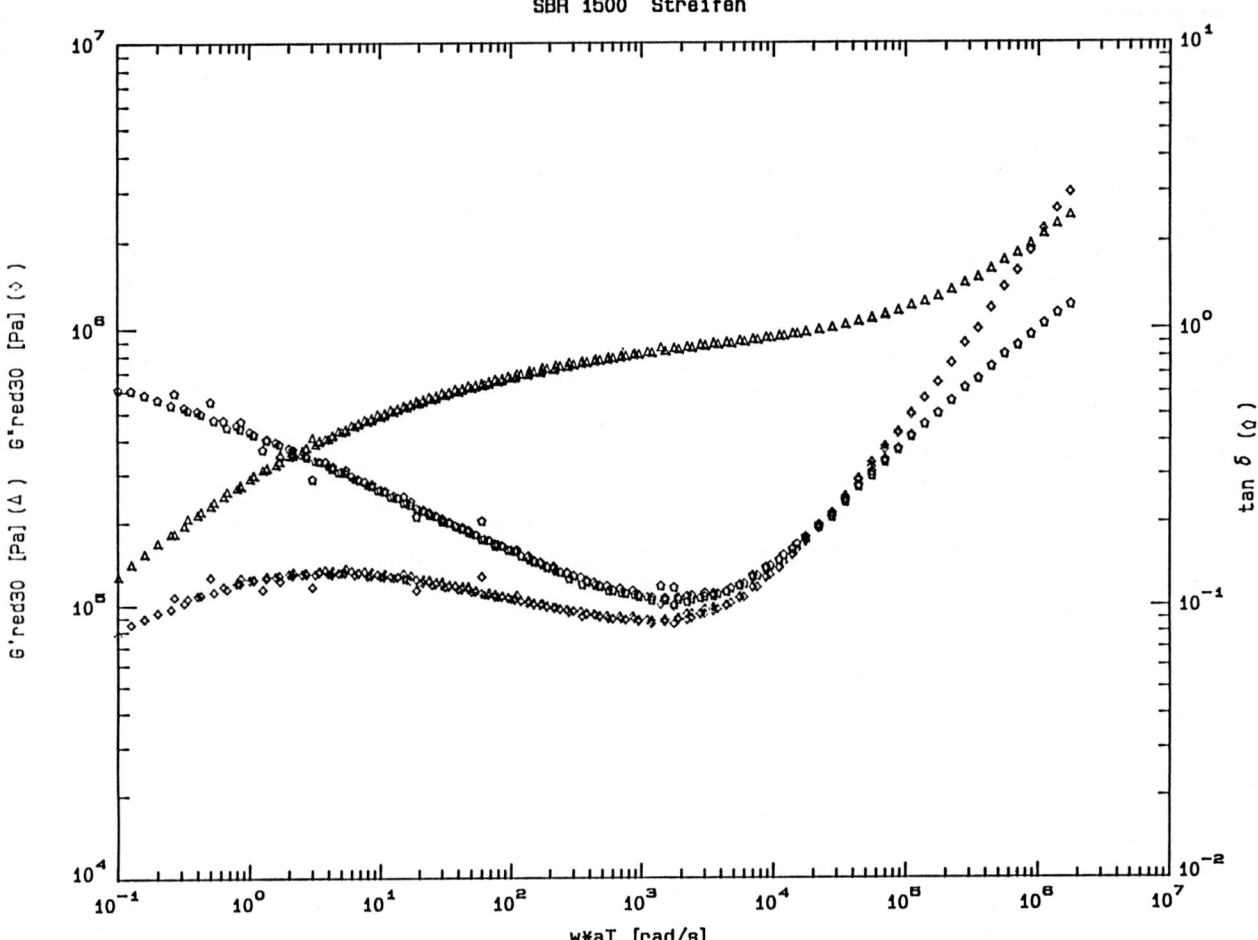

Fig. 4. Master curves of $G'(\omega)$, $G''(\omega)$ and $\tan\delta$ for SBR 1500

Table 3. Compound recipes

Compound No.	1	2	3	4	5	6	7	8	9
Rubber type	NR	Buna CB 10	IIR	SBR 1500	SBR 1516	S-SBR A	S-SBR B	S-SBR C	S-SBR D
Component [phr]									
Rubber	100	100	100	100	100	100	100	100	100
Zinc oxide	3	—	5	—	—	—	—	—	—
Stearic acid	2	—	3	—	—	—	—	—	—
Carbon black N-339	50	50	50	50	50	50	50	50	50
Aromatic oil	—	5.0	—	5.0	5.0	5.0	5.0	5.0	5.0
Sulphur	2.2	1.8	2.0	1.8	1.8	1.8	1.8	1.8	1.8
CBS[1])	0.5	1.4	—	1.4	1.4	1.4	1.4	1.4	1.4
DPG[2])	0.5	0.5	—	0.5	0.5	0.5	0.5	0.5	0.5
TMTD[3])	—	—	3.0	—	—	—	—	—	—
MBTS[4])	—	—	1.0	—	—	—	—	—	—
t_{95} [min] (Rheometer, 150°C)	10	5	60	40	70	20	65	30	55
ML (1 + 3) [units] compound, 100°C	57	100	100	79	74	80	85	67	63

[1]): N-cyclohexylbenzothiazole-2-sulphenamide. — [2]): Diphenylguanidine. — [3]): Tetramethylthiuram disulphide. — [4]): Dibenzothiazole disulphide.

We measured the temperature dependence of the dynamic mechanical properties $E^*(T) = E'(T) + iE''(T)$ of the filled vulcanizates. The temperature profile was determined over a temperature range from $-70°$ to $25°C$. The measurements were carried out in tension using an dynamic mechanical tester under conditions of constant deformation. The testing was performed at 10 Hz on specimens with a cylindrical shape. The amplitude of the sinusoidal deformation was small to ensure linear viscoelastic response.

It is generally accepted that $\tan\delta$ determined at 10 Hz and at a temperature fairly far above the T_g of rubbers of interest prove to be the significant material property that relates the energy dissipated per cycle in a rolling tire [8]. Therefore, we use $\tan\delta$ (25°C) to assess the rolling resistance. The relation between RR and $\tan\delta$ is based on an internally used equation. The corresponding laboratory values of the RR of small specimens show good correlations to tire road tests. Correlations between RR and $\tan\delta$ are generally used in tire industry to assess the tire rolling resistance from laboratory values of the viscoelastic properties. Clearly, these correlations cannot be valid generally. They differ between different companies with different equipment for measuring the RR of tires during drum or actual road tests.

As in the case of WSR, the rating number was introduced, and an improvement in RR compared to SBR 1500 was indicated by RR < 100.

We end this section with some final remarks. Although molecular considerations in the transition zone are expected to be applicable only to the unfilled bulky polymers we tested cured samples loaded with 50 phr carbon black. This is borne out by the experience that lightly crosslinking and filling does not change the wet skid ranking between the polymers. The short-range nature of transition-region motions are dominant in all cases. Road test results produced a similar order in wet skid rating [5].

Results

1) Figure 3 shows that the glass-transition temperature T_g is not sufficient to allow a complete description of the wet skid behavior. A (linear) correlation between increasing WSR with increasing T_g is only observed for cis-1,4-polybutadiene, natural rubber and emulsion styrene-butadiene copolymers (SBR 1500, SBR 1516). Butyl rubber as well as the solution SBR rubbers, which contain vinyl groups of different amount, show large deviations.

The different WSR properties between NR and IIR can be phenomenologically explained by the different widths, magnitudes and locations of the loss curves. The sharpness in the loss maximum correlates with the slope of the relaxation spectra in the transition zone. Increasing steepness of the spectra compresses the transition zone into a narrower region of the frequency scale, increasing entanglement density compresses it into a narrower range of magnitudes [2]. On a molecular level, the different WSR properties between NR and IIR can be explained by the higher mobility of the rubber backbone. This higher mobility appears as a decrease in the effective local viscosity surrounding a given chain segment rather than an "internal" viscosity of the chain segment itself [2].

Obviously, the vinyl structure causes reinforced skid rating properties in the case of S-SBR's [8, 9]. Vinyl groups more influence the intramolecular chain dynamics, whereas the effects of low frequency intermolecular dynamics remain nearly un-

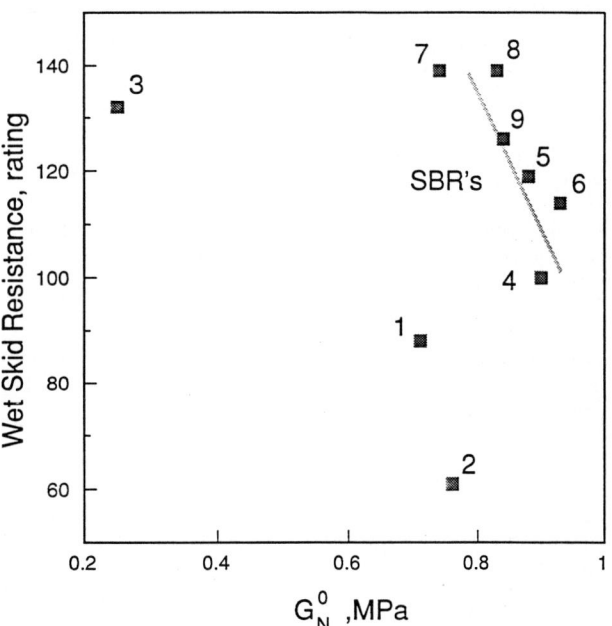

Fig. 5. Wet skid resistance as a function of the plateau modulus G_N^0

changed. Therefore, the RRs of S-SBR's do not change remarkably within a range where skid resistances widely differ (Fig. 1).

Again, the WSR properties of S-SBR's can be explained by the mobility of the polymer backbones. This mobility depends on the weight fraction of butadiene and styrene and the proportions of cis, trans, and vinyl microstructure in the butadiene moiety [2].

2) Figure 5 shows that the plateau modulus G_N^0 alone — as well as the glass temperature T_g — is not sufficient to allow a complete description of the wet skid behavior. A somewhat better correlation is observed for the styrene-butadiene polymers alone. The modulus G_N^0 is related to the local segment number density of the polymer [10, 11]. This density can be varied by the vinyl and/or styrene content. Decreasing values of the moduli lead to lowering of the onset frequency ω_{tr} which locates the boundary on the frequency scale between the plateau and transition zone [2]. At frequencies above ω_{tr}, viscoeleastic losses will be prominent. A lowering of ω_{tr} leads to a broadening of the frequency scale which determines the skid behavior. This leads to an improved WSR.

Discussion and conclusions

1) Both T_g and G_N^0 of the polymers give only poor correlation to the values of the WSR. A similar conclusion is valid for the RR (Figs. 6, 7). Obviously, the viscoelastic properties of the whole transition zone seem to be a necessary condition influencing WSR (i.e., the mobility of chain contours between entanglements is the key factor). A further indication of this statement is seen in Fig. 8 where WSR shows a good correlation to the average loss tangent $\overline{\tan\delta(T)}$ of the vulcanizate. The averaging is performed within that region which correlates with the wet skid behavior (Fig. 2). The corresponding frequency region depends on the length scales of the asperities of testing surface (asphalt in our case) and the tester sliding speed. This region can be considered to range from approximately $10^3 \ldots 10^5$ Hz [12] and is located within the glass-transition zone. A temperature-time conversion rule was applied in order to convert, approximately, the frequency region to a temperature region [18]. The shift factor was of the WLF-type and the reference temperature was $T_g + 50\,°C$. The frequency of the dynamical tester was 10 Hz; and the skid measurements were carried out at 20°C. The corresponding approximated temperature regions are given in Table 1. The averages $\overline{\tan\delta}$ were estimated from the cor-

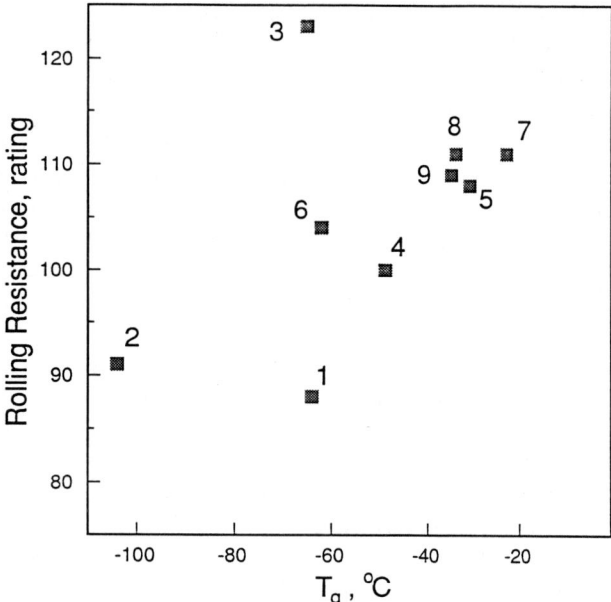

Fig. 6. Rolling resistance as a function of the glass-transition temperture T_g

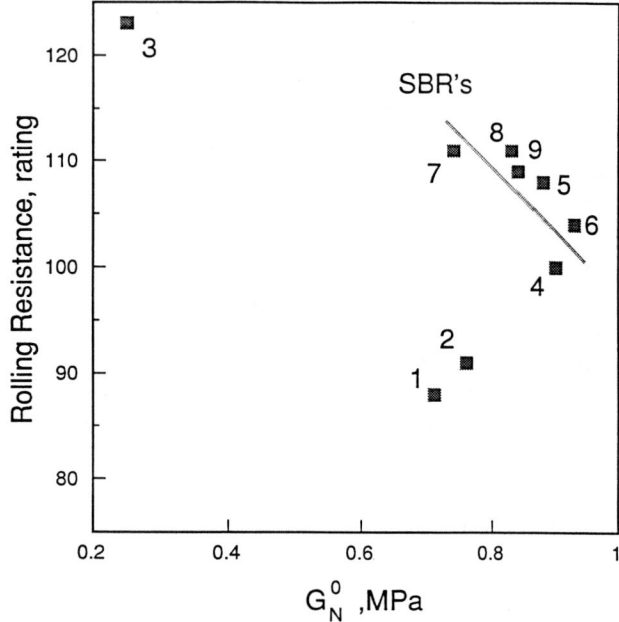

Fig. 7. Rolling resistance as a function of the plateau modulus G_N^0

responding $\tan\delta(T)$-data of the temperature sweeps (dynamical tester) within these regions.

2) Unfortunately, no simple molecular model can predict the real viscoelastic behavior in the glass

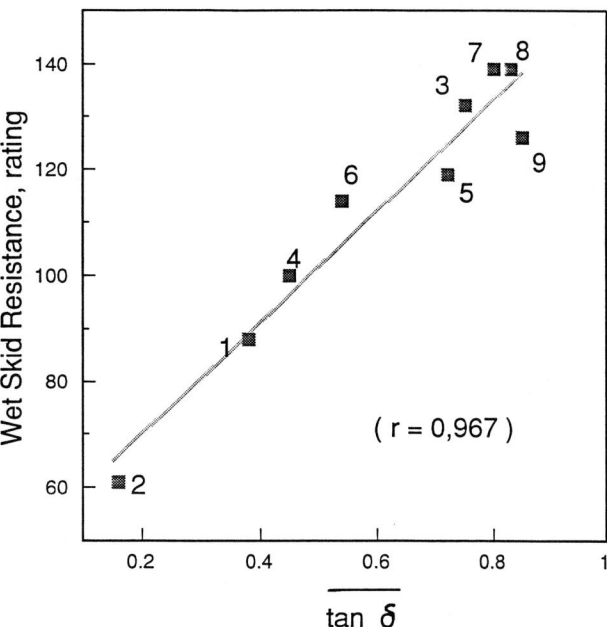

Fig. 8. Wet skid resistance as a function of an averaged loss tangent tan δ

transition zone. It is known that the sharpness in the loss maximum emphasizes the failure of the Rouse-model [13] to provide a detailed description of the properties in the transition zone [2]. This model predicts $G'(\omega) \sim G''(\omega) \sim \omega^{1/2}$, and tan δ = 1 independent of frequency in this region. Otherwise, a drop in tan δ at high frequencies and at low frequencies is observed. The former is associated with the entrance into the glassy zone; whereas the latter is associated with the entanglements. More precisely, the hypothesis can be stated that the high frequency dependence of the loss curve is related to (intramolecular) local chain dynamics and the low frequency dependence to inter-molecular correlations. We use therefore a modified Rouse-model which considers a tree-like network of Gaussian chains (Kloczkowski et al. [14]). This model takes into account the motion of the junctions together with the motion of the sub-chains dividing each chain between the junctions. We apply this model in our short-time region by identifying the junctions with the entanglement points (Fig. 9). The corresponding expression for the frequency dependent modulus is given in the appendix. To test the co-operative bead-spring model, we calculate the damping curves for several polymers which are well characterized by the following parameters (Table 4): monomeric friction coefficient ζ_0; molecular mass of the monomeric unit M_{mon}; characteristic ratio C_∞; statistical segment length l_{stat}; plateau modulus $G_N^0 = \rho RT/M_e$ (M_e — molecular weight between two successive entanglements); number of monomer units within a statistical segment z. The primary relaxation time of a single subchain is given as follows: $\tau_0 = z\zeta_0 l_{stat}^2/3k_BT$.

Figure 10 shows that the model predicts excellent wet skid properties for butyl rubber (IIR) and high-vinyl butadiene rubber (1,2-BR), but poor skid properties for HCBR (cis-1,4-BR). Further, the experimentally observed ranking can be predicted:

IIR > 1,2 BR > SBR 1500 > NR > cis-1,4-BR .

The results led us to the following final conclusions: The high-frequency dependence ($\omega \gg \omega_0$) of the loss curve is related to the local chain dynamics, whereas the large scale Rouse-like motion of many chains units of length M_e dominates in the low frequency region ($\omega \ll \omega_0$). The frequency ω_0 marks the position of the maximum in the frequency scale. In the surrounding of the

Table 4. Molecular parameters which enter the tree-like bead-spring model

	$\log \zeta_0$ [Ns/m^2]	M_{mon} [g/mol]	G_N^0 [MPa]	l_{stat} [Å]	z	M_{stat} [g/mol]	M_e [g/mol]	τ_0/z [s]
IIR	−7.16	56	0.28	10.5	4	224	8.110	5.6 E−06
1,2-BR	−7.11	54	0.62	11.7	5	270	2.055	6.3 E−06
SBR 1500	−9.11	65	0.90	10.6	6	393	2.544	6.3 E−08
NR	−9.41	68	0.44	8.8	4	272	3.825	3.2 E−08
cis-1,4-BR	−9.75	54	0.76	9.6	5	270	2.970	1.4 E−08

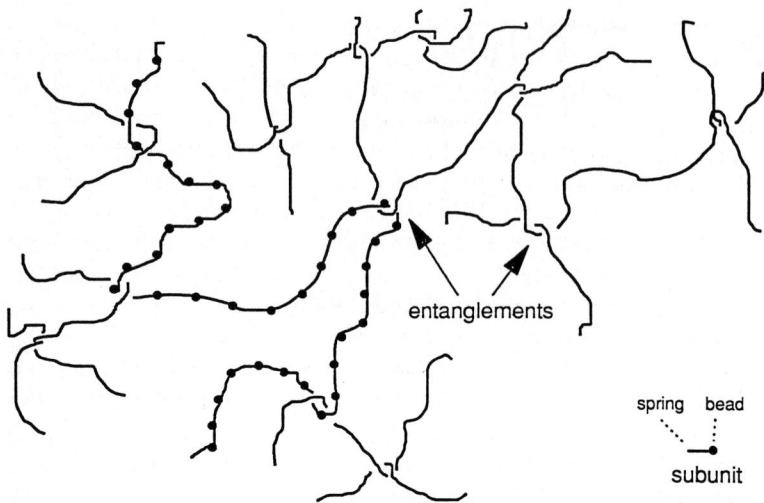

Fig. 9. Tree-like entanglement network model

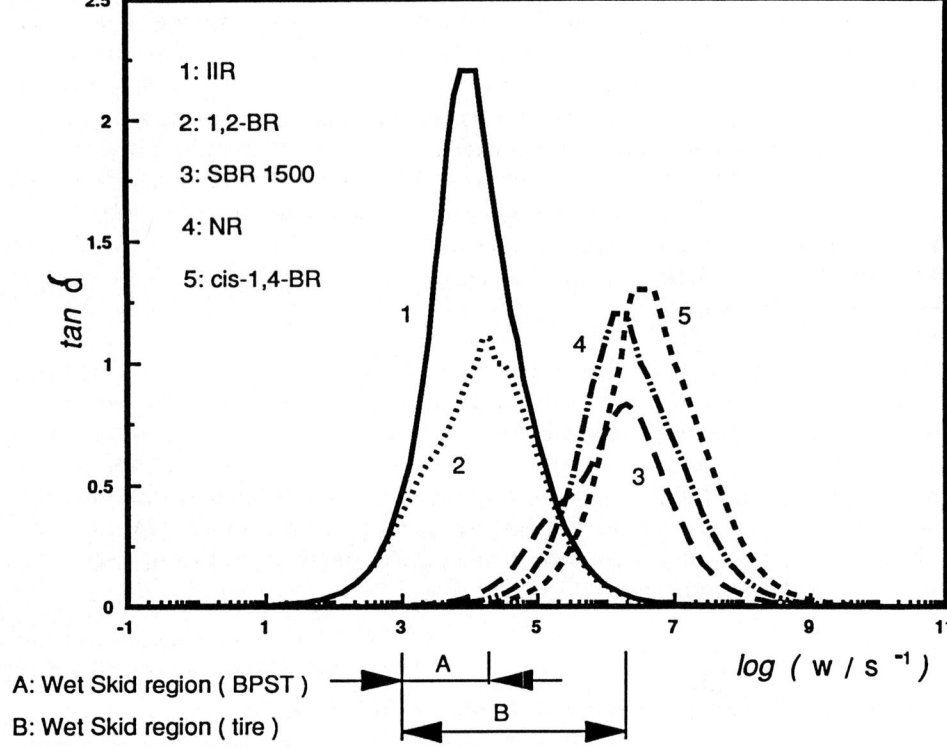

A: Wet Skid region (BPST)
B: Wet Skid region (tire)

Fig. 10. Damping curves for several polymers predicted by the tree-like entanglement network model

loss maximum act both intra- and intermolecular correlations. It is interesting to note that similar conclusions explain the shape of the dielectric relaxation function at the glass-transition of polymers [15].

Appendix

The KMF-model (Kloczkowski, Mark, Frisch [14]) of network dynamics considers the motion of junctions together with the motion of the subchains (statistical segments) dividing each chain between the junctions. We apply this model in the short-time region by replacing the junctions by the entanglements.

We start from the expression of the shear relaxation modulus $G(t)$ which is defined as an integral of the relaxation spectrum $H(\tau)$:

$$G(t) = G_N^0 + \int_{-\infty}^{+\infty} H(\tau) \exp(-t/\tau) d\ln\tau \quad (1)$$

Fourier-transforming of Eq. (1) leads to the expression for the complex shear modulus $G^*(\omega) = G'(\omega) + iG''(\omega)$:

$$G'(\omega) = G_N^0 + A \int_{-\infty}^{1} dx \frac{\omega^2}{[2/\tau_0(1-x)]^2 + \omega^2} \xi(x;n,4) \quad (2)$$

$$G''(\omega) = A \int_{-\infty}^{1} ds \frac{2\omega/\tau_0(1-x)}{[2/\tau_0(1-x)]^2 + \omega^2} \xi(x;n,4) \quad (3)$$

$$\xi(x;n,\phi) = \frac{U'_n(x)+(\phi-2)U'_{n-1}(x)-(\phi-1)U'_{n-2}(x)}{[4(\phi-1)-\{U_n(x)+(\phi-2)U_{n-1}(x)-(\phi-1)U_{n-2}(x)\}^2]^{1/2}} \quad (4)$$

The loss factor follows from $\tan\delta = G''(\omega)/G'(\omega)$. $U_n(x)$ is the Chebyshev polynomial of the second kind. The recursion formulae are

$$U_0(x) = 1$$
$$U_1(x) = 2x \quad (5)$$
$$U_{n+1}(x) = 2xU_n(x) - U_{n-1}(x), \quad n = 1,2,\ldots$$

The derivative $U'_n(x) = dU_n(x)/dx$ can be expressed by the Chebyshev polynomial of the first kind:

$$U'_n(x) = (x^2-1)^{-1}[(n+2)T_{n+1}(x) - U_{n+1}(x)], \quad (6)$$
$$(|x| < 1)$$

$$T_0(x) = 1$$
$$T_1(x) = x$$
$$T_{n+1}(x) = 2xT_n(x) - T_{n-1}(x), \quad n = 1,2,\ldots$$

Further, the following representations of the polynomials are used:

$$T_n(x) = \begin{cases} \cos(n \arccos x), & x \in [-1,1] \\ \cosh(n \arccosh x), & x \geq 1 \\ (-1)^n \cosh(n \arccosh(-x)), & x \leq 1 \end{cases} \quad (7)$$

$$U_n(x) = \frac{\sin((n+1)\arccos x)}{\sin(\arccos x)}, \quad -1 \leq x \leq +1. \quad (8)$$

The quantity ϕ denotes the functionality which is equal to 4 in our case; n is the number of subchains and $A = \nu k_B T$ (ν — number of beads per unit volume). The primary relaxation time τ_0 has been introduced in the discussion and conclusions section.

The integrals in Eqs. (3) and (4) are defined only for the bands $|y| < 1$ with (physical) positive slopes of $y(x)$; otherwise, it is zero. The function $y(x)$ is given as follows:

$$y(x) = [U_n(x) + (\phi-2)U_{n-1}(x) - (\phi-2)U_{n-2}(x)]/2\sqrt{\phi-1}. \quad (9)$$

The molecular parameters, which enter the calculation, are given in Table 4.

Acknoweledgements

All experimental assistance provided by C is gratefully acknowledged. Special thanks are due to Dr. T. Alshuth (Deutsches Institut für Kautschuktechnologie e.V., Hannover) for measuring the viscoelastic curves of SBR 1500 and to Messr. Kendziorra for his assistance in the calculations. The author wishes to thank Continental AG for permission to publish this article.

References

1. Nordsiek KH, Wolpers J (1990) Kautschuk & Gummi. Kunststoffe 43:755
2. Ferry JD (1980) Viscoelastic Properties of Polymers. John Wiley & Sons, Inc, NY
3. Aggarwal SL, Hargis IG, Livigni RA, Fabris HJ, Marker LF (1986) In: Lal J, Mark JE (eds) Advances in elastomers and rubber elasticity. Plenum Press, pp 17—36
4. Rahalkar RR (1989) Rubb Chem Tech 62:246
5. Gargani L, De Ponti P, Bruzzone M (1987) Kautschuk & Gummi. Kunststoffe 40:935
6. Aharoni SM (1986) Macromolecules 19:426
7. Wu S (1989) J Pol Sci: Part B: Polym Phys 27:723
8. Bond R, Morton GF, Kroll LH (1984) Polymer 25:132
9. Duck EW, Locke JM (1968) J Inst Rubber Ind 2:223
10. Graessley WW, Edwards SF (1981) Polymer 22:1329
11. Heinrich G, Straube E, Helmis G (1988) Adv Poly Sci 85:33
12. Kawakami S, Hirakawa H, Misawa M (1989) Int Polym Sci Technol 16:41

13. Rouse PE (1953) J Chem Phys 21:1272
14. Kloczkowski A, Mark JE, Frisch HL (1990) Macromolecules 23:3481
15. Schönhals A, Schlosser E (1989) Coll Polym Sci 267:125
16. Colby RH, Rubinstein M, Viovy JL (1992) Macromolecules 25:996
17. Heinrich G, Straube E (1984) Acta Polymerica 35:115
18. Heinrich G (1992) Kautschuk Gummi Kunstst 45:173
19. Heinrich G, Rennar N, Stähr J (1992) Kautschuk Gummi Kunstst 45:442

Author's address:

Dr. G. Heinrich
Continental AG
Materials Research
Postfach 169
D-3000 Hannover 1, FRG

Relaxation in permanent networks

V. Kraus, H.-G. Kilian, and W. v. Soden*)

Abteilung Experimentelle Physik, Universität Ulm, FRG
*) Abteilung Angewandte Physik, Universität Ulm, FRG

Abstract: Large-strain relaxation in permanent networks (natural rubber, PMMA) is explained by applying thermodynamics of irreversible processes and using the van der Waals network model. The memory function is shown to be strain-invariant. Independent Onsager processes characterized by the relaxation-time distribution are altogether coupled in the same manner to the global level. The macrosocpically nonlinear response, including necking in the glass-transition regime is discussed.

Key words: van der Waals networks; relaxation; irreversible thermodynamics; large deformations

Introduction

Studying relaxation phenomena in permanent networks should deliver solid information about the dynamics of macromolecular chains linked in a network. Chains are not able to perform large scale diffusion processes like reptation. If we want to explain relaxation, we are confronted with the question how time-dependent processes in networks with a broad chain-length distribution run off. The underlying processes are in any case controlled by cooperation of chains across junctions. What about universal features? This could be found out by describing macroscopic relaxation phenomena in terms of a classical thermodynamics of irreversible processes [1, 2]. Hope is engendered that the characteristics can be identified when the description of the overall behavior essentially under large deformations is made quantitative.

One might suspect that deformation-induced deviations from equilibrium are different for chains of different length. This is expected if the constraints, which induce deviation from equilibrium are continuously distributed, thus being of "continuum-like" local origins. But at a large enough distance from the glass transition range, segmental Platzwechsel (Rouse-motions) are so rapid in comparison with the experimental time window that they allow to adjust "internal equilibrium" (within well-defined small enough, but possibly time-dependent volume elements). Every interpretation of relaxation in networks has therefore to discuss how processes on the global level and relaxation mechanisms are interrelated.

In the limits of equilibrium the global level of networks is well defined. The stress-strain behavior is fully described by means of the van der Waals modification of the Gaussian network [3—6]. It will be shown that it is essential for any clear characterization of the relaxation mechanisms including large strains, that the equilibrium stress-strain behavior itself is nonlinear.

Consequently, even for a linear treatment of the elementary relaxation, the macroscopic response must in principle be nonlinear. Limitations to the linear regime of relaxation have extensively been studied by experts like Ferry [7], Pechold [8], MacKnight [9] — to mention only a few of these authors. The results point in the direction suggested above. But most of the interpretations attribute nonlinear amplitude effects to nonlinear elementary relaxation phenomena. It will be demonstrated in this paper that a different answer can be given by investigating relaxation under very large strains.

One of the few attempts to tackle large strains (known to us) goes back to Tschoegl [10]. He proposed an empirical power law for describing simple extension of networks under different strain rates. We have the advantage of knowing the Gibbs function of a van der Waals network. By extending the Gibbs function by introducing additional internal variables ("hidden variables"), the correct description of equilibrium stress-strain behaviour is à priori guaranteed. The fundamental question of whether relaxation is of the classical Onsager type or not can convincingly be answered only by a full interpretation of adequate experiments. To this purpose, stress-strain cycles at different constant strain rates were carried out at various temperatures. What we would like to know is how the stress-strain patterns, as shown in Fig. 9, come about.

Equilibrium state of reference

We define a van der Waals modification of the Gaussian network [3, 4]. Finite-chain extensibility and global interaction are accounted for by the two van der Waals parameters λ_m and a. We use the van der Waals equation of state in the mode of simple extension (nominal force $f(\lambda)$)

$$f(\lambda) = G \cdot D(\lambda) \cdot \left(\frac{1}{1-\eta} - a\Phi^{1/2}(\lambda)\right), \quad (1)$$

with

$$\phi(\lambda) = \frac{1}{2}(\lambda^2 + 2/\lambda - 3), \quad (2)$$

$\eta = \sqrt{\Phi/\Phi_m}$, $\Phi_m = \Phi(\lambda_m)$ and $D(\lambda) = \lambda - \lambda^{-2}$. The quality of the fit to experiments is demonstrated by Figs. 1 and 2. It is important that the van der Waals

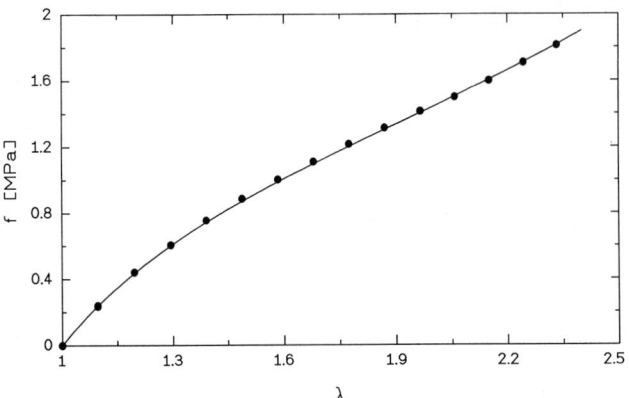

Fig. 2. Quasistatic stress-strain curve of crosslinked PMMA at 140°C. The parameters according to Eq. (1) are $\lambda_m = 5.6$, $a = 0.44$ and $G = 0.93$ MPa

network is characterized in all types of strain by the same fundamental parameters λ_m, a and M_u [6]. These parameters practically do not depend on temperature. The maximum strain λ_m, describes the maximum chain extensibility in networks with finite chain lengths. The second parameter a is a phenomenological parameter which characterizes global interaction across junctions [3, 11]. M_u is the molecular weight of the stretching invariant unit. It is now possible to explicitly formulate the strain-energy function [4]

$$W(\lambda) = -G\left\{2\Phi_m[\ln(1-\eta) + \eta] + \frac{2}{3}a\Phi^{3/2}\right\}$$

$$= G \cdot w. \quad (3)$$

The shear modulus G of a permanent network [12, 13] is given by

$$G = \frac{\rho RT}{M_u \lambda_m^2}, \quad (4)$$

where R is the gas constant, T the absolute temperature and ρ the density of the polymer. $W(\lambda)$ describes the work exchanged during isothermal deformation. But, to keep temperature constant during deformation, it is in principle necessary that heat is exchanged with a heat bath in a well-defined manner. To describe this, we use the differential of the Gibbs free energy of a van der Waals network

$$dg = -sdT + Vdp + fd\lambda. \quad (5)$$

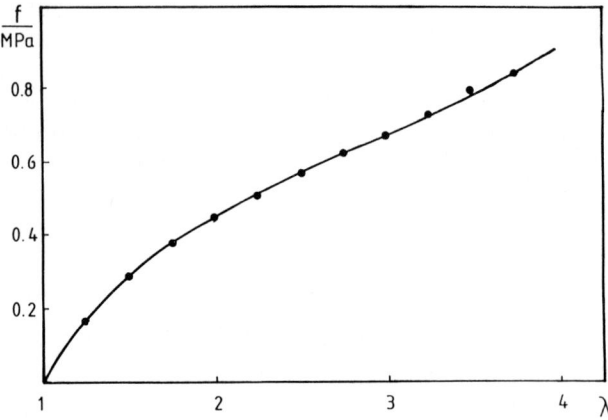

Fig. 1. Quasistatic stress-strain curve of natural rubber at 90°C. The fundamental parameters are $\lambda_m = 10.5$, $a = 0.3$ and $G = 0.38$ MPa

It was shown elsewhere [14] that thermoelasticity and the strain-induced increase in volume are altogether fully described if the adequate thermodynamic coefficients in the unstrained system are known (thermal expansion, isothermal compressibility, temperature-coefficient of non-isoenergetic rotational isomers). Every strain-induced effect onto intrinsic condensed matter properties of networks is small enough to allow a linear approach.

The extended Gibbs functions

At a small distance to equilibrium where local homogeneity in the sense of a continous system is guaranteed, it is possible to apply the thermodynamics of irreversible processes. Most elegant is to extend the Gibbs function by introducing a set of orthogonal hidden variables $\{\xi_i\}$. These variables are the extensive conjugates of de Donders affinities A_i [1, 2]. They are defined to disappear at equilibrium; they therefore provide a direct measure of the "distance to equilibrium" of each of these à priori independent internal processes. The extended Gibbs function at constant pressure reads

$$dg = -sdT + Vdp + fd\lambda - \sum_i A_i d\xi_i . \tag{6}$$

This Gibbs function is a complete differential defining a stable system. It is supposed that the mechanic strain-energy of networks is comprised of two approximately independent energy forms. On one hand, this is the mode of compression (or dilation) and on the other hand, changes in shape are due to shear-processes. The last term in equation (6) describes exchange of energy due to time-dependent internal processes. The equations of state are given by

$$f = \left(\frac{\partial g}{\partial \lambda}\right)_{p,T,\xi_i} , \tag{7}$$

$$s = -\left(\frac{\partial g}{\partial T}\right)_{p,\lambda,\xi_i} , \tag{8}$$

$$A_i = -\left(\frac{\partial g}{\partial \xi_i}\right)_{p,\lambda,T,\xi_K \neq \xi_i} . \tag{9}$$

These equations are interrelated by the necessary conditions of integrability. This includes the very important assumption that during relaxation internal equilibrium must at any moment be fully achieved. Hence, strain-induced thermoelastic and volumetric effects are relaxation dependent.

Keeping in mind that relaxation is accompanied by a well-defined intrinsic entropy production, irreversible deformation is, in principle, nonisothermal. It is therefore a crucial question of how far a quasi-isothermal treatment of relaxation of networks under large strains and at different strain rates is a good approximation. At the moment, we rest content with treating relaxation as a quasi-isothermal and isobaric process.

The constitutive equation

We formulate an isobaric and isothermal constitutive equation for systems where the Gibbs function is known. Because the thermodynamic limit is included, the formulation is unique and definite. In networks strain-energy is exchanged at the global level. The system has then to achieve the equilibrium state by intrinsic processes (basic relaxation mechanisms as defined by the complete set of hidden variables ξ_i) which are linked with the global level in a well-defined manner under isothermal conditions. Each of the mechanisms is coupled to the global level in the same manner (scalar and isotropic coupling) [3, 14]:

$$(g)_T = \frac{1}{2} f_1 w(\lambda) + \sqrt{w(\lambda)} \sum_i f_{12}^{(i)} \xi_i + \frac{1}{2} \sum_i f_2^{(i)} \xi_i^2 . \tag{10}$$

The Onsager ansatz

The relaxation mechanism will obey the Onsager relationship [15, 16, 17]

$$\dot{\xi}_i = a_i A_i , \tag{11}$$

where a_i are phenomenological kinetic coefficients. Hence, we suppose that the relaxation rate is controlled by the momentary distance to equilibrium characterized by the set of affinities. Now, it is a crucial point, in which way the elementary relaxation mechanisms enter into the formulation of the Nachwirkungsintegral?

The solution

A single relaxation process

In a representation with a single relaxation process [14], we are led to the equations of state

$$f = \frac{1}{2} w'(f_1 + f_{12} w^{-1/2} \xi) \tag{12}$$

$$-A = f_{12} w^{1/2} + f_2 \xi . \tag{13}$$

Together with the Onsager equation, we arrive at the Thomson-Poynting differential equation [18]

$$\dot{f} + \tilde{\tau}_\varepsilon^{-1} f = \frac{G w'(\lambda)}{2\tilde{\tau}_\varepsilon} (1 + \tilde{\tau}_\sigma \dot{\lambda}) \qquad (14)$$

with the relaxation times

$$\tilde{\tau}_\varepsilon^{-1} = \tau_\varepsilon^{-1} + \dot{\lambda} \left(\frac{w'}{w} - \frac{w''}{w} \right);$$

$$\tilde{\tau}_\sigma^{-1} = \tau_\sigma^{-1} \frac{4w'}{(w')^2}, \qquad (15)$$

and

$$G = \frac{1}{2} \left(f_1 + \frac{f_{12}^2}{f_2} \right), \qquad (16)$$

and its solution

$$f(t) = \frac{G w'(t)}{2}$$

$$\cdot \left[1 + \frac{\Delta \tau}{\tau_\varepsilon} \left(1 - \frac{1}{\tau_\varepsilon} \Psi \right) \right],$$

$$\Psi = \int_0^t e^{-\frac{(t-t')}{\tau_\varepsilon}} \left(\frac{w(t')}{w(t)} \right)^{1/2} dt'. \qquad (17)$$

As a consequence of having the elementary relaxation processes coupled to the global level, it comes about that the macroscopic relaxation times depend on strain via the potential $w(\lambda)$ and its derivatives. A pronounced dependence on strain is conceivable if the strain-energy $w(\lambda)$ is nonlinear. This yields, of course, to a nonlinear macroscopic response. This nonlinear behavior is originated with the nonlinear equilibrium stress-strain behavior, which influences the Nachwirkungsintegral in its dependence on strain.

Because we know the strain-energy function of a van der Waals network, the entropy production can be calculated when the sample is stretched under a constant strain rate. The calculation, as depicted in the sketch in Fig. 3, clearly demonstrates a nonlinear energy dissipation.

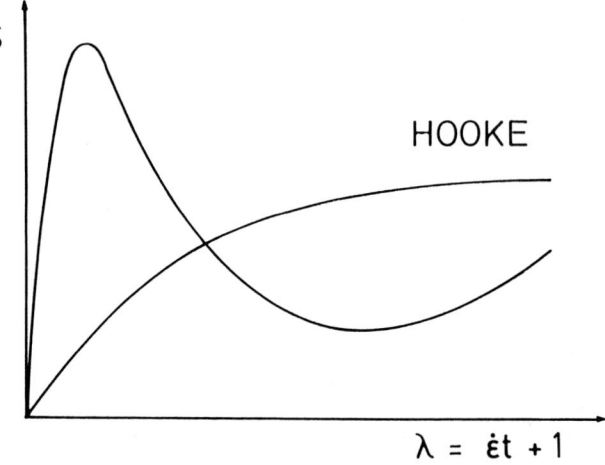

Fig. 3. Entropy production of a van der Waals-network with a single relaxation time compared with a linear Hookian system

A relaxation-time distribution

Nothing in principle must be changed when proceeding to systems with a relaxation-time distribution. We arrive at the general soluton [3, 19]

$$f(t) = G w'(t) \left\{ 1 + \frac{\Gamma}{G} \left(1 - \frac{2}{\Gamma} \Psi \right) \right]$$

$$\Psi = \int_0^t m(t - t') \left(\frac{w(t')}{w(t)} \right)^{1/2} dt' \qquad (18)$$

with $w'(t) = \partial_\lambda w(\lambda(t))$ and $\lambda(t) = \dot{\varepsilon} t + 1$ where $\dot{\varepsilon}$ is the strain rate. The memory function

$$m(t - t') = \sum_i \frac{h_i}{\tau_i} e^{-\frac{(t-t')}{\tau_i}} \qquad (19)$$

contains the normalized relaxation-time spectrum $h_i(\tau_i)$ (Figs. 4, 7). $\Gamma = G_g - G$ is the relaxation strength and G_g the modulus, attained at high frequencies ($G_g = G(\omega \to \infty)$). Equation (18) expresses the superposition of the various relaxation mechanisms as the most prominent identification of the Onsager approach. The memory function should be strain invariant and independent of the type of strain. These universal features rely heavily on the fact that the mechanisms are altogether coupled in the same manner with the global level (10).

The analytical form of the stress-strain relationship shows classical features: The macroscopically

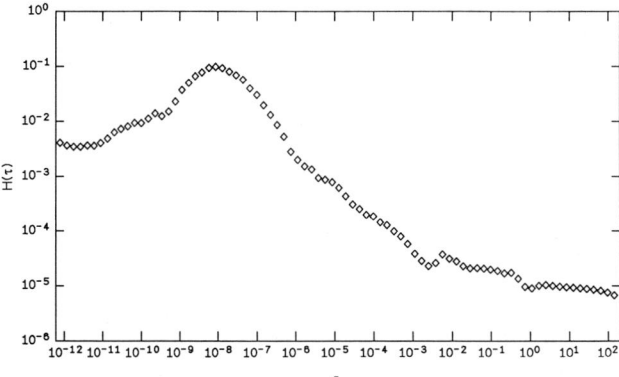

Fig. 4. Relaxation-time spectrum of NR. The corresponding WLF-parameters are $T_S = 274.61$ K, $B = 4.57$ and $C = 101.57$

nonlinear response displays a fading memory that is of utmost importance in systems with a relaxation time distribution. Only those relaxation times matter which compare with the time window of the experiment.

The discussion of stress-strain cycles under constant strain rates allows us to prove the utility of our approach. This becomes essential at large strains where finite chain extensibility and global interaction come into play. It is a question of substantial interest to find out whether it is sufficient to know the fundamental parameters of the van der Waals network and the phenomenological relaxation-time spectrum for describing relaxation in strained networks — even in the high temperature tail of the glass-transition.

Experimental

Natural rubber (from Degussa, crosslinked with 1.35 wt. % DCP) and permanently crosslinked PMMA (from Röhm) are the two examples we refer to. For the stress-strain experiments, we used a tensile tester (Zwick 1445) with a heatable sample chamber. The natural rubber sheets (thickness: 2 mm) were cut into samples with a length of 80 mm and a width of 10 mm. The PMMA was cut into dogbone-shaped samples with a length of 50 mm and a width of 4 mm (thickness: 1 mm).

Natural rubber

Master curves of the complex shear modulus are deduced from small-amplitude shear-relaxation experiments. The relaxation-time spectrum (Fig. 4) was then obtained by applying the Schwarzl-Stavermann relations [7]. The fundamental parameters of the network have been deduced from the fit to the quasi-static stress-strain pattern measured at 90 °C (Fig. 1). Here, no hysteresis is measurable, even under large strain rates.

The calculation fits the stress-strain cycles at room-temperature fairly well without any parameter adjustment. The temperature-shift of the relaxation spectrum was deduced from the WLF-equation [20]

$$\log a_T = -\frac{B(T - T_S)}{C + T - T_S}. \qquad (20)$$

Nonlinearity of the stress-strain cycles is more pronounced at lower temperatures (Figs. 5, 6), where, due to the shift of the spectrum, a significant fraction of shorter relaxation-times enters the experimental time window. The description is surprisingly good, except in the small-strain regime (Fig. 6). In the largest part the stress-strain cycles seem to run as quasi-isothermal processes. The discrepancies at small strains increase with the strain rate or at lower temperatures.

PMMA

Again, we determined the fundamental network parameters first (Fig. 2). The calculation fits the stress-strain curve fairly well far above the glass-

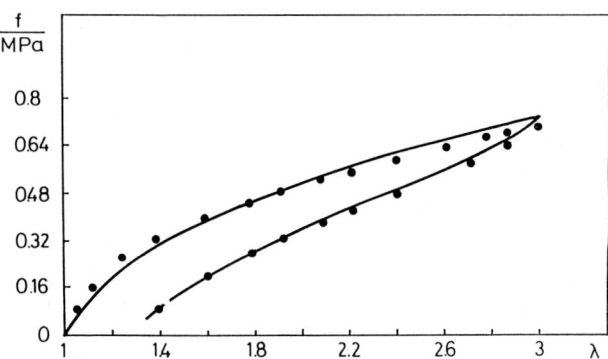

Fig. 5. Stress-strain cycles of NR at -50 °C at the strain rate $\dot{\varepsilon} = 1 \cdot 10^{-3}$ s^{-1}

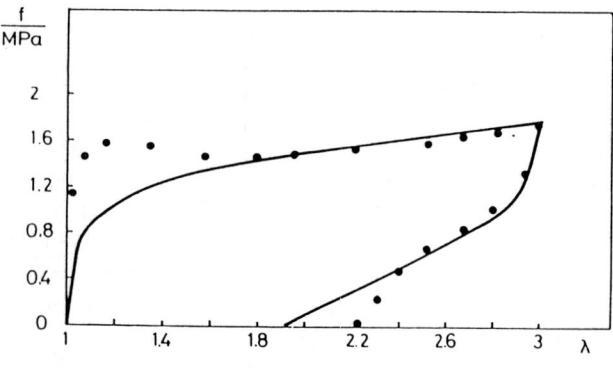

Fig. 6. Stress-strain cycles of NR at −60°C at the strain rate $\dot\varepsilon = 1 \cdot 10^{-2}$ s^{-1}

Fig. 9. Stress-strain cycles of PMMA at T 3 120°C. The strain rates are $2 \cdot 10^{-2}$ s^{-1} (a), $7 \cdot 10^{-3}$ s^{-1} (b), $4 \cdot 10^{-3}$ s^{-1} (c) and $6 \cdot 10^{-4}$ s^{-1} (d)

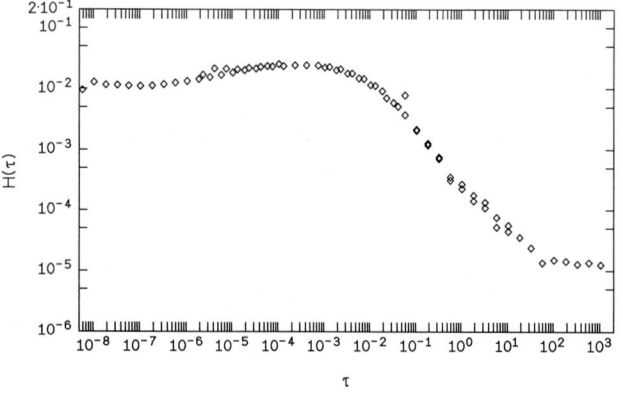

Fig. 7. Relaxation-time spectrum of PMMA at $T = T_S = 144.05$°C. The WLF-parameter are $B = 6.2$ and $C = 65$

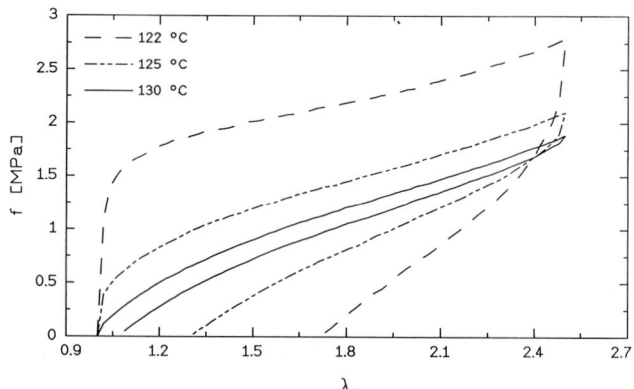

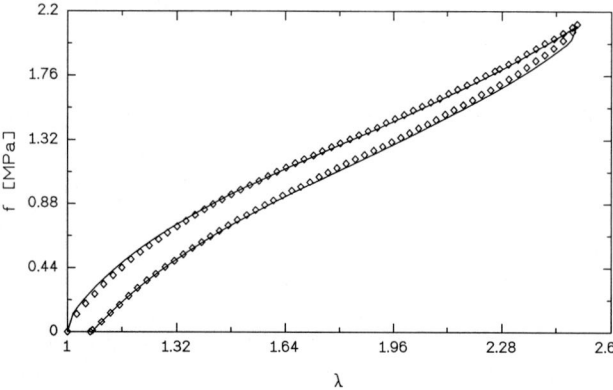

Fig. 8. Stress-strain cycle of PMMA at $T = 140$°C. The strain rate is $\dot\varepsilon = 1.3 \cdot 10^{-2}$ s^{-1}. The solid line is computed with the spectrum of Fig. 7

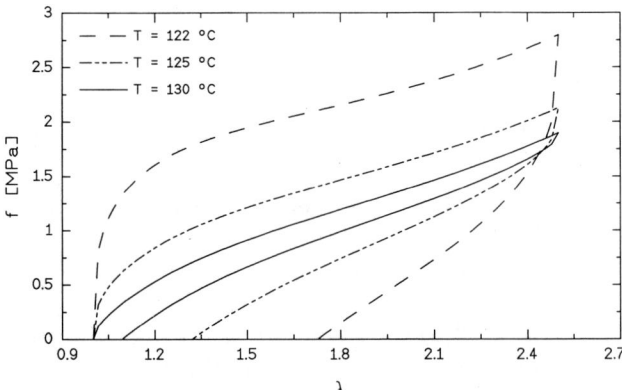

Fig. 10. Stress-strain cycles of PMMA at a strain rate $\dot\varepsilon = 6 \cdot 10^{-4}$ s^{-1} at the temperatures as indicated. a) Experiment, b) isothermal calculation

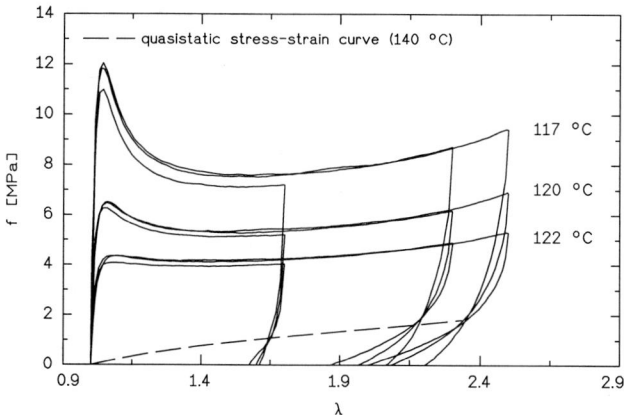

Fig. 11. PMMA stress-strain cycles at temperatures as indicated with a strain rate $\dot{\varepsilon} = 7 \cdot 10^{-4}$ s^{-1}, drawn to different maximum strains

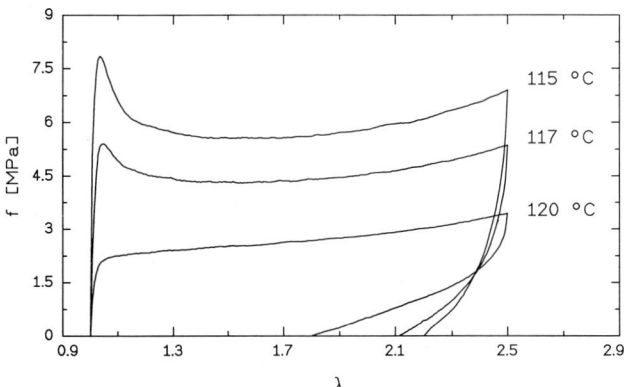

Fig. 12. PMMA stress-strain cycles at a strain rate $\dot{\varepsilon} = 7 \cdot 10^{-4}$ s^{-1} at the temperatures as indicated

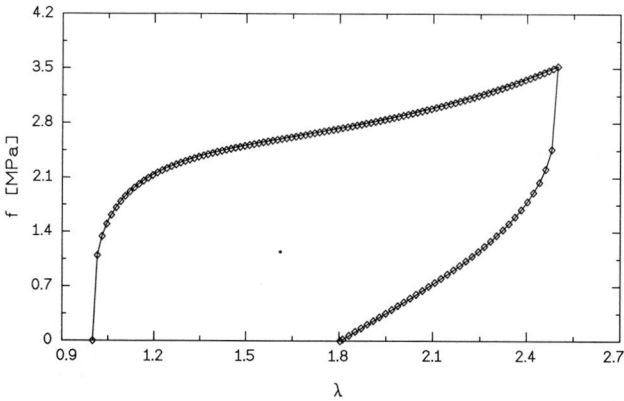

Fig. 13. Isothermal calculation of stress-strain cycles of PMMA. Solid line: $T = 117.75\,°C$, $\dot{\varepsilon} = 7 \cdot 10^{-5}$ s^{-1}, sqares: $T = 121.85\,°C$, $\dot{\varepsilon} = 7 \cdot 10^{-4}$ s^{-1}

transition temperature (Fig. 8). The relaxation-time spectrum was deduced from measurements of small strain shear deformations (Fig. 7). According to Fig. 9 a very pronounced maximum develops at temperatures in the high-temperature tail of the glass-transition (at temperatures above the inflection point of the dynamic modulus $G'(w)$). the quasi-static glass-transition temperature as determined by DSC is about 113.4 °C.

It is an important finding that analogous modifications of stress-strain pattern are observed when cycles under constant strain-rate are performed at different temperatures. Exactly the same stress-strain cycle is computed for an an adequate couple of variables T and $\dot{\varepsilon}$ (Fig. 13). We come to the statement that stress-strain cycles of PMMA and rubber-networks display a unique frequency-temperature relationship as a manifestation of linear relaxation behavior up to large strains. This is also documented by the phenomenon that stress-strain cycles for different velocities at the same temperature or at different temperatures at the same strain rate run altogether through a single point during unloading (Figs. 9, 10a, 11, 12), provided that they were drawn to the same maximum strain.

This very peculiar phenomenon is exactly what is obtained by our calculation (Fig. 10b) by using the relaxation-time spectrum as shown in Fig. 7.

The dashed line in Fig. 11 is the equilibrium stress-strain curve. Hence, if the same maximum strain $\lambda_{\max}$ is enforced, unloading stress-strain curves of rubbers cross the equilibrium curve at the same macroscopic strain. This demonstrates that the nonlinear macroscopic stress-strain relaxation is made by the superposition of a set of linear Onsager processes. The relaxation velocity of each à priori independent mechanism is regulated by its individual distance from equilibrium. The driving force is zero when the equilibrium curve is met. To have the whole set of the relaxation mechanism uniquely linked to the global level must be the reason why any relative deviation from equilibrium is the same for each elementary process.

This also makes understandable that each of the relaxation mechanisms in the frequency window of interest passes the equilibrium curve at the same time and at the same macroscopic strain. This is a demonstration that the distance of the momentary stress-strain curve from the equilibrium curve characterizes, in a quasi-stationary regime, the relative deviation of the whole set of relaxation mechanisms. It is now evident that the nonlinear behavior is the consequence of having each of the mechanism linked in the same scalar and isotropic manner to the nonlinear state of reference. Each of the relevant mechanisms is affected in a different but well-defined manner, like being "synchronized" from the beginning. Each one seems to be forced to the same à priori distance from equilibrium.

The nonisothermal process

As mentioned above, entropy production occurs during deformation. Because of being coupled to the heat bath which is the system itself, the sample is heated up during deformation. If now internal equilibrium is locally established within a time scale which is very much shorter than the experimental window, we should be in the position to describe nonisothermal stress-strain cycles. If we adjust the shiftfactor a_T at each point of the stress-strain curve, this leads to the representation of experiments as shown in Fig. 14. We see in Fig. 15 that the shiftfactor has to be continuously lowered down to a stationary value in the large strain regime. This is in principle accompanied with a rise of temperature in the neck. We in fact measured temperature patterns of this shape (Fig. 16) with an infrared-camera for PMMA at 75°C and for polycarbonate at room temperature. It is therefore explained why our isothermal calculation was successful in the large strain regime. Heating of the sample by orientational entropy and dissipation is in addition also behind the very pronounced nonlinear stress-strain behavior in the small strain regime.

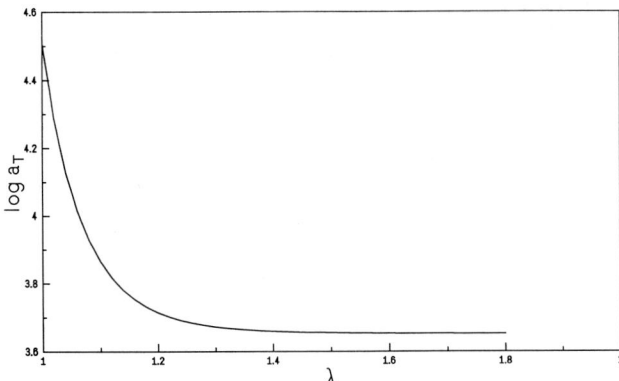

Fig. 15. Strain-induced lowering of the shiftfactor a_T

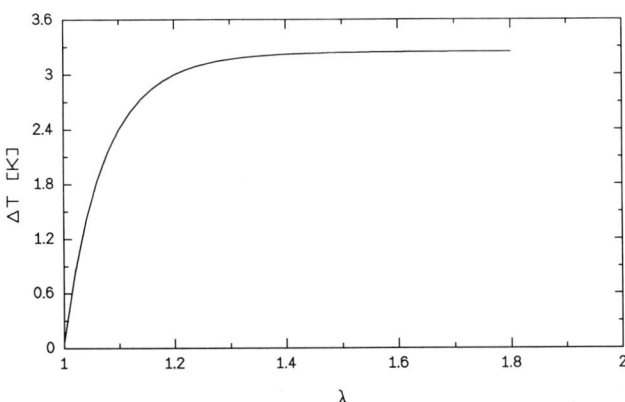

Fig. 16. Assumed rise of temperature during extension of a PMMA sample at a surrounding-temperature of 117°C in the high temperature tail of the glass-transition

The heterogeneous deformation

At the lowest temperatures investigated here, we observe that heterogeneous deformation forms a more or less pronounced neck. We nevertheless succeed in correctly describing the macroscopic stress-strain curve. The neck is very sharp at low temperatures or at large strain rates at a defined temperature in the glass-transition range.

Necking in the glassy state was interpreted by ourselves as a strain-induced transition [21] where "coexistence" of two differently strained phases seems to be the appropriate model for understanding this phenomenon. This situation is in the present case much more complicated. From a discussion of the stability conditions, it comes out that single relaxation mechanisms, each one characterized by its variable τ_k, may violate the condition of

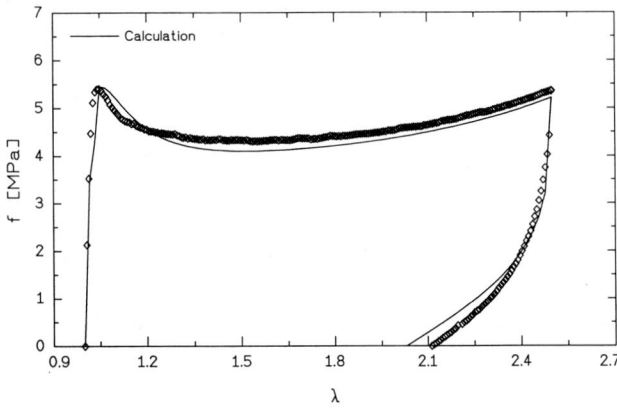

Fig. 14. Fit of a PMMA stress-strain curve at $T = 117°C$ (Fig. 12) with the temperature profile of Fig. 16

stability. The system is then forced to go into a new non-neighbored state of increased extension. The internal variables are constant during the transformation. A delicate situation arises because of the running relaxation. One has to consider the process in the hyperspace of the variables $(T, \lambda, \{\xi_k\})$. Any solid discussion has to represent the time-dependent modifications of the stability in a system of very large dimensions. Can we understand that we correctly calculate the stress-strain pattern independent of whether the deformation is homogeneous or heterogeneous? One of the evident reasons might be that, according to the Maxwellian construction, the work exchanged in the transition region must be the same for the actual transition and the quasi-continuous calculation. Because of having a large set of ξ_k's the course of the transition may run off macroscopically quasi-continuously.

Despite the qualitative discussion, we are confronted with the interesting consequence that the conditions of stability in a relaxing network can be violated also due to the intrinsic heating and stress activation that strongly modify relaxation of each of the relaxation processes. The system has then to undergo a time-dependent transformation into nonneighbored homogeneous states of deformation. These processes display features of a classical phase transition. Phenomenologic criteria, as for a phase transition, were proven to characterize the necking process in cold drawn polycarbonate [21]. Because every selfconsistant description has also to explain this low temperature behavior in the glassy state, the above considerations are supported in principle.

Effects of vitrification

We learned that the relaxation behavior up to the inflection point of the dynamic shear modulus $G'(\omega)$ ($\omega \cdot \langle \tau \rangle = 1$) is related to the whole relaxation-time spectrum. This spectrum is shifted with temperature as demanded by the WLF-procedure. This seems to be changed substantially in constant-rate stress-strain cycles in the range below the above limit. Fig. 17 shows that, on unloading, the characteristics of the pattern are changed in principle. The stress-strain curves pass through the equilibrium curve at different strains and no longer cross over. For sufficiently large strain rates the remnant strain at $f(t) = 0$ is the largest for samples with the highest asymptotic stress at $\lambda = \lambda_{max}$.

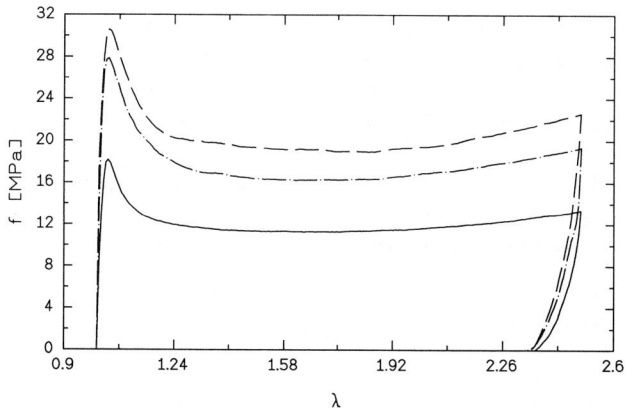

Fig. 17. Stress-strain cycles of PMMA carried out at temperatures below the inflection point of the dynamic shear modulus $G(\omega)$ ($\dot{\varepsilon} = 7 \cdot 10^{-4}\,\mathrm{s}^{-1}$). (—: $T = 105\,°\mathrm{C}$, —·—: $T = 95\,°\mathrm{C}$, ----: $T = 90\,°\mathrm{C}$)

Conclusive remarks

In the temperature range above the actual glass-transition, we have achieved a full description of deformation and relaxation in permanent networks. The global level can be considered as a weakly interacting van der Waals conformation gas-network, characterized by the fundamental parameters M_u, λ_m and a. The Gibbs function of this system is well defined. The relaxation process is describable within the framework of thermodynamics of irreversible processes. The Gibbs function is extended to deliver the appropriate constitutive equation. The relaxation-time distribution defines the memory function as mode-independent strain invariant (Onsager approach). Nonlinear viscoelasticity is a consequence of linking the relaxation mechanism to the nonlinear global level. Entropy production in the glass-transition range above the inflection point of $G'(\omega)$ implies a measurable heating during deformation. Unstable non-equilibrium states are behind the heterogeneous macroscopic deformation. They may be understood as a generalized type of phase-transition-like process.

Acknowledgement

We want to thank Röhm, Darmstadt, for preparing the PMMA sheets. Also we want to thank the Deutsche Forschungsgemeinschaft for support. Discussions with Dr. B. Heise and Dr. J. Koenen are gratefully acknowledged.

References

1. Meixner J, Reik HG (1963) Thermodynamik der irreversiblen Prozesse, Springer, Berlin
2. deGroot SR, Mazur P (1962) Non-equilibrium Thermodynamics, North Holland, Amsterdam
3. Ambacher H, Enderle HF, Kilian HG, Sauter A (1989) Coll Polym Sci 80:209
4. Enderle HF, Kilian HG, Vilgis T (1984) Colloid Polym Sci 262:696
5. Kilian HG (1980) Polym Bull 3:151
6. Kilian HG (1981) Polymer 22:209
7. Ferry JD (1980) Viscoelastic Properties of Polymers, J. Wiley & Sons Inc., New York
8. Pechhold W, Sautter E, Soden W, Stoll B, Grossmann HP (1979) Macromol Chem Suppl 3:247
9. Aklonis JJ, MacKnight WJ (1983) Introduction to viscoelasticity of polymers, 2nd Ed., Wiley & Sons, New York
10. Tschoegl NW (ed) (1989) The phenomenological theory of linear viscoelastic behaviour, Springer, Berlin Heidelberg
11. Vilgis T, Kilian HG (1986) Colloid Polym Sci 264:137
12. Kilian HG (1985) Colloid Polym Sci 263:30
13. Kilian HG, Unseld K (1986) Colloid Polym Sci 264:9
14. Kilian HG, Vilgis T (1984) Colloid Polym Sci 262:691
15. Onsager L (1931) Phys Rev 37:405
16. Onsager L (1931) Phys Rev 38:2265
17. Meixner J (1954) Z Naturforsch 9a:654
18. Keller H (1977) Irreversible Thermodynamik, de Gruyter, Berlin
19. Enderle HF (1988) Finite Viskoelastizität von Van der Waals-Netzwerken, PhD thesis Universität Ulm
20. Williams ML, Landel RF, Ferry JD (1955) J Amer chemSoc 77:3701
21. Koenen JA, Kilian HG, Heise B (1989) J Polym Sci: B: Polym Phys 27:1235

Authors' address:

Prof. Dr. H.-G. Kilian
Abt. Experimentelle Physik
Universität Ulm
Albert-Einstein-Allee 11
D-W-7900 Ulm, FRG

Do local motions influence rheological properties near the gelation threshold?

D. Lairez, M. Adam, E. Raspaud, J.R. Emery[1]), and D. Durand[1])

Service de Physique de l'Etat Condensé, DRECAM, C.E.-Saclay, Gif-sur-Yvette, France
[1]) U.R.A. C.N.R.S. n° 509 & 807, Université du Maine, Le Mans, France

Abstract: Generally speaking, the law: $G^* \sim G_\infty \varepsilon^t f(i\omega/\omega^*)$ with $\omega^* = \omega_0 \varepsilon^{s+t}$, describes the viscoelastic properties near the gelation threshold. G^* is the complex elastic modulus, ω^* is the frequency associated with the longest relaxation time, and ε is the relative distance to the gelation threshold. G_∞ and ω_0 are linked to the local properties of the viscoelastic medium. At low frequency ($\omega < \omega^*$), before the threshold, the viscosity diverges as $(G_\infty/\omega_0)\varepsilon^{-s}$ while after the threshold, the elastic modulus emerges as $G_\infty \varepsilon^t$. At higher frequency ($\omega > \omega^*$), before and after the threshold, one expects the complex modulus to be ε independent and to obey the frequency power law: $G'' = G' tg(\pi u/2) = G_\infty (\omega/\omega_0)^u$, with $u = t/(s+t)$. — Static properties of branched polymers near the gelation threshold are percolation-like, while dynamical properties and, more precisely, the s and t exponents values measured by different authors are very scattered. — We suspect that the frequency which describes local properties of polymers is not a constant ω_0, but the frequency ω_g associated with the glass transition. The frequency ω_g decreases with increasing connectivity, which would lead to ε dependent prefactors of power laws and to non-universal apparent exponents. This could explain the apparent non-universality observed for the rheological properties near the gelation threshold.

Key words: Gelation; dynamical properties; viscosity; power law prefactor; polyurethane

Introduction

Generally speaking [1], it is possible to describe the viscoelastic properties near the gelation threshold using Eq. (1):

$$G^* = G_\infty \varepsilon^t f(i\omega/\omega^*) , \quad (1)$$

with

$$\omega^* = \omega_0 \varepsilon^{s+t} , \quad (2)$$

where ε is the relative distance to the gelation threshold; G_∞ and ω_0 are linked to the microscopic properties (very high frequency) of the viscoelatic medium; ω^* is the characteristic frequency associated with the slowest relaxation process.

Equation (1) allows us to describe the frequency behavior of the material near the gel point. For $\omega < \omega^*$ and before the gel point ($p < p_c$), the material has a ω^1 dependence as a viscous medium. Thus, $f(i\omega/\omega^*)$ is equal to $i\omega/\omega^*$, and Eq. (1) expressed in terms of viscosity ($\eta^* = G^*/i\omega$) becomes:

$$\eta^* = \eta_\infty \varepsilon^{-s} \quad (3)$$

with

$$\eta_\infty = G_\infty/\omega_0 .$$

After the gel point ($p > p_c$) the material behaves like an elastic medium for which the modulus is independent of frequency. The function $f(i\omega/\omega^*)$ is equal to a constant and, therefore, Eq. (1) leads to:

$$G^* = G_\infty \varepsilon^t . \quad (4)$$

Equations (3) and (4) show that the exponents s and t govern the divergence of the viscosity and the emergence of the elastic modulus near the gel point, respectively.

In the high-frequency domain, where $\omega^* < \omega$, a power law is expected for the complex function $f(i\omega/\omega^*)$ which is equal to $(i\omega/\omega^*)^u$. This leads to:

$$G^* = (G_\infty/\omega_0^u)\omega^u e^{iu\pi/2}\varepsilon^{t-(s+t)u} . \qquad (5)$$

This frequency range corresponds to a solicitation of the sample at a frequency higher than the characteristic frequency ω^* associated with the largest cluster. Therefore, the modulus is independent of this cluster radius ξ and thus of the relative distance to the gel point ε. Consequently, using Eq. (5), one can establish the expression:

$$u = t(s + t) . \qquad (6)$$

To summarize, in this frequency range ($\omega^* < \omega$), one expects to find:

— a particular value for u, equal for the real and imaginary parts of the modulus and compatible with s and t values.
— a ratio $G''/G' = \mathrm{tg}(u\pi/2)$ compatible with s and t values.
— G' and G'' independent of ε.

Considering experimental results reported in the literature, static properties of polymer clusters near the gelation threshold are well described by the percolation model [2]. On this basis, two main theoretical approaches predict different exponent values for s, t and, thus, u. The Rouse model [3—5], which assumes no hydrodynamic interaction between polymeric clusters, predicts $s = 1.33$, $t = 2.67$, ad $u = 0.67$. Percolation theory and electrical analogy [6] postulate that, if the elasticity has a purely entropic nature, the complex elastic modulus is the analog of the conductivity of a network with randomly distributed resistors and capacitors. Simulations performed in three dimensions predict $s = 0.75$, $t = 1.95$, and $u = 0.72$ [7].

Experimental results reported in the literature, concerning various chemical systems and performed by different teams, can be classified into two categories:

— gelation carried out without solvent: exponents values are widely scattered and dynamical behavior seems to be non-universal: $0.7 < s < 2$ and $3 < t < 4$ [4, 5, 8—10].

— gelation carried out in presence of solvent: dynamical behavior seems to be percolation-like: $0.8 < s < 1$ and $1.9 < t < 2$ [10—13].

However, it is convenient to note that the mean value of the exponent u is found to be $u = 0.71 \pm 0.04$, whatever the chemical system.

In this paper, results concerning polyurethane systems are reviewed and discussed with respect to the above theoretical predictions and these literature results.

Experimental results

Material

The chemical system studied is the polycondensation of triol monomers with diisocyanate monomers [14]. Condensation reactions occur between OH groups of the triols and NCO groups of the diisocyanate to give urethane bonds. Such a system presents a sol/gel diagram which depends on the chemical reaction extent (a) and on the stoichiometric ratio (r). The chemical reaction extent is defined as the ratio of the number of formed bond to the total number of possible bonds, while the stoichiometric ratio is defined as the inital ratio of NCO groups number to OH groups number. We studied two particular ways to cross the sol/gel transition:

— reaction bath: the stoichiometric ratio is equal to 1 and, therefore, all OH groups can find an NCO group to react with, provided one waits long enough. The gel point is reached at a critical value a_c of the chemical reaction extent to which corresponds the time t_c.

—quenched samples: in order to obtain chemically stable samples, NCO groups are in a minority and the chemical reaction is carried out until complete reaction of these groups ($a = 1$). In this case, the gel point is approached by means of the stoichiometric ratio and is reached at the critical value r_c.

In the case of the reaction bath, the degree of connectivity p is governed by a and the relative distance to the gel point ε is defined by:

$$\varepsilon = |a - a_c|/a_c . \tag{7}$$

Near the gel point, ε can be approximated to:

$$\varepsilon = |t - t_c|/t_c . \tag{8}$$

In the case of quenched samples the relative distance to the gel point is defined by:

$$\varepsilon = |r - r_c|/r_c . \tag{9}$$

Static properties of quenched samples near gelation threshold

On quenched samples, the growth process of polymer clusters was investigated by light scattering and small-angle neutron scattering experiments and found to be well described by percolation theory. Near the gel point, the weight average molecular weight (M_w) [15], the mass distribution curve [16] and the fractal dimension of polymer clusters in the bulk state (without solvent) [17] behave respectively as:

$$M_w \sim \varepsilon^{-\gamma} \text{ with } \gamma = 1.71 \pm 0.06 \tag{10}$$

number of clusters of mass $M \sim M^{-\tau}$

$$\text{with } \tau = 2.20 \pm 0.04 \tag{11}$$

$$D_p = 2.50 \pm 0.06 , \tag{12}$$

while percolation simulations performed in three dimensions find: $\gamma = 1.74$, $\tau = 2.20$ and $D_p = 2.50$, respectively [18]. As static properties are well described by the percolation model, it is tempting to use it to describe mechanical properties.

Dynamical mechanical analysis

Dynamical rheological measurements were performed on quenched samples at 40 °C in a conventional cone-and-plate rheometer (Carrimed controlled stress rheometer) in the frequency range from 10^{-3} to 10 Hz [1]. The high-frequency part of the spectra ($\omega^* < \omega$) shows a power law dependence of both the real and imaginary parts of the modulus with an exponent u in good agreement with the value predicted by percolation theory:

$$u = 0.71 \pm 0.03 . \tag{13}$$

Moreover, a Cole-Cole representation of the elastic modulus, $G'' = f(G')$, is linear with a slope $\text{tg}(\delta_c)$ which is consistent with the previous exponent value:

$$\text{tg}(\delta_c) = \text{tg}(u\pi/2) = 1.89 \text{ leading to } u = 0.69 . \tag{14}$$

This first result is consistent with the theoretical prediction of Eq. (5). At these rheological frequencies, the observed modulus value is found [19] to be constant only in the very small ε range [5.4×10^{-4}, 7.3×10^{-3}]. The reason may be the low frequency used for these measurements. In order to increase the range of ε for which $\omega^* < \omega$, high-frequency mechanical measurements were performed on the same polyurethane samples.

An ultrasonic spectroscopy method, based on Fourier analysis of longitudinal ultrasonic pulses sent through the material [20], allows access to the longitudinal modulus M^* which is the sum of pure dilatation (K^*) and pure shear (G^*). If we assume, as it is expected, that the bulk component (K^*) has relaxed at high frequency and if velocity dispersion is neglected, the exponent value u can be deduced from the absorption spectrum:

$$a/\omega \sim M'' \sim G'' \sim \omega^u . \tag{15}$$

The exponent value u found experimentally [21] is consistent with rheological measurements, but no modulus plateau as a function of ε was observed. In the same way, ultrasonic measurements performed in the reaction bath do not show this expected modulus plateau. Recent high-frequency measurements performed on other reported systems [22] present the same behavior.

Zero shear viscosity and elasticity

Using a magnetic sphere viscometer, the zero shear viscosity was measured. Previous measurements performed in the reaction bath and reported elsewhere [10] allow us to deduce the following exponent values, independent of the temperature and of the solvent concentration:

$$s = 0.78 \text{ and } t = 3.2 . \tag{16}$$

On quenched samples prepared at different relative distances to the gel point, results are quite

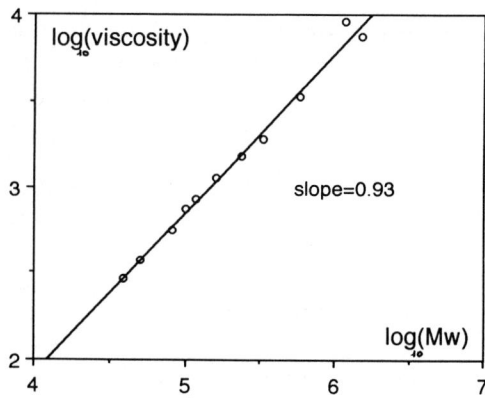

Fig. 1. Power law dependence of the viscosity versus the average molecular weight. The slope of the straight line corresponds to the "apparent exponent" s/γ and found to be 0.93 at 30 °C

different. Figure 1 shows, on a log-log scale, the evolution of the viscosity with the weight average molecular weight M_w (determined by light scattering). The slope of the straight line, equal to the ratio s/γ, is found experimentally to be 0.93 at 30 °C. Knowing the γ value, one obtains:

$$s = 1.6 \ . \tag{17}$$

In addition to this discrepancy, between the results obtained on reaction bath and quenched samples, a temperature dependence of the exponent s of quenched samples is found (see Fig. 2). Thus, we do not, properly speaking, measure an exponent, but rather an "apparent exponent". Such a behavior may be explained by a non-constant power law prefactor which would be dependent on connectivity extent. More precisely, we suspect that the frequency which describes local properties of polymers is not a constant ω_0, but rather the frequency ω_g associated with the glass transition.

In order to verify this hypothesis, preliminary dielectric measurements were performed at different temperatures in the frequency domain corresponding to the glass transition relaxation ($10^{-1} < \omega$ (Hz) $< 10^6$). Figure 3 shows the master curve thus obtained from one quenched sample. The Williams-Landel-Ferry (WLF) shift factor a_T [23], determined by superposition of different temperature spectra, allows us to determine, for this sample, the temperature dependence of ω_g. This temperature dependence accounts for the viscosity temperature behavior, as Fig. 4 shows. Such a WLF behavior for the viscosity may be expressed by the equation [23]:

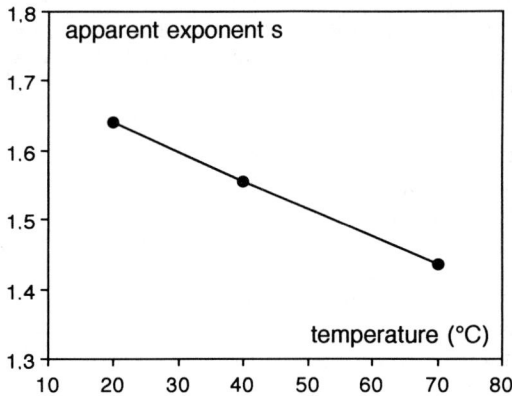

Fig. 2. Temperature dependence of the "apparent exponent" s for quenched samples

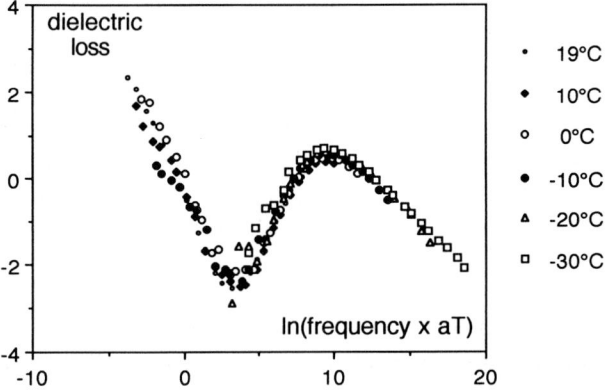

Fig. 3. Dielectric master curve for one sample at $\varepsilon = 0.01$. The WLF shift factor is obtained by superposition of the relaxation peak

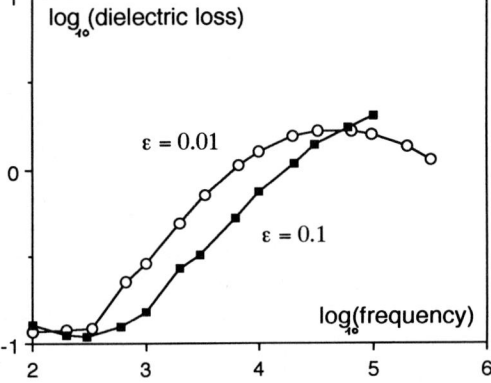

Fig. 4. Temperature dependence of the viscosity and the frequency ω_g for one sample at $\varepsilon = 0.01$. The product of the viscosity by ω_g is independent of the temperature

$$\eta = \eta_\infty e^{B/R(T-T_g)}, \quad (18)$$

where T_g is the temperature at which η becomes infinite, R is the gas constant, and B the activation energy. For the sample at the relative distance to the gelation threshold $\varepsilon = 0.01$, it is found $\eta_\infty = 7\,10^{-3}$ Poise and $B = 15$ kJ. One has to note that this temperature dependence does not account for the viscosity temperature variation of the other samples having different degrees of connectivity. This means that the parameters T_g, B and, therefore, ω_g, which describe local properties of polymers, do not remain constant with the connectivity extent. To illustrate this aspect, Fig. 5 shows dielectric spectra at 7°C for two quenched samples whose relative distances to the threshold are 10^{-1} and 10^{-2}. The frequency ω_g decreases with increasing connectivity. To first approximation[1]), the difference between the relaxation frequency ω_g of these two samples is of the order of:

$$\Delta \log_{10}(\omega_g) \cong 0.6. \quad (19)$$

Using Eq. (3), and assuming a constant value for G_∞ and replacing ω_0 by ω_g, the "apparent exponent" s can be expressed as:

$$-\Delta \log(\eta)/\Delta \log(\varepsilon) = s + \Delta \log(\omega_g)/\Delta \log(\varepsilon). \quad (20)$$

The preliminary results obtained in this study, with respect to this equation, are summarized in Table 1.

In addition to the temperature dependence of the "apparent exponent" s, Fig. 6 shows its concentration dependence. When a small amount of solvent is added to polymers, the "apparent exponent" s decreases and its temperature dependence vanishes.

---

[1]) For the sample $\varepsilon = 0.1$, the maximum of the dielectric loss was not observed experimentally but was deduced by a xy translation of the peak obtained from the sample $\varepsilon = 0.01$.

Table 1. Comparison between the "apparent exponent" s found in the bulk and the theoretical expected values. The discrepancy between the former and the percolation value may be due to the ε variation of ω_g

"apparent exponent" s $-\Delta \log(\eta)/\Delta \log(\varepsilon)$ $= s + \Delta \log(\omega_g)/\Delta \log(\varepsilon)$	exponent s theoretical value:	$\Delta \log(\omega_g)/\Delta \log(\varepsilon)$
$1.4 < \ldots < 1.7$	percolation: 0.75 Rouse: 1.33	0.6

Discussion and conclusion

The above experimental results and preliminary dielectric results are consistent with the assumption that the relevant frequency related to the local motion of the polymer corresponds to the relaxation frequency (ω_g) associated with its glass transition and is ε dependent. A determination of the exponent s, which does not take into account this fact, leads to an apparent exponent value which includes a quantity $\Delta \log(\omega_g)/\Delta \log(\varepsilon)$. This ratio decreases with increasing the temperature and with solvent

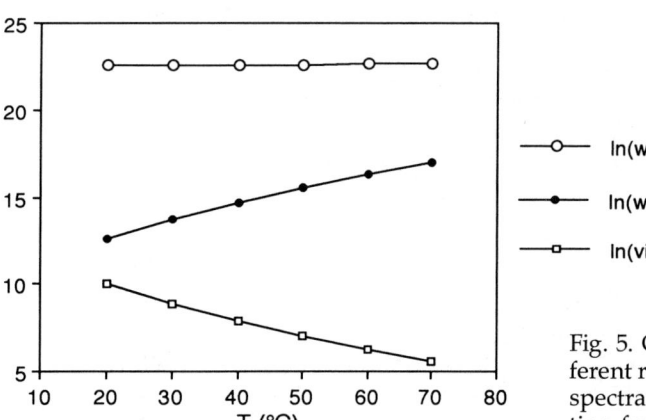

Fig. 5. Comparison of the dielectric spectra of two samples at different relative distance to the threshold ($\varepsilon = 0.1$ and $\varepsilon = 0.01$). The spectra are obtained at 7°C and the difference between the relaxation frequency ω_g is approximately 0.6 decade

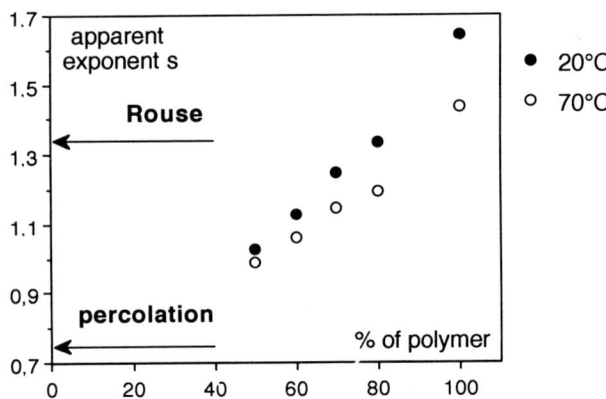

Fig. 6. Concentration dependence of the "apparent exponent s". With addition of solvent the exponent decreases and its temperature dependence vanishes

addition (ω_g tends to the solvent value whatever the relative distance to the threshold is). It is important to note that the ω_g behavior does not act on static properties of the polymeric system. The parameter which is affected by these non-constant local motion properties is the persistence time rather than the persistence length.

The "universal" value of exponent u uncountered in the literature can be explained by the fact that prefactors play no part, whatever the method of its determination:

— u can be directly measured by frequency analysis of a sample at a given relative distance to the gel point (ε), consequently, the ε dependence of the prefactor is irrelevant.
— u is the ratio $t/(s + t)$ where the s and t exponent values result from measurements using samples at different ε. The ε dependence of prefactors vanishes when exponent u is calculated by this method.

In the frequency range $\omega^* < \omega_g$ theory predicts a modulus independent of ε. The ε dependence of ω_g may explain why this behavior is never observed. With such an analysis, one may understand the apparent non-universality of the dynamical properties in the bulk, and the percolation-like behavior in presence of solvent. In this paper we focused our interest on the viscosity, but the problem of the emergence of the elastic modulus after the gel point remains to be considered using the same framework.

Acknowledgements

We thank J. P. Pascault and G. Seytre for dielectric measurements.

References

1. Durand D, Delsanti M, Adam M, Luck JM (1987) Europhys Lett 3:297
2. Adam M (1991) Makromol Chem Symp 45:1
3. de Gennes PG (1978) C R Acad Sci 286B:131
4. Rubinstein M, Colby RH, Gillmor JR (1989) Polym Preprint (Washington) 30:81
5. Martin JE, Adolf D, Wilcoxon JP (1988) Phys Rev Lett 61:2620
6. de Gennes (1980) J Physique-Colloq 41:C3
7. Clerc JP, Giraud G, Laugier JM, Luck JM (1990) Adv Physics 39:191
8. Adolf D, Martin JE, Wilcoxon JP (1990) Macromolecules 23:527
9. Colby RH, Rubinstein M (1991) Society of Rheology Meeting, October
10. Adam M, Delsanti M, Durand D (1981) Pure and Appl Chem 53:1489
11. Allain C, Salomé L (1990) Macromolecules 23:981
12. Axelos MAV, Kolb M (1990) Phys Rev Lett 64:1457
13. Adam M, Aimé JP (1991) J Physique II, October
14. Adam M, Delsanti M, Durand D (1985) Macromolecules 18:2285
15. Adam M, Delsanti M, Munch JP, Durand D (1987) J Physique 48:1809
16. Bouchaud E, Delsanti M, Adam M, Daoud M, Durand D (1986) J Physique 47:1273
17. Adam M, Lairez D, Boué F, Busnel JP, Durand D, Nicolaï T (1991) Phys Rev Lett 67:3456
18. Stauffer D, Coniglio A, Adam M (1982) Adv Polym Sci 44:103
19. Durand D, Adam M, Delsanti M, Munch JP (1988) In: Universalities in condensed matter, Jullien R, Peliti R, Rammal R, Boccara N (eds) Springer-Verlag, Berlin
20. Emery JR, Tabellout M (1990) Revue Phys Appl 25:243
21. Tabellout M (1989) PhD, Le Mans, France
22. Hodgson DF, Amis EJ (1990) Macromolecules 23:2512, (1991) J Non-Crystalline Solid:1
23. Ferry JD (1980) In: Viscoelastic properties of polymers. J. Wiley & Sons, New York

Authors' address:

Didier Lairez
SPEC-DRECAM
C.E.-Saclay
F-91191 Gif-sur-Yvette Cedex, France

On the dynamics of moderately and lightly crosslinked polymer networks

J.-U. Sommer

Fachbereich Physik, Institut Physik III, Universität Regensburg, FRG

Abstract: The relaxation behavior of networks, where the entanglement density is comparable with the crosslink density, can be explained by a broad distribution of the lengths of the network chains. A stretched exponential behavior is obtained with a fractional exponent of 1/3. — In the case of lightly crosslinked systems the influence of entanglements on the dynamics of a tagged network chain is investigated. It is assumed that the entanglement dynamics can be viewed as a cooperative stochastic process. A mode coupling ansatz is ued which yields nonuniversal scaling behavior of the characteristic relaxation time of the form: $\tau_N = \tau_0 \cdot N^\gamma$, where the exponent γ depends on the averaged entanglement distance. Moreover, the relaxation function can be represented by a stretched exponential function: $g(t) \propto e^{-(t/\tau_N)^\beta}$. The fractional exponent β also depends on the entanglement density.

Key words: Polymer networks; long-time relaxation; mode-coupling approximation

Introduction

One of the most interesting features of polymer networks is their long-time relaxation behavior. For lightly crosslinked systems, relaxation processes are observable with times of 10^5 [1—3]. In contrast to polymer solutions and polymer melts, there exist no well established models describing the dynamics of polymer networks.

Ferry [4] proposed the idea by analogy to deGennes' [5] reptation concept, that dangling ends in the network[1]) are responsible for the long-time dynamics of lightly crosslinked polymer systems. It was assumed that the dangling ends can disentangle from the network topology; thus, the effective physical crosslink density $v_{phys}(t)$ decreases with increasing time t by this process. Curro and Pincus [6] have used reptation dynamics for one star-arm [7] and derived a simple algebraic relaxation behavior for the dynamical modulus, where the exponent is proportional to the crosslink density. In their model it is not possible to eliminate the crosslink density by rescaling time and modulus [8]. Moreover, from this point of view the dynamics of the regular network chains was completely neglected.

The opposite approach was taken by Grassley [9], who investigated only the junction point motion in a tree-like structure as a generalization of the Rouse model [10] for branched structures. Entanglement restrictions were ignored; thus, the model was confined to highly crosslinked networks for short time scales.

A further approach was given by Heinrich [12], who considered a reptationlike process of junction points in a highly entangled network structure.

To distinguish between the purely network chain dynamics[2]) and the influence of entanglement restrictions, it is nessesary to define two corresponding classes of networks. As in the case of melts with low molecular weight N ($N < N_c$), we will define a class of networks with low average chain length where the crosslink density is comparable to or even greater than the entanglement density, i.e.,

---

[1]) Chains which have only one junction point to the network structure.

[2]) By analogy to the single chain dynamics of dilute polymer solutions and polymer melts with low molecular weight.

where entanglements are negligible. In this paper, we will call such networks moderately crosslinked[3]). In this case, the relaxation or retardation spectra shows only one peak or relaxation region [13].

If the crosslink density is much less than the entanglement density, we will call the network lightly crosslinked. Here, the relaxation or retardation time spectra show two characteristic peaks. The additional relaxation region is characterized by a long-time relaxation process which can be interpreted as the influence of the entanglement constraints. If we associate this long-time relaxation region with a characteristic timescale, i.e., the position of the peak maximum τ'_N, a very sensible dependence of τ'_N is observable from the averaged network chain length[4] N [2]. It can be written in the form:

$$\tau'_N \sim N^\gamma. \tag{1}$$

There are indications that γ is a nonuniversal quantity depending on the specific sample [13] with values between 3. and 12..

Moderately crosslinked systems

We assume that, in this case, the fundamental relaxation process is the Rouse dynamics of the network strands. Moreover, a major difference from polymer solutions or melts arises due to the distribution of chain lengths of the macroscopic sample.

Uncrosslinked systems are well described by a Gaussian distribution function of the chain lengths. The parameters of the distribution function are N_0, the averaged chain length, and σ, the width of the distribution. A small value of σ implies a narrow distribution and, consequently, the monodisperse value N_0 represents a physical quantity.

Let us consider a stochastical crosslink process to establish a network starting from a melt. The distribution of network chain lengths is equal to the distribution of gaps between two successive crosslink points along one tagged chain in the melt. Equation (2) displays a good approximation for this distribution:

$$p(N) = p_0 e^{-p_0 N}, \tag{2}$$

where p_0 is the probability to crosslink one segment along the melt chain. The averaged chain lentgh is given by: $\langle N \rangle = p_0^{-1}$. For such a distribution there exists no consistent way of taking only the monodisperse ensemble expansion. Moreover, there is no central or "characteristic" value of the distribution. In the case of long-time relaxation, we have to take the a posteriori average as follows:

$$g(t, p_0) = \sum_{N=1}^{N_0} e^{-p_0 N} p_0 g(t, N), \tag{3}$$

where $g(t, N)$ is the reduced modulus of one network chain of length N and $g(t, p_0)$ is the reduced modulus of an ensemble of network chains with the crosslink density p_0. The reduced modulus is defined as: $g(t) = G(t) - G(\infty)$. N_0 describes the length of the chains in the melt. For times comparable and even greater than the Rouse-time[4]) $t_R = N^2$, $g(t, N)$ is well described by the expression [14]:

$$g(t, N) \simeq (4\pi t)^{1/2} e^{-t/N^2}, \tag{4}$$

$g(t, p_0)$ can now be obtained from (3) by a saddle-point estimation to the first order. The result is [15]:

$$g(t, p_0) \simeq p_0^{1/3} t^{-1/3} 2^{1/6} 6^{-1/2} e^{-C(tp_0^2)^{1/3}}, \tag{5}$$

where C is a numerical constant of order 1.

For long times the distribution of chain lengths yields a stretched exponential behavior with an universal exponent of 1/3. Note that a single chain (or monodisperse ensemble) obeys a simple exponential behaviour for long times (see Eq. (4)).

Lightly crosslinked systems

The dynamics of networks changes dramatically when the crosslink density becomes smaller then a critical value[5]) v_c. The second relaxation region mentioned in section 1 comes into play, connected with a strong modification of long-time relaxation functions [13]. Between the two relaxation regions the so-called entanglement plateau is observable [4]. It is the author's assumption that the long-time relaxation process is now governed by cooperative motions of the entanglements.

---

[3]) Here is a difference from the more chemically motivated definition for the term "moderately crosslinked".

[4]) Here we use an appropriate microscopic time scale.
[5]) v_c is not connected with the connectivity behaviour of the network. It is defined only by aspects of network dynamics.

The idea of our model is as follows: consider a given chain segment within an entanglement constraint. Obviously its dynamics must be strongly correlated to the dynamics of the other segments belonging to this constraint during times much longer than the microscopic fluctuation time of a nonentangled segment. Without going into details of the interaction of the segments within an entanglement, we assume that these constraints produce a correlated stochastic force environment for the tagged segment. Hence, we distinguish between correlated $\xi^{(2)}$ and uncorrelated $\xi^{(1)}$ stochstic forces acting on the chain segments.

The corresponding equation of motion now reads:

$$\mu \dot{r}_i(t) + h \int_0^t \phi_{ij}(tt') \dot{r}_j(t') dt' + k M_{ij} r_j$$
$$= \xi_1^{(1)}(t) + \xi_i^{(2)}(t) , \qquad (6)$$

where:

$$\frac{h}{M_0 \mu} \phi_{ij}(tt') = \langle \xi_i^{(2)}(t) \xi_j^{(2)}(t') \rangle \delta_{ij} . \qquad (7)$$

M_0 is given by the ratio k/μ and represents the inverse microscopic time constant.

We now use a mode-coupling ansaatz [16] of the form:

$$\phi(t) = \lambda_2 g^2(t) + \lambda_3 g^3(t) + \ldots . \qquad (8)$$

The motivation for (8) is given in [16]. The main point is the approximation of many particle correlators appearing in (7) by simple products of the one particle correlator of the relevant observable. $g(t)$ corresponds here to the relevant one-particle correlator.

A solution of (6) and (8) can be given by using a slightly modified one-chain dynamics [14]. For the time-scale of interest this dynamics is in good agreement with the usual Rouse model. An expression for the relaxation function:

$$g(z) = \int_0^\infty e^{-zt} g(t) dt \stackrel{\text{def}}{=} LT[g(t)](z) \qquad (9)$$

can be given in the form:

$$g(z) = \cfrac{1}{z + \cfrac{1}{1 + g(z)(1 - \varepsilon) + \lambda_2 LT[g^2(t)](z) + \ldots}} , \qquad (10)$$

with $\varepsilon = 2/N$ and N is the averaged chain length of the network strands. Equation (10) possesses just the form of the usual mode-coupling equation investigated by Goetze [16]. Because of the strong nonlinearities, Eq. (10) cannot be solved exactly. In principle at least, a numerical solution can be achieved.

Nevertheless, an interesting scaling solution is obtained around some plateau value f_c. In this scaling region all contributions of λ_i become irrelevant for $i \geq 3$ [16]. A small parameter δ determines the effective timescale as follows:

$$g(t) \geq f_c: \quad \tau_N = |\delta|^{-\frac{1}{2a}} \qquad (11)$$

$$g(t) \leq f_c: \quad \tau'_N = |\delta|^{-\frac{1}{2a} - \frac{1}{2b}} = |\delta|^\gamma , \qquad (12)$$

with:

$$\delta = -\frac{2}{N}\left(1 - \frac{N}{2N_c^2}\right) . \qquad (13)$$

The following definition is assumed here:

$$f_c \stackrel{\text{def}}{=} \frac{1}{2 N_c} . \qquad (14)$$

The exponents a and b are related to the plateau f_c:

$$\frac{\Gamma^2(1-a)}{\Gamma(1-2a)} = \lambda = 1 - f_c = \frac{\Gamma^2(1+b)}{\Gamma(1+2b)} . \qquad (15)$$

The parameter $\varepsilon = 2/N$ is already fixed and the only assumption we have to make is to take f_c as the entanglement plateau of the network. Equation (14) gives the definition of the corresponding entanglement spacing. It is noteworthy that the factor 1/2 in (14) is arbitrary[6]. With this single assumption, it is possible to completely describe the dynamic around the plateau:

on scale τ_N: $\quad g(t) = f_c + A t^{-a} \qquad (16)$

on scale τ'_N: $\quad g(t) = f_c - B t^b . \qquad (17)$

Von Schweidler's law is displayed by (17). For times even greater than τ'_N a good fit exists to a stretched exponential behavior of the form [16]:

$$g(t > \tau'_N) \sim e^{-\left(\frac{t}{\tau'_N}\right)^\beta} . \qquad (18)$$

---
[6]) It corresponds to the case of free fluctuating knots.

The exponents a, b, γ and β are now related to the plateau as follows:

$$a \simeq b \simeq \left(\frac{6}{\pi^2} f_c\right)^{\frac{1}{2}} \sim N_e^{-\frac{1}{2}} \qquad (19)$$

$$\gamma \simeq \frac{1}{a} \simeq \frac{1}{b} \simeq \left(\frac{\pi^2}{3} N_e\right)^{\frac{1}{2}} \sim N_e^{\frac{1}{2}} \qquad (20)$$

and

$$\beta \simeq \frac{\ln(2)}{\ln(2N_e)} . \qquad (21)$$

Characterizing the long-time relaxation region a nonuniversal exponent γ is predicted which scales with the square root of the entanglement spacing. The (often used) KWW-fit (18) also has a nonuniversal exponent β which scales inversely to the logarithm of the entanglement spacing.

In summary, it can be noted that the mode-coupling approximation (8) recovers the main features which are experimentally observable. Moreover, it predicts a significant dependence of the characteristic exponents on the entanglement plateau value and thus gives a hint for further dynamical experiments on lightly crosslinked rubbers.

References

1. Chasset R, Thirion P (1965) Proceedings of the Conference on Physics of Non-Crystalline Solids, Ed. by J. A. Prins, North Holland Publishers Co.
2. Plazek DJ (1966) J Polym Sci A4:745
3. Sperling LH, Tobolsky AV (1968) J Polym Sci A6:259
4. Ferry JD (1970) Viscoelastic Properties of Polymers, 2nd Ed., J Wiley
5. deGennes PG (1971) J Chem Phys 55:572
6. Curro JG, Pincus P (1982) Macromolecules 16:559
7. Doi M, Edwards SF (1989) The Theory of Polymer Dynamics, Clarendon Press, Oxford
8. McKenna GB, Gaylord R (1988) Polymer 29:2027
9. Grassley G (1980) Macromolecules 13:372
10. Rouse PE (1953) J Chem Phys 21:1272
11. Heinrich G, Vilgis T (1992) Macromolecules 25:404
12. Havranek A (1988) Progr Trents Rheology II:203
13. Sommer J-U (in preparation)
14. Sommer J-U (1991) J Chem Phys 95:1316
15. Goetze W (1989) Bifurcation senarios for glass transition In: Conference Proceedings of the "V[th] International Symposium on Selected Topics in Satistical Mechanics", Dubna USSR
    Goetze W (1986) Dynamical Phenomena near the Liquid Glass Transition In: Proceedings of the 6th Conference on "Liquid and Amorphous Materials", Passodella Mendola, Martin Nijhoff, Dortrecht

Author's address:

Jens-Uwe Sommer
Fachbereich Physik
Institut Physik III
Universität Regensburg
D-W-8400 Regensburg, FRG

A new approach to polymer networks including finite chain extensibility, topological constraints, and constraints of overall orientation

G. Heinrich and W. Beckert[1]

Continental AG, Materials Research, Hannover, FRG
[1]) Institut für Technologie der Polymere, Dresden, FRG

Abstract: A new statistical theory of polymer networks is presented. In contrast to the Gaussian chain model, the idea of the theory is to include the effects of limited chain extensibility, stiffness, configurational constraints of each network strand, and constraints of overall orientation. The theory uses the Ronca/Yoon worm-like chain model in spherical approximation. The configurational constraints are modeled by virtual tubes. The partition function of the network is formulated in single-chain approximation. The model describes, in a consistent manner, the fact that, at large extensions, the entropy decreases, and when the polymer is stretched out there is no further configuration left which the polymer can occupy. The elastic free energy and the stress-strain relationship are derived. The results show, for example, the typical upturn behavior in the Mooney-plot. The location of the upturn depends on the length of the network chains. Furthermore, it is shown how the action of configurational constraints becomes weaker in the case of predominance of shorter network strands and finite chain extensibility. The extension of the theory to the effects of constraints of overall orientation is discussed.

Key words: Networks; finite chain extensibility; worm-like chains; topological constraints; constraints of overall orientation

Introduction

The phantom network models are unsatisfactory in that they fail to take into account the entanglements and the limited chain extensibility (LCE). The LCE depends on the chain flexibility and chain length.

The network model developed in this paper uses the RONCA/YOON-worm-like chains in spherical approximation which makes the chain mathematically more tractable without violating the requirement of finite extensibility [1]. This worm-like chain model (WLCM) closely represents the short-range statistics of real chains and allows the evaluation of the orientation-dependent configurational partition function. The topological constraints are modeled by configurational tubes which describe — in a mean field manner — the effects of entanglements as well as configurational packing effects [2]. The configurational part of the partition function of the whole network is formulated in single-chain approximation.

We derive the elatic free energy and the stress-strain relationship of the networks. The theory predicts in a natural way the typical "upturn" behavior in the Mooney-plot of the stress-strain curve. The exact location of the upturn depends on the length of the network chains. Furthermore, it will be shown how the action of configurational constraints becomes weaker in the case of predominance of shorter network strands and finite chain extensibility.

The case of LCE and overall orientational constraints is discussed shortly, and will be investigated in a separate paper.

The worm-like chain model with limited chain extensibility and topological constraints

The configurational probability density function of a free worm-like chain with contour length L is given by [1]

$$p[r(s)] = N^* \exp\left\{-f/(2k_BT) \int_0^L \left(\frac{\partial t(s)}{\partial s}\right)^2 ds\right\} \quad (1)$$

(N^* — normalization),

where f is the bending force constant. The geometrical constraints of real chains associated with the existence of a maximum limiting curvature may be incorporated by developing the tangent vector $t(s) = \partial r(s)/\partial s$ in a Fourier series truncated at an arbitrary cut-off wavelength L_c, i.e.,

$$t(s) = a_0 + \sum_{m=1}^{m_c=L/L_c} A_m \cos(m\pi s/L), \quad (2)$$

where a_0 and A_m are random variables. In Eq. (2), π/L_c represents the maximum allowed curvature, independent of temperature and chain length. The requirement of finite extensibility is expressed by the condition (spherical approximation)

$$\int_0^L t^2(s)ds = L, \quad (3)$$

which replaces the unitary condition ($|t(s)| = 1$) of the classical worm-like chain ($L_c = 0$).

The polymer network is a system of long chains, where neighboring molecules permeate regions occupied by other chains in the system. Overlapping chains physically entangle with each other and restrict the relative motion of large segments of neighboring chains. Therefore, we now consider the statistics of confined worm-like chains. The confinement is described using the picture of harmonic-like topological constraints. Physically, the picture rests on the assumption that entanglements do not act at specific locations along the chains. Rather, neighboring chains are thought to form a tube that restricts lateral movement to the neighborhood of a reference (tube-axis) configuration $R(s)$.

The configurational probability density function of a constrained WLCM with LCE is given by

$$p^*[r(s) \mid R(s)] = N^* \exp\left\{-\frac{f}{2k_BT} \int_0^L ds \left(\frac{\partial t(s)}{\partial s}\right)^2 \right.$$
$$\left. - w_0^2 \int_0^L ds(r(s) - R(s))^2\right\}, \quad (4)$$

where $w_0^2 = l/d_0^4$ denotes the strength of the configurational tube constraints [2]. The mean lateral tube dimension is denoted by d_0 and l is the length of the basic chain units. The probability density of the reference configuration $R(s)$ is given by Eq. (1) when replacing $t(s)$ by $u(s) = \partial R(s)/\partial s$.

From Eqs. (1)—(4), the partition function for a given value of the end-to-end chain vector,

$$r = \int_0^L t(s)ds = La_0,$$

and a given configuration $R(s)$ is expressed as

$$Z(r, \{B\}) = \int_{-\infty}^{+\infty} \prod_m dA_m \exp\left\{-\frac{f\pi^2}{4Lk_BT}\sum_m m^2 A_m^2\right.$$
$$\left. - w_0^2 \frac{L^3}{2(2\pi)^3} \sum_m \frac{1}{m^2}(A_m - B_m)^2\right\}$$
$$\delta(\sum A_m^2 - 2(1 - r^2/L^2)), \quad (5)$$

where

$$u(s) = b_0 + \sum_m B_m \cos(m\pi s/L), \quad a_0 = b_0. \quad (5a)$$

The δ-function expresses the condition in Eq. (3).

After Fourier representation of the δ-function and integration over the Fourier coefficients A_m, we obtain

$$Z(r, \{B\}) = \frac{1}{2}\left(2\pi \frac{LT}{L_c T^*}\right)^{3/2} m_c^{-1} a^{-1}$$
$$\cdot \int d\omega e^{-i\omega}\left[\prod_m (m^2 - i\omega/a)^{-3/2}\right]$$
$$\cdot \left[\prod_m \left(1 + \frac{w_0 L^3}{2(2\pi)^3 c^2 m^2 M}\right)^{-3/2}\right]$$
$$\cdot \exp\left\{-Mc^2\left[1 - \frac{M}{M + \frac{w_0^2 L^3}{2(2\pi)^3 c^2 m^2}}\right]B_m^2\right\} \quad (6)$$

with

$$k_B T^* = \pi^2 f / 2L_c, \quad a = 2c^2(1 - r^2/L^2),$$

$$M = m^2 - \frac{i\omega}{2c^2} \frac{1}{1 - r^2/L^2}, \quad c^2 = \frac{L_c T^*}{2TL}.$$

In the high temperature limit $T \gg T^*$, the dominant contribution to the integration over ω is given at very large values of the argument. In this limit the partition function becomes

$$Z(r, \{B\}) \approx \frac{1}{2} \left(2\pi \frac{LT}{L_c T^*}\right)^{3/2 \cdot m_c - 1} \cdot a^{3/2 m_c - 1}$$

$$\cdot \int_{-\infty - i\varepsilon}^{+\infty + i\varepsilon} d\omega \, e^{-i\omega} (-i\omega)^{-\frac{3}{2} m_c}$$

$$\cdot \left[1 - \frac{1}{32\pi} \frac{w_0^2 L^3 (1 - a_0^2)}{-i\omega}\right]$$

$$\cdot \exp\left\{-\frac{w_0^2 L^3}{2(2\pi)^3} \prod_m \frac{1}{m^2} B_m^2\right\}. \quad (7)$$

The network model

Our network model rests on the following physical picture: Before crosslinking, the polymer melt consists of free fluctuating chains which are subjected to the θ-theorem of Flory [3]. The permanent crosslinking points reduce the configurational degrees of freedom and the system undergoes a transition from an annealed to a quenched state. Actual constrained configurations $r(s)$ of the network chains are located to the vicinity of the reference configuration, $R(s)$, immediately after a sudden crosslinking reaction. Therefore, we have to perform quenched averages under the condition of a frozen configuration $R(s)$. The problem in calculating the free energy is then the averaging over a random structure $\{B\}$. The most advanced way to perform the averages is the replica trick, well described and applied to conventional rubbers in the paper of Deam and Edwards [4]. We instead use the classical and simple way of a three-chain network, and start from the free energy per chain with end-to-end distance r:

$$F(\underline{\lambda} r) = \int p(\{B\}) F^*(\{B\}, \underline{\lambda} r; w_{0i}(\lambda_i)) d\{B\}, \quad (8)$$

$(i = 1, 2, 3)$

with

$$F^*(\{B\}, r) = -k_B T \log Z(r, \{B\}), \quad (9)$$

and

$$p(\{B\}) = \prod_m p_m = \prod_m (b_m/\pi)^{3/2} \exp(-b_m B_m^2). \quad (10)$$

Equation (8) contains the assumption of pure affine transformation of the chain's end-to-end distance: $r \to \underline{\lambda} r$. Generalization to the case of free fluctuating crosslinks is not a major problem (in this case, only the mean end-to-end distance transforms in an affine manner). The deformation dependence of the lateral tube dimension d_0 is considered according to [2] and is given as follows: $d_i = d_0 \lambda_i^{1/2}$.

The algebra, which will be presented elsewhere in more details, started with a Fourier representation of the δ-function (Eq. (5)) and integration over the coefficients A_m. We note that the integrand has singularities only in the half-plane $\text{Im}\,\omega < 0$. Therefore, the free energy has a singularity for $r > L$ by the Cauchy theorem. Hence, the model always satisfies the condition of finite chain extensibility.

The normalized stress-strain relationship (Mooney-plot) for an incompressible ($\lambda_x = \lambda$, $\lambda_y = \lambda_z = \lambda^{-1/2}$) network consisting of v_c network chains per unit volume is given as follows:

$$\Phi \equiv \frac{\sigma(\lambda)}{v_c k_B T (\lambda - \lambda^{-2})} = \Phi_c + \Phi_t, \quad (11)$$

where $\sigma(\lambda) = 1/V \cdot dF(\lambda)/d\lambda$, and

$$\Phi_c \approx \frac{1 - N^{-1}((\lambda - \lambda^{-3})/(\lambda - \lambda^{-2}))}{1 - N^{-1}(\lambda^2 + 2/\lambda) + N^{-2}(2\lambda + \lambda^{-2}) + N^{-3}}, \quad (12)$$

$$\Phi_t \approx a[(1 - \lambda^{-3/2}) - 3N^{-1}(\lambda^2 - \lambda^{-5/2}) + b(\lambda^2 - \lambda^{-5/2})]/(\lambda - \lambda^{-2}). \quad (13)$$

The quantities a, b take the form

$$a \approx (lL/d_0^2)N/3, \quad b \approx NT/T^*, \quad (14)$$

with $k_B T^* = \pi^2 f / 2L_c$. For $b = 0$, the case of the classical worm-like chain with $L_c = 0$ and $T d\ln C_\infty / dT = -1$ follows. C_∞ is the characteristic ratio r^2/Nl^2 for chains comprising N bonds of length l. The limiting case $a = 0$, $N \to \infty$ yields the classical phantom network results.

Figure 1 shows some selected results. It is seen how the location of the upturn depends on the

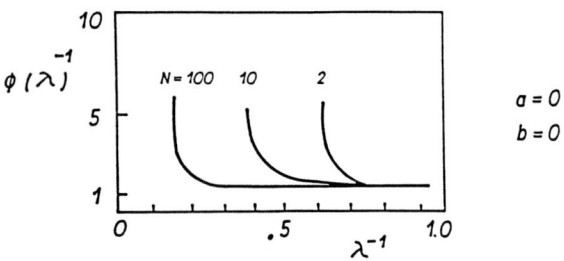

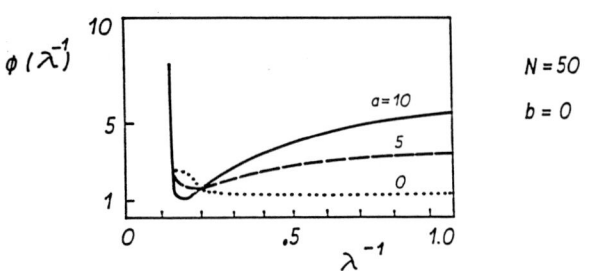

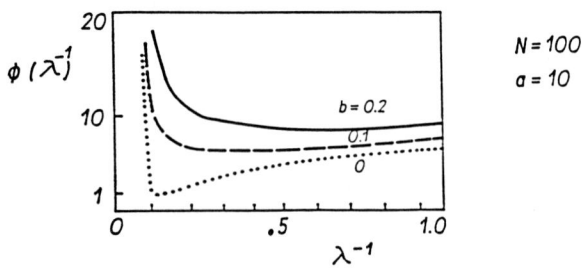

Fig. 1. Dependence of the Mooney-stress on the parameters of the chains: N — number of the basic units per chain; a — strength of the configurational constraints (Eq. (14)); b — stiffness parameter (Eq. (14)), the case $b = 0$ corresponds to the classical worm-like chain

length of the network chains. Furthermore, it is shown how the action of configurational constraints becomes weaker in the case of predominance of shorter network strands and finite chain extensibility. Similar conclusions were obtained by Kovac and Crabb [5] using modified Gaussian chain statistics and following the three-hard-tube approach of Gaylord [9].

Limited chain extensibility and constraints of overall orientation

So far, we have only considered chains without any interactions other than the entanglements. The worm-like chains allow the evaluation of the orientation-dependent configurational partition function in a natural way [1]. It is known that nematic effects always reduce the free energy of deformation [6]. In this case, the many chains partition function contains a δ-function which expresses the condition of overall constraints [1], $\gamma = \langle \sin^2 \psi \rangle$:

$$\delta \left(\sum_k \int_0^L (t_{kx}^2(s_k) + t_{ky}^2(s_k)) \, ds_k - N_c L \gamma \right). \quad (15)$$

N_c denotes the number of polymer chains (k) forming states of different overall orientation, and ψ is the angle between the tangent vector t and the orientation axis of the sample.

Solutions of the problem can be obtained after some approximations [7]:

i) the many-chain partition function is reduced to an effective one-chain partition function by Fourier representation of the δ-function;
ii) the deformation dependence of the end-to-end vectors is assumed to be affine.

Then, analytical expressions of the free energy can be derived in the two cases of:

i) not too short chains using steepest descent approximation ($m_c \gg 1$);
ii) weak stiffness operating in the high temperature limit ($T \to \infty$).

The calculations and results will be presented in a separate paper [7]. We note that some derived stress-strain relations agree with corresponding computer simulation results of short-chain polyethylene networks [8].

Conclusions

The network thoery presented in this paper incorporates the effects of finite-chain extensibility, stiffness, topological constraints, and constraints of overall orientation. Although analytical solutions are found only in some limiting cases, the theory more closely represents the real chain properties in comparison to the Gaussian phantom networks. Clearly, the single-chain approximation is an oversimplification of the statistical mechanics of polymer networks. Nevertheless, the model may be used as a route to an improved statistical mechanics of real networks with quenched disorder.

References

1. Ronca G, Yoon DY (1982) J Chem Phys 76:3295
2. Heinrich G, Straube E, Helmis G (1988) Adv Pol Sci 85:33
3. Flory PJ (1953) Principles of Polymer Chemistry. Ithaca, New York, Cornell Univ Press
4. Deam RT, Edwards SF (1976) Phil Trans Roy Soc London A 280:317
5. Kovac J, Crabb CC (1986) Macromolecules 19:1744
6. Jarry JP, Monnerie L (1979) Macromolecules 12:316
7. Heinrich G, Beckert W, in preparation
8. Mark JE, Curro JG (1983) J Chem Phys 79:5705
9. Gaylord RJ (1982) Polym Bull 8:325

Authors' address:

Dr. G. Heinrich
Continental AG
Materials Research
Postfach 169
D-3000 Hannover 1, FRG

Formation of networks — a lattice model for kinetic growth processes

M. Schulz

Fachbereich Physik, Technische Hochschule Merseburg, FRG

Abstract: Starting from a microscopical master equation for simple kinetic processes on a lattice (diffusion, aggregation, decomposition), an exact field theoretical representation for the gelation process was predicted. As a result of the renormalization group theory, one gets the critical dimension $d_c = 6$ and the same universality as in the random percolation for the diffusion limited gelation process and new universality class (also with $d_c = 6$) for the reaction limited gelation.

Key words: Gelation; percolation theory; renormalization; group approximation

Introduction

The structure of networks is determined by the local kinetic (diffusion, aggregation, decomposition) of the growth process. The different methods for a theoretical description have one central problem, the connection between time evolution and space correlations. A special question is the critical behavior near the sol-gel transition. Here, local effects are irrelevant and only the long-range correlations determine the critical region. Therefore, it is reasonable to use a lattice theory for the following calculations. The advantages of such a mathematical description are the simultaneously numerical and analytical treatment of the problem and relative simple basic equations. This paper gives only an analytical description of the gelation problem. The simple lattice theory for describing the sol-gel transition is the well known percolation theory [5]. In the bond-percolation version of this theory all lattices sizes are randomly occupied with a given probability p. In the course of the kinetic growth theory p is a function of the time $p = p(t)$, which follows, for instance, from effective kinetic equations. Near the sol-gel transition the linear relation $p = p_c + a\,|t - t_c|$ reflect the identity between time and probability $|p - p_c| \sim |t - t_c|$ as parameter of the growing system. The critical point (p_c, t_c) is nonuniversal (it is determined from the lattice structure, kinetic constants, etc.), but the average gyration radius, the degree of polymerization, and the gel fraction behave as universal scaling laws

$$\xi \sim |p - p_c|^{-\nu} \sim |t - t_c|^{-\nu} \tag{1}$$

$$M^{(2)} \sim |p - p_c|^{-\gamma} \sim |t - t_c|^{-\gamma} \tag{2}$$

$$P_{gel} \sim |p - p_c|^{\beta} \sim |t - t_c|^{\beta}. \tag{3}$$

A restriction of the percolation theory is the neglect of the structure (the bonds on different lattice sides i, j are noncorrelated $\langle b_i b_j \rangle = \delta_{ij}$ and all lattice sites are occupied by monomers a priori) as a result of the chemical kinetic.

Model

A more realistic model, which contains kinetic and diffusion effects, started from a randomly distributed configuration of monomers $\{m_i\}$, where m_i is the number of monomers at the lattice site i. The monomers realized jumps between neighbored lattice sites (diffusion), and they build clusters as a result of collisions with other monomers or clusters. The cluster configuration is given by $\{c_i\}$ (number of cluster elements at the lattice site i). The formation of bonds is connected with the kinetic of the monomer-cluster configuration, but not in an unique manner. The degrees of freedom for the bond formation are determined by the aggregation reaction between monomers and clusters and also by the formation of bonds between clusters (cluster-cluster aggregation) and in the interior of a cluster (cyclization effect). The last two effects are possible only in the case that the two lattice sites, which are neighbored by a bond, are occupied by cluster-elements ($c_i \neq 0$). Formally, this reaction kinetic can be described by a probability distribution $P(c,b,t)$, which characterized the dynamic of the lattice configurations. The solution of the full master-equation for $P(c,b,t)$ is possible by using reasonable approximations. Writing $P(c,b,t)$ as a product of the probability $P(c,t)$ and the conditional probability $P(c,t \mid b,t)$, it is possible to use the following approximations:

- The probability $P(c,t)$ is a measure for the configuration $c = \{c_i\}$ of cluster elements, which is determined by aggregation and diffusion processes of monomers and cluster elements.
- The probability $P(c,t \mid b,t)$ is the measure of the bond configuration by fixing the configuration of the cluster elements c. This probability can be approximated using the following assumptions:
  - $P(c,t \mid b,t)$ vanishes for all configurations b in which one or more bonds are neighbored to a lattice site with $c_i = 0$.
  - $P(c,t \mid b,t)$ has an essential statistical weight only for a small region around the average number of bonds $\bar{N}_b$.
  - all configurations for a fixed number of bonds have the same statistical measure. This is a reasonable approximation near the sol-gel transition, because the detailed structure of the bonds can be neglected.

With the bond probability $p_b(t)$ (averaged ratio between all occupied bonds and all allowed bonds), it follows that

$$P(c,t \mid b,t) = \Theta(c,b) p_b(t)^{N_b}(1 - p_b(t))^{N_c - N_b},$$

where $\Theta(c,b) = 0$ if a bond of the configuration $b = \{b_i\}$ is not allowed, and N_c is the number of all possible bonds for a fixed configuration c.

That means that the average of a statistical value $X(c,b)$ is a quenched average of a value $\bar{X}(c,t)$ with the distribution function $P(c,t)$ and

$$\bar{X}(c,t) = \text{tr}_b\, P(c,t \mid b,t) X(c,b).$$

Using the approximation for $P(c,t \mid b,t)$, we get, instead of (4),

$$\bar{X}(c,t) = \frac{1}{Z} \sum_{\{\omega\}} e^{-H} X(b,c,t), \qquad (5)$$

with the Hamiltonian [1] for the well known Potts-model [11] on a dilute lattice [2]:

$$H = -J(t) \sum_{\langle ij \rangle} c_i c_j [\delta_{\sigma_i \sigma_j} - 1] + \omega \sum_i [\delta_{\omega_i 1} - 1]. \quad (6)$$

In this representation $\sigma_i = 1, 2, \ldots, s$ is the state of a spin at the lattice site i and $\langle ij \rangle$ means all neighbored pairs of lattice sites. The kinetic of the bond formation determines the interaction constant $J(t)$ because of the connection $e^{-J(t)} = 1 - p_b(t)$. Hence, the determination of the scaling behavior near the sol-gel transition decays in two separate problems:

— determination of the time-dependent probability $P(c,t)$ or relevant functions, which describe the averaged structure of the configurations c as a function of time, diffusion and chemical kinetic coefficients
— solution of the quenched averaged percolation problem near the sol-gel transition.

Time-dependent static structure factor

From the following calculations it results that a central point of the quenched average percolation problem is the knowledge of the time-dependent static structure factor (the Fourier-transformed one-time density correlation function). The analytical investigations start from a simple kinetic model which contains all important properties of the gelation process. Note that, in principle, there exist enough possibilities for a numerical (see, for example, the cluster-cluster-aggregation [7, 8]) and also analytical enlargements. The reaction scheme is based on monomers which can jump from a lattice site to a neighbored point and motionless cluster elements. These fixed clusters approximately realized the situation that the real diffusion coefficient for clusters with m elements is proportional to m^{-1}, e.g., clusters with a high mass are nearly fixed. To consider the excluded volume effect a virtual correcture reaction (decomposition of cluster-elements at lattice sites with $c_i \geq 2$) is useful. Using the reaction scheme:

- diffusion $m_i \to m_j$ $D_0 D_{ij}$
- dimer formation $m_i + m_j \to c_i + c_j$ $K_0 D_{ij}$
- cluster growth $m_i + c_j \to c_i + c_j$ $K_0 D_{ij}$
- correction $2c_i \to m_i + c_i$ g_0.

($D_{ij} = 1$ if i and j neighbored, otherwise $D_{ij} = 0$), we get a one step master equation [6] for the probability distribution $P(m, c, t)$. This equation with discrete variables $\{m_i\}$ and $\{c_i\}$ was transformed in a general Focker-Planck equation by using the Poisson-transformation

$$P(m,c,t) = \int_{\tilde{G}} d^2 a \prod_i \left[\frac{(a_i^1)^{m_i} e^{-a_i^1}}{m_i!} \frac{(a_i^2)^{c_i} e^{-a_i^2}}{c_i!} \right] f(a,t) . \quad (7)$$

The integration region $\tilde{G}$ is the full complex plane. (In this case, it is possible to construct a Focker-Planck equation with a positive definite kernel in the second order differential part). The function $f(a,t)$ has the same Markovian character as the original probability $P(m,c,t)$, but it is not necessarily a positive function. As a result of this Poisson transformation the evolution equation for the pseudo-probability $f(a,t)$ becomes

$$\frac{\partial}{\partial t} f(a,t) = [\partial_i^a \partial_j^b m_{ijkl}^{abcd} a_k^c a_l^d - \partial_i^a A_i^a(a)] f(a,t) . \quad (8)$$

Equation (8) is the field theoretical representation of the molecular lattice dynamics. The coefficients m_{ijkl}^{abcd} are functions of the kinetic coefficients K_0 and g_0 only, the functions $A_i^a(a)$ are second-order polynoms of the field a. From (8) it is possible to calculate the free propagator $G_{ij}^0(t - t')$ as the linear response function around the stationary mean field solution. Using this free propagator, we obtain the perturbation theory in a straightforward manner. The construction of a polynomial perturbation theory (which can be present in a Feynman diagram series) is the decisive advantage of the Poisson representation. The mean field representation is a sufficient solution of the Focker-Planck Eq. (8) for the problem of percolation. Using the connection identity,

$$\langle c_i c_j \rangle = \langle a_i^2 a_j^2 \rangle + \delta_{ij} \langle a_i^2 \rangle , \quad (9)$$

it follows for the static structure factor at the time t that

$$S(q) = c_1(t) + \frac{c_2(t)}{M_0 + q^2} . \quad (10)$$

Here, the "mass" M_0 is the parameter of the system, which determined the character of the static structure factor. This "mass" is given by the ratio $M_0 \sim K_0 D_0^{-1}$, e.g., for the case of a reaction limited network formation process it follows $M_0 \to 0$ and, therefore, $S(q) = c_1(t) + c_2(t) q^{-2}$, otherwise, in the case of diffusion-limited gelation it is $S(q) \sim$ const.

Quenched averaged percolation problem

The main idea for the calculation of the quenched averaged percolation problem is the identity of the general Potts-model and the percolation problem in the limit $s \to 1$, where s is the number of spin states. We start our consideration from a quenched site dilute lattice, for which the lattice sites are randomly occupied with Potts spins. The Potts Hamiltonian for this quenched site-diluted model is given by (15). From this Hamiltonian follows (in a well known way [1]) the characteristic values of the percolation theory which are, in the case of a dilute lattice, a functional of the lattice structure. After averaging over the actual structure c, we get for the average number of clusters N_{Cl}, the average cluster mass P_{gel} and the weight-average molecular weight $M^{(2)}$ (all values per unit volume):

$$N_{Cl} = \langle F(c,\omega) \rangle_c |_{\omega=0} + \bar{c} - 1 \quad (11)$$

$$P_{gel} = \frac{\partial}{\partial \omega} \langle F(c,\omega) \rangle_c |_{\omega=0} + \bar{c} - 1 \quad (12)$$

$$M^{(2)} = \frac{\partial^2}{\partial \omega^2} \langle F(c,\omega) \rangle_c |_{\omega=0} + \bar{c} - 1 , \quad (13)$$

with the structure dependent free energy:

$$F(c,\omega) = \lim_{N \to \infty} \frac{1}{N} \frac{\partial}{\partial s} \ln \sum_{\{\sigma\}} e^{-H} \quad (14)$$

($\bar{c}$: concentration of occupied lattice sites).

Analogously, the quenched averaged pair connectedness P_{ij} can be obtained by introducing an inhomogeneous field ω_i at each site i:

$$P_{ij} = \frac{\partial^2}{\partial \omega_i \partial \omega_j} \langle F(c,\omega) \rangle_c |_{\omega=0} . \quad (15)$$

Writing the Potts spins states in (4) as a set of s vectors e_a^a in a $s-1$ dimensional space [3, 4] ($a = 1, ..., s$ and $a = 1, ..., s-1$), we have

$$e_a^\alpha e_a^\beta = s\delta^{\alpha\beta} - 1,$$

which reflect the full symmetry of the model; the partition function Z becomes a functional integral with the new continuous field variables Φ_i^a ($a = 1, ..., s-1$, $i = 1, ..., N$ lattice site, $\omega = 0$):

$$Z[c] = \int D\Phi$$

$$\cdot \exp\left\{-\frac{1}{2}\sum_{ij,a}\Phi_i^a M_{ij}\Phi_j^a + \sum_{i,a} V(c_i\Phi_i^a)\right\}, \quad (16)$$

and the structure dependent potential:

$$V(c_i\Phi_i^a) = \ln\sum_{\{a\}} \exp\{c_i\Phi_i^a e_a^a\}$$

$$= c_i \frac{1}{2}\Phi_i^2 - c_i \frac{1}{6} Q_{abc}\Phi_i^a\Phi_i^b\Phi_i^c$$

$$- c_i \frac{1}{24} T_{abcd}\Phi_i^a\Phi_i^b\Phi_i^c\Phi_i^d. \quad (17)$$

Using $c_i = \bar{c} + \xi_i$ and writing the tensor $\Phi M \Phi$ in terms of local values ($\Phi^2, (\nabla\Phi)^2$), it follows a simple diagram theory. For example, the two-point Green function $G(q, \xi)$ becomes a diagram series, which is shown in Fig. 1. Each wave line characterized in this representation an external field ξ. The quenched average over this external field with the given probability $P(c,t)$ generated the Fourier transformed correlation functions $\langle \xi_q \xi_{q'} ... \xi_{q^n}\rangle$. A simple algebraic investigation shows [10] that a separation of this function is possible with

$$\langle \xi_{q^1}\xi_{q^2}...\xi_{q^n}\rangle$$

$$= \sum_{\text{combinations}}\prod_{\text{pairs}}[S(q_i)\delta(q_i - q_j)] + o\left(\frac{1}{V}\right). \quad (18)$$

$S(q)$ is the static structure factor. Therefore, the quenched average over the diagrams produces the sum over all combinations of different pair connections between the ξ lines. The weight of such a structure line is $S(q)$. Generally, the behavior of $G_0(q)$ and $S(q)$ in the region $q \to 0$ determines the scaling law near the percolation threshold [9]. In principle, the free propagator behaves in this case (massless theory) as q^{-2} and the structure factor may be a general power function $S(q) \sim q^{-a}$. (Note

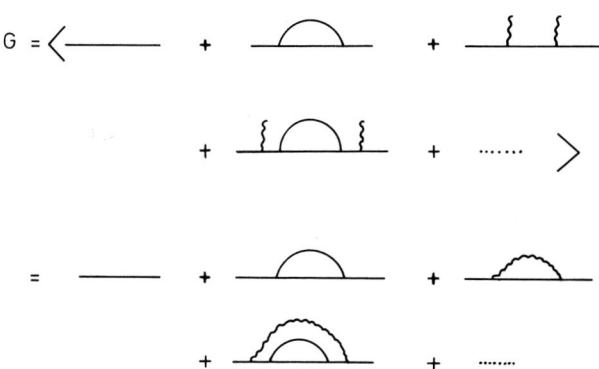

Fig. 1. Diagram expansion and quenched average of the two-point Green function G

that, in the case of diffusion limited network formation $a = 0$, in the case of reaction-limited network formation a part of the structure factor with $a = 2$ exists). Under this consideration, it follows for the divergence degree of a vertex function $\Gamma^{(E)}$ with N external legs that

$$\Gamma \sim p^{2n_4 + (2-a)l_a + (L-1)(d-6) - 2E},$$

where p is a given momentum scale, n_4 the number of vertices with four legs, l_a the number of structure lines, and L the number of loops. For fixed L the divergence degree have the maximal value in the following cases.

- $2 > a$: $n_4 = 0$ $l_a = 0$
- $2 = a$: $n_4 = 0$ $l_a = $ arbitrary
- $2 > a$: $n_4 = 0$ $l_a = l_a^{\max}$.

That means, if $a < 2$, the structure of the c-distribution is irrelevant. On the other hand, it follows that, in this case, the scaling law of the simple percolation theory is always valid for the region $a < 2$, i.e., all critical exponents have the same values as i the simple percolation theory, only the nonuniversal critical point p_c will be changed. Another situation is given with $a = 2$. Now, it follows a set of new relevant diagrams, connected by internal structure lines, which changes the typical behavior near the sol-gel transition. In this case, the problem contains relevant dimensionless interaction constants (u, v) which describe the Φ^3 interaction and the $\xi\Phi^2$ interaction. Using the renormalization group theory in a straightforward manner, it follows for the ε-expansion around the critical dimension $d_c = 6$ that the fixed point for the simple percolation theory ($u^* = \neq 0$, $v^* = 0$) will be unstable and a new stable fixed point ($u^* \neq 0$, $v^* \neq 0$) was formed. The critical exponents, which follow from the ε-expansion, are

$$\eta = 0 \tag{19}$$

$$v = 2 + \frac{1}{8}\varepsilon + \frac{1}{32}\varepsilon^2 . \tag{20}$$

Using these equations, the other critical exponents follow by the well-known scaling relations. Unfortunately, the extrapolation from $\varepsilon \ll 1$ to the interesting region $\varepsilon \sim 3$ has a high speculative character, because of the strong Taylor coefficients in the ε-expansion, but the tendency of the scaling exponents can be seen. It follows that:

d	Classical	Simple percolation	Percolation with $S \sim q^{-2}$
η	0	$-\frac{1}{21}\varepsilon + o(\varepsilon^2)$	$o(\varepsilon^3)$
v	$\frac{1}{2}$	$\frac{1}{2} + \frac{5}{84}\varepsilon + o(\varepsilon^2)$	$\frac{1}{2} + \frac{1}{8}\varepsilon + o(\varepsilon^2)$
β	1	$1 - \frac{1}{7}\varepsilon + o(\varepsilon^2)$	$1 + o(\varepsilon^3)$
γ	1	$1 + \frac{1}{7}\varepsilon + o(\varepsilon^2)$	$1 + \frac{1}{4}\varepsilon + o(\varepsilon^2)$
α	-1	$-1 + \frac{1}{7}\varepsilon + o(\varepsilon^2)$	$-1 + \frac{1}{4}\varepsilon + o(\varepsilon^2)$
δ	2	$2 + \frac{2}{7}\varepsilon + o(\varepsilon^2)$	$2 + \frac{1}{4}\varepsilon + o(\varepsilon^2)$
σ	$\frac{1}{2}$	$\frac{1}{2} + o(\varepsilon^2)$	$\frac{1}{2} - \frac{1}{16}\varepsilon + o(\varepsilon^2)$
τ	$\frac{5}{2}$	$\frac{5}{2} - \frac{1}{14}\varepsilon + o(\varepsilon^2)$	$\frac{5}{2} - \frac{1}{16}\varepsilon + o(\varepsilon^2)$

Conclusions

As the main result of the last considerations, it follows that the behavior near the sol-gel transition is relatively stable against a change of the static structure factor and, therefore, of a change of the network formation process. In the case that the structure factor is a power law $S(q) \sim q^{-a}$ with $a < 2$ for $q \to 0$ (long-wave limit; small-angle region of the x-ray scattering), we get unchanged critical exponents as in the case of the random percolation for the sol-gel transition. Clearly, a change of the nonuniversial critical probability exponents follows in the case of $S(q) \sim q^{-2}$ — this is, from the view of the kinetic theory, the situation of the reaction-limited network formation. Because of the small convergence radius in the ε-expansion an additional numerical simulation is desirable.

References

1. Coniglio A, Stanley HE, Klein W (1979) Phys Rev Lett 42:518
2. Wu FY (1982) Rev Mod Phys 54:235
3. Zia RKP, Wallace DJ (1975) J Phys A: Math Gen 8:1495
4. Amit DJ (1976) Phys A: Math Gen 9:1441
5. Stauffer D, Coniglio A, Adam M (1982) Adv Polym Sci 44:103
6. Gardiner CW (1983) Handbook of Stochastic methods, Springer, Berlin
7. Meakin R (1984) J chem Phys 81:4637
8. Jullien R, Botet R, Kolb M (1984) J Phys Lett (France) 45:L211
9. Schulz M (1992) J Stat Phys 67:1109
10. Schulz M, Handrich K (1991) Phys Stat Sol (b) 163:55
11. Potts RB (1952) Proc Camb Phil Soc 48:106

Author's address:

Michael Schulz
TH Merseburg
Fachbereich Physik
D-O-4200 Merseburg, FRG

Gelation and 1,1-transition in three-dimensional condensation and chain polymerization

P. G. Babayevsky

Department of Composite Technology, Division of Composite Physics and Chemistry,
Tsyolkovsky Institute of Aviation Technology, Moscow, Russia

Abstract: The data on chemical analysis, gel fraction, and TBA experiments have been used to identify chemical and physical transformations during the isothermal cure of three model reactive systems: trimethylolphenol, epoxy, and unsaturated ester, through condensation, addition, and free radical chain polymerization, respectively. For the first two systems cured by stepwise, statistically homogeneous reaction, TBA technique revealed physical gelation identified as 1,1-transition which proceeded and superimposed on chemical gelation and vitrification. In the unsaturated ester cured through heterogeneous (microgel) polymerization, microgelation and vitrification are associated with a space-filling packing of rubbery or glassy microgels, respectively.

Key words: Trimethylolphenol; epoxy and unsaturated ester systems; isothermal cure; torsional braid analysis; gel fraction; fractional conversion; gelation; 1,1-transition; vitrification

Introduction

During the isothermal cure, reactive monomer, oligomer or prepolymer systems undergo transformation from the liquid to solid (rubbery or glassy) state as a result of the proceeding of complicated and varied chemical and physical processes. Chemical processes involve irreversible change in molecular composition and topology of the reactive systems, and physical processes — reversible, by temperature increase — change in chain conformation, molecular packing and phase structure. Physical processes during the isothermal cure are caused by chemical reactions and in their turn make an influence on the cure mechanism and kinetics. At some degree of the cure these interdependent processes cause qualitative (phase and relaxation) transitions: gelation, vitrification, phase separation, etc.

According to the present theories [1] gelation is a percolation phase transition associated with irreversible molecular topology transformation, the incipient formation of an infinite polymer network (spanning molecular cluster in terms of percolation theories) by growth, branching and chemical crosslinking of polymer chains, and relaxation transition from liquid to sol-gel rubber. Similar, but, reversible transformations known as transition from A to B stage in thermosetting resins, 1,1-transition in polymer melts, gelation in polymer solutions, and melt-rubber transition in block and graft copolymer elastoplastics can be caused by physical processes: formation of labile chain entanglements and crystalline, liquid crystalline or amorphous glassy microphase domains as polymer network "crosslinks" [2]. The same physical processes are involved in isothermal vitrification of the reactive systems (corresponding to the glass transition temperature T_g, rising to the cure temperature T_{cure}).

The objective of this work was to identify isothermal chemical and physical transition of three monomer reactive systems cured by different mechanisms of reactions: step-growth (condensation and addition) and chain (free radical) polymerization, through their dynamic mechanical behavior and solubility using torsional braid analysis (TBA) and gel fraction (GF) data, respectively, in wide ranges of cure temperature and degrees of conversion.

Experimental

The monomer cured by polycondensation (trimethylolphenol, TMP) was synthesized in the laboratory [3] as a low viscous liquid with 50 mol% of methylol groups. As a reactive system cured through additive polymerization the stoichiometric mixture of a diglycidyl ether of bisphenol A (Russian analog of Shell Epon 828 epoxy resin, M.w. 390—410, epoxy equivalent 195) with m-phenylene diamine (DGEBA-mPDA) was used. The mixture was prepared by stirring mechanically for 3 min at 70°C and stored in dessicator in a freezer until needed. Triethylene glycol dimethacrylate (trade mark — TGM-3) with 1 wt% of tert-butyl perbensoate (TEGDMA-tBPB) was employed as a system cured through free radical chain polymerization. The chemical formulae of the reactants are shown in Fig. 1.

TBA experiments were performed [4] using as specimens loose, heat-cleaned glass braids impregnated by the liquid systems studied. The specimens, after having been dried in vacuum at room temperature, were cured isothermally in the analyzer at different temperatures while monitoring changes in relative rigidity ($1/p^2$) and mechanical damping index ($1/n$). The similar specimens, after having been cured to a given degree of conversion, were extracted from the TBA analyzer and were used to determined gel fraction and fractional conversion. The GF data for all the systems studied were determined by boiling in acetone for 3 h in a Soxhlet extraction column. Fractional conversion of TMP was determined by chemical analysis of the residual methylol groups [5] of DGEBA-mPDA and TEGDMA-tBPB by differential scanning calorimetry (Perkin-Elmer DSC-4). Reaction heats of the partially cured specimens were obtained from a temperature scan at 5°C/min from 20° to 250°C.

Results and discussion

Critical to a choice of the cure temperatures of the systems studied are the values of their glass transition temperatures at the onset of gelation (T_{gg}) and in the fully cured state ($T_{g\infty}$). T_{gg} values corresponded approximately to $\tilde{T}_{cure}$, at which damping peaks of gelation and vitrification coincide, and are about 130°C for TMP and 70°C for DGEBA-mPDA. $T_{g\infty}$ for the fully cured DGEBA-mPDA and TEGDMA-tBPB systems are 160°C and 130°C,

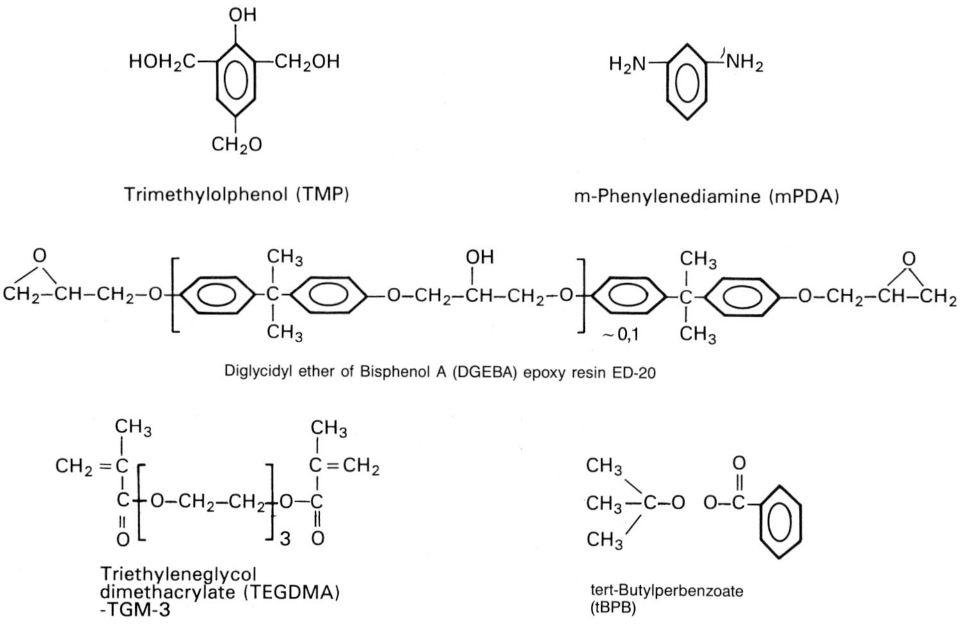

Fig. 1. Chemical formulae of the reactants

respectively. The values of T_{gg} for TEGDMA-tBPB and $T_{g\infty}$ for TMP could not be determined.

As the most interesting ranges of T_{cure} were chosen: for TMP, $T_{cure} < T_{gg}$ and $T_{gg} < T_{cure} < T_{g\infty}$; for DGEBA-mPDA, $T_{cure} < T_{gg}$, $T_{gg} < T_{cure} < T_{g\infty}$ and $T_{cure} > T_{g\infty}$; and for TEGDMA-tBPB, $T_{cure} < T_{g\infty}$ and $T_{cure} < T_{goo}$. The experimental results obtained for the systems studied at different T_{cure} in these ranges are summarized in Fig. 2 and Table 1.

The data for the systems cured by condensation and addition polymerization (TMP and DGEBA-mPDA) show that the onset of insolubility corresponds to some critical fractional conversions (a_i) close to those calculated for statistically homogeneous gelation. TBA curves at $T_{gg} < T_{cure} < T_{g\infty}$ have two distinct damping peaks: the smaller one at low fractional conversion (a_s) is usually associated with chemical gelation and the larger one at higher fractional coversion (a_1) is associated with vitrification [4]. At $T_{cure} < T_{gg}$ instead of the smaller damping peak a shoulder is observed on the vitrification peak and at $T_{cure} < T_{g\infty}$ the vitrification peak disappears. A comparison of TBA with GF data for the systems (Fig. 2a, b; Table 1) shows that, at low T_{cure}, $a_s < a_i$ and a_s approach to a_i only at $T_{cure} > T_{g\infty}$. It seems to be obvious that, at low T_{cure}, smaller damping peaks on TBA curves are associated with relaxation transformations similar to 1,1-transition in polymer melts as a result of the incpient formation of an infinite network through polymer chain entanglements or microphase separation [2] before the onset of chemical gelation. Gillham was the first who proposed the possibility of 1,1-transition during the reactive systems cure as a pregel phenomenon [6] and he determined a difference for the time of gelation from TBA and GF data for an epoxy/diamine system [7].

For the TEGDMA-tBPB system cured through free radical chain polymerization the gel fraction is

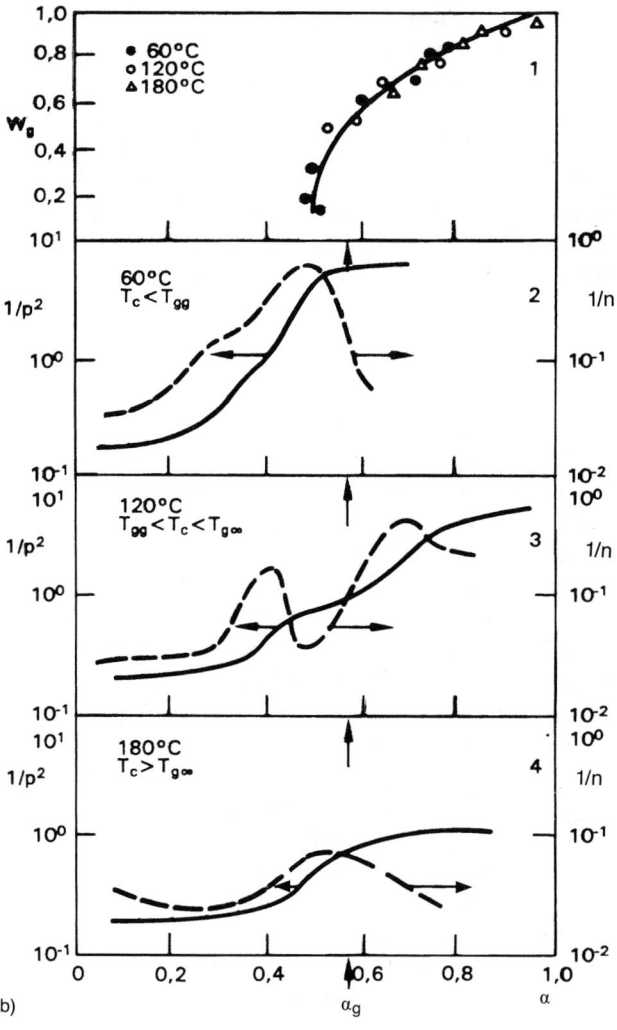

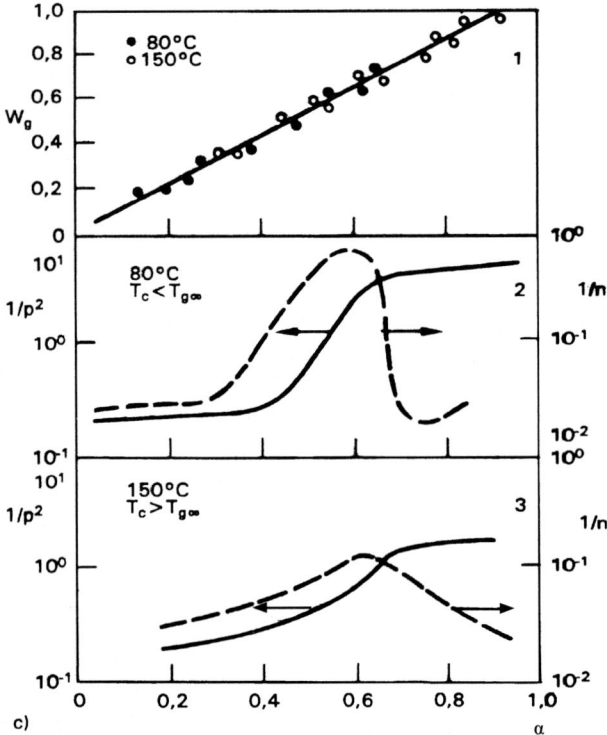

Fig. 2. Gel fractions (1) — W_g and TBA parameters (2—4): relative rigidity $1/p^2$ (——) and damping index $1/n$ (———) as a function of the degree of conversion α for isothermal cure at different temperatures of TMP (a), DGEBA-mPDA (b), and TEGDMA-tBPB (c), a_g: statistically calculated degree of conversion at the gel point

Table 1. Fractional conversions correspond to the onset of insolubility (a_i), smaller (a_s) and larger (a_l) damping peaks for three model reactive systems at different temperatures of cure

System	T_{cure} (°C)	a_i	a_s	a_l
TMP	120	0.46	0.28	0.37
	180		0.29	0.65
DGEBA-mPDA	60	0.52	0.30	0.47
	120		0.38	0.67
	180		0.51	—
TEGDMA-tBPB	80	—	—	0.62
	150	0.62		—

to the filling factor for random packing of spherical particles.

Conclusions

By comparing the torsional braid analysis and gel fraction data as a function of fractional conversion for three model reactive systems during the isothermal cure through three-dimensional condensation and chain polymerization at different temperatures, it is shown that, in homogeneous condensation polymerization, physical gelation identified as 1,1-transition proceeds and superimposes on chemical gelation. In heterogeneous (microgel) free radical chain polymerization both macrogelation and vitrification are associated with microgel agglomeration.

References

1. Stauffer D, Coniglo A, Adam A (1982) Adv Polymer Sci 44:103—158
2. Boyer R (1980) J Macromol Sci — Phys 18:461—553; 563—651
3. Martin R (1951) US Patent 2579329 Chem Abstr (1952) 46:33286
4. Babayevsky P, Gillham J (1973) J Appl Polymer Sci 17:2067—2077
5. Analytical Chemistry of Polymers, Kline G (ed) Interscience Publishers, New York, London
6. Gillham J (1979) Polymer Eng Sci 19:676—682
7. Enns J, Gillham J (1982) J Appl Polymer Sci 28:2567—2591
8. Dušec K, Prins W (1969) Adv Polymer Sci 6:1—102
9. Berlin A, Korolev G, Kefeli T, Sivergin Yu (1983) Acrylic Oligomers and Materials, Chimia, Moscow (in Russian), pp 47—110

Author's address:

Prof. P. G. Babayevsky
Dostoyevsky str. 3
103030 Moscow, Russia

directly proportional to the fractional conversion and only one damping peak (large or small) is associated with vitrification (at $T_{\text{cure}} < T_{g\infty}$) and with gelation (at $T_{\text{cure}} > T_{g\infty}$), respectively, and is observed on TBA curves at both ranges of T_{cure} studied (Fig. 2c). This can be explained by the heterogeneous (microgel) mechanism of the cure [8, 9]. At the beginning of the cure the system consists of a dispersion in the monomer of swollen microgel ("popcorns") with high degree of vinyl group conversion. The volume of the microgels being in the glassy state at $T_{\text{cure}} < T_{g\infty}$ and in the rubbery state at $T_{\text{cure}} > T_{g\infty}$ is increasing during the cure, and a space-filling packing of the glassy and rubbery particles is associated with vitrification and macrogelation, respectively. This is confirmed by the fact that fractional conversions corresponding to the transformations (a_s and a_l at $T_{\text{cure}} < T_{g\infty}$ and $T_{\text{cure}} > T_{g\infty}$, respectively) are both about 0.62 and close

Effect of thermal history on amylose gelation

J. L. Doublier[1]), I. Coté[2]), G. Llamas[1]), and G. Charlet[2])

[1]) INRA-LPCM, Nantes, France
[2]) Département de Chimie and CERSIM, Université Laval, Québec, Canada

Abstract: Amylose is known to dissolve in water at temperatures above 130°C. The literature unanimously reports that, provided the polymer content exceeds about 1%, any solution will form a gel upon cooling. The phenomenon is attributed to liquid-liquid phase separation followed by crystallization. In the present work, gels were prepared using carefully controlled thermal histories in sealed glass tubes. The crucial parameter is the dissolution temperature T_d, i.e., the maximum temperature at which the solution was heated. If $T_d < 160°C$, a gel forms on cooling while amylose crystals precipitate out of solutions heated at higher temperatures. The effect of T_d on the final macroscopic state of the system is independent of the polymer concentration (between 1% and 12%), the final temperature (between 2° and 45°C) or the cooling rate (between rapid quenching and 1°/h). Moreover, the thermal history effect was thermoreversible and unrelated to polymer degradation. Gels prepared using different T_d were investigated by oscillatory shear measurements. The kinetics of gelation were strongly dependent upon T_d, confirming that the thermal history should be carefully controlled in such studies. The observations suggest that amylose gelation involves crystallization and requires the presence of nuclei, whose number is determined by T_d. Heating at low temperature ($<160°C$) leaves a large number of nuclei in the solution. Upon cooling, a gel forms because the nuclei induce rapid crystallization, with any given chain involved in different crystals. In contrast, dissolution at high T_d may yield complete melting of the nuclei, which induces a slower but more complete crystallization on cooling.

Key words: Amylose; thermal history; rheology; gelation

Introduction

Amylose, a (1—4) linked α-glucan, is the linear component of starch and is considered to be responsible for much of the gelation of this macromolecular system. This polysaccharide is normally not soluble in water at room temperature. When dispersed in a hot neutral aqueous medium, solutions are unstable and set to gels, upon cooling, provided the concentration is high enough. Despite several recent studies [1—6], the mechanisms underlying gelation of amylose are still not well understood. At present, the phenomenon is thought to be a liquid phase separation, with crystallization of amylose chains occurring subsequently within the polymer-rich phase [1, 2]. There remains, however, some debate on the details of the process, particularly the crystallization mechanism, the effect of molecular weight, the value of the critical concentration for gelation, etc. Part of the disagreement may arise from differences in the procedures employed, particularly to solubilize amylose. Three methods are mainly used. A first procedure consists of dissolving amylose in alkali (i.e., 1 M KOH) followed by neutralization [6—8]. This method has been extensively employed in light-scattering studies. Evidence for complete solubilization has been given on fresh solutions from the absence, in the Zimm plots, of a downward curvature at low angle. However, depending upon concentration and molecular weight, aggregation could be seen from the development of turbidity after several hours [10—12]. A second, elegant method consists of regenerating amylose solutions from the amylose-n-butanol complex by heating to 90°C, butanol then being removed in a hot nitrogen stream [1, 2]. Again, light-scattering experiments showed that fresh amylose solutions were free of aggregates, suggesting complete solubilization [1]. The third method involves heating a suspension of amylose in water at temperatures between 130 to 170°C [3—5]. In [3], it is reported that heat treatment at 160—170°C for 10—20 min yields optically clear solutions without any degradation. The above procedures have been shown to result in amylose gelation or precipitation depending on molecular weight and concentration. However, no comparison between the different preparation methods was ever conducted. The original purpose of the present work was to ascertain the effect of the solubilization temperature on the properties of gels prepared by heating neutral aqueous solutions, as a first step in a comparative study of amylose gels prepared by various routes. The results, however, show that the gelation mechanism is likely to be different than previously purported. The final macroscopic state of a system prepared by cooling aqueous amylose solution is not always a gel, but depends on the solution thermal history.

Materials and methods

Three potato amylose samples, S1, S2, and S3, were used. S1 and S2 were from Sigma Chemicals and differed by their viscosity-average molecular weight (M_v) which was of 10^5 and $2.3\ 10^5$, respectively, as assessed from intrinsic viscosity measurements in KOH (0.2 M) according [13]. S3 was from AVEBE and had a M_v of $3.9\ 10^5$.

Preparation of samples

Amylose and water were sealed together in a glass tube which was then heated for 20 min in an oil bath at a constant dissolution temperature (T_d) chosen between 150° and 172°C. These solutions were then cooled at 25°C, either by rapid quenching in water or by slow cooling at a constant rate. The evolution of the macroscopic state of the system was then followed by direct visual observation or either of the following two methods.

Turbidimetry

The development of turbidity at 25°C was monitored by measuring the absorbance of the system at 640 nm as a function of time, using a UV-visible spectrophotometer.

Rheology

Rheological measurements were performed at 25°C in oscillatory shear experiments on the Rheometrics Fluids Spectrometer (RFS2) using the cone-and-plate fixture (5 cm diameter, 0.04 rad angle). Measurements were first performed for 15 h at 1 rad/s and at a strain amplitude of 0.1 which was confirmed to lie within the limits of linear viscoelasticity; then a mechanical spectrum between .01 and 100 rad/s was recorded.

X-ray diffraction

An amylose-water mixture containing 6% by weight of sample 1 was heated at 168°C for 20 min, then quenched at room temperature. After a few days, the resulting precipitate was recovered by filtration and dried under vacuum. The powder was then densely packed into a quartz capillary (Charles Supper Co., Natick, Massachussetts; I.D. 1.5 mm). A wide-angle diffraction spectrum was obtained using nickel-filtered Cu Kα radiation (λ = 0.154187 nm) produced by a Rigaku RU-200BH rotating anode x-ray generator operated at 50 kV and 160 mA. Collimation of the incident beam was achieved through a Soller slit and a 2.0-mm pinhole. All scans were recorded in the symmetrical transmission geometry, the normal to the sample surface being positioned at an angle θ with respect to the incident beam while the diffracted intensity was measured at 2θ by a scintillator equipped with a pulse-height analyser.

Results and discussion

Heating at any dissolution temperature (T_d) between 150° and 172°C yielded optically clear solutions. Cloudiness appeared slowly on cooling after a lag time which depended upon concentration and T_d. At a given concentration, the higher the T_d, the slower the development of opacity. Moreover, the final macroscopic state of the dispersion also depended upon T_d. This is illustrated in Fig. 1

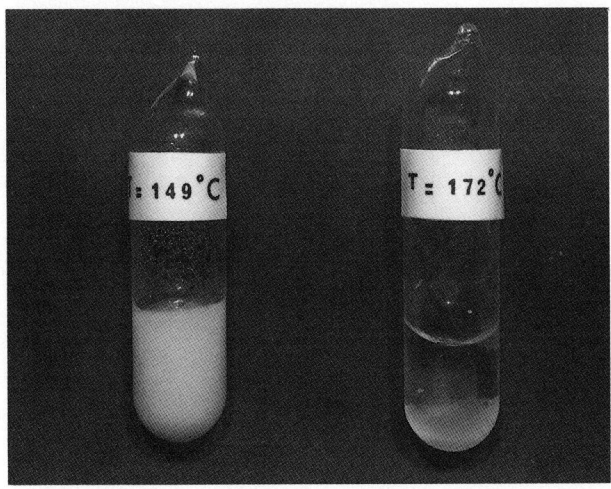

Fig. 1. Final state of amylose-water dispersions heated at $T_d = 149\,°C$ (left tube) and $T_d = 172\,°C$ (right tube) (concentration: 2%)

which shows the final state of two 2% dispersions of S1 dissolved for 20 min at either $T_d = 149$ or 172 °C and subsequently quenched at room temperature. A gel was obtained for $T_d = 149\,°C$, whereas a precipitate appeared when $T_d = 172\,°C$. We obtained similar figures at different concentrations between 1 and 12%, irrespective of the sample investigated (S1, S2 or S3) which suggests that the phenomenon does not depend strongly on concentration and molecular weight. Furthermore, varying the final temperature between 2° and 45°C, as well as the cooling rate between a rapid quenching and 1 °C/h yielded similar results.

To our knowledge, this is the first report of such a drastic effect of thermal history on amylose gelation. The literature unanimously reports that only very dilute solutions, or solutions of very low molecular weight give rise to precipitation instead of gelation. In those cases, gelation is prevented because the polymer chains are, respectively, too scarce or too short to allow for the onset of a macromolecular network. Gidley [5] reported that amylose aqueous dispersions treated at 160–170 °C give gels for molecular weight higher than $5 \cdot 10^4$ g · mol^{-1} and concentration higher than 1% by weight.

On the other hand, it may be argued that the effect of thermal history described above is a consequence of a degradation of the polymer at high temperature. The dissolution at 172 °C could induce depolymerization to an extent that would bring hydrolyzed amylose molecules to a molecular weight below the critical value for gelation. Such a partial depolymerization has, for instance, been reported in solutions heated at temperatures higher than 170 °C for times longer than 30 min [3]. The effect of thermal history proved to be completely reversible, which precludes polymer degradation. When an aqueous suspension of the precipitate obtained after dissolution at 172 °C (the second tube in Fig. 1 for instance) is heated at 150 °C, a gel (very similar to that obtained in the first tube in Fig. 1) forms on cooling.

The nature of the precipitate deserves further comment. Figure 2 shows the diffraction pattern of the precipitate recovered from a 6% solution prepared at $T_d = 168\,°C$. This diagram is of a B-type recrystallized amylose [5]. The amylose gels obtained by heat treatment at 150° or 160°C did not display any well-resolved pattern, and the amorphous halo produced by the water present in large excess obscured the crystalline peaks, if there were any. The diffraction pattern of Fig. 2 is exceptionally sharp in comparison with those reported in the literature for amylose precipitates [5]. This suggests that cooling from $T_d > 168\,°C$ leads to amylose crystals of a large size and/or high perfection. This may therefore provide a route to prepare highly crystalline amylose of a high molecular weight.

The kinetics of the phase separation in amylose-water mixtures were investigated by turbidimetry, in light of the above effect of thermal history. The results are illustrated in Fig. 3 for 1.5% solutions

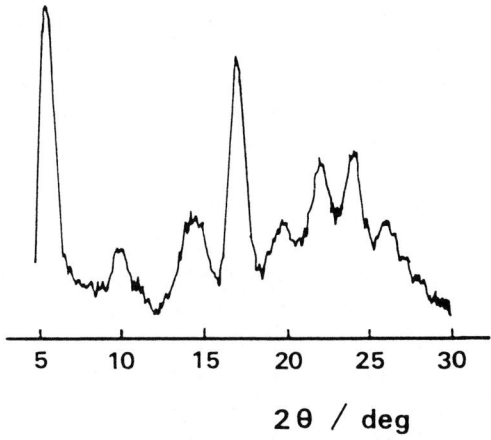

Fig. 2. X-ray diffraction pattern of a precipitated amylose ($T_d = 168\,°C$)

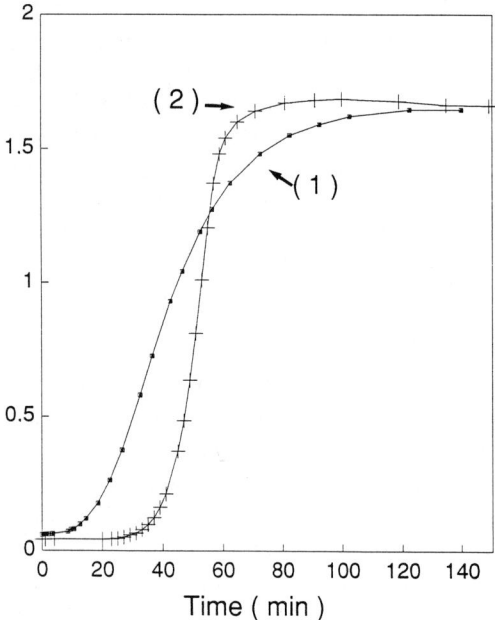

Fig. 3. Evolution of turbidity at 25°C for amylose-water dispersions after dissolution at T_d = 152°C (curve 1) and T_d = 172°C (curve 2)

heated either at 152° or 172°C. When T_d = 152°C the dispersion remained clear for about 8 min before opacity progressively developed, whereas cloudiness appeared only beyond 30 min for T_d = 172°C. Cloudiness likely results from the formation of particles large enough to scatter the incident light. The results of Fig. 3 suggest that large aggregates were formed much earlier when the system was heated at 152°C than when the dissolution was performed at 172°C. This could be explained in terms of the number of nuclei present in the solution after cooling from T_d. Dissolution at 172°C would induce extensive melting of the amylose crystals, while heating at a lower T_d could leave a large number of nuclei in the solution. In the latter case, aggregation would proceed much faster upon cooling. In effect, crystallization would be so fast that amylose chains could participate in more than one aggregate. At high enough concentration, this would generate a three-dimensional network responsible for gelation. Subsequent crystallization would be restricted. In contrast, amylose would take a much longer time to phase separate from a solution where most nuclei were previously molten by heating at higher T_d. Crystal growth would oc-

cur in rare, well-separated areas of the solution, therefore resulting in separated crystals of a larger size and higher perfection. This model for amylose gelation is consistent with DSC measurements performed on recrystallized amylose [14]. The authors reported a major endotherm at temperatures between 150°C and 145°C, with a maximum at 137°C. The thermogram also showed a smaller endotherm centered around 157°C. Although further studies are required to confirm this single result, the existence of two distinct melting ranges supports the above model. While most of the amylose would be dissolved in water upon heating at a given temperature below 150°C, according to the present work, some crystals, possibly of a different polymorphic nature or morphology, would not melt until a higher temperature.

The effect of T_d on the rheological properties of a rapidly cooled 1.5% dispersion of sample S2 aqueous amylose systems was also investigated. Figure 4 illustrates the variation of the storage modulus G' at 25°C as a function of time. Three traces are shown corresponding to T_d = 152°, 160°, and 172°C, respectively. The trace exhibited for T_d = 152°C is consistent with previously reported data at low concentration [4, 6, 8]. It is characterized by a lag period which lasts ~33 min, followed by a rapid increase in G', whereas G'' remained low. At longer times, a slower G' increase is observed. Gels prepared at T_d = 160°C resulted in a similar plot. The final G' value was of the same order (~30 Pa)

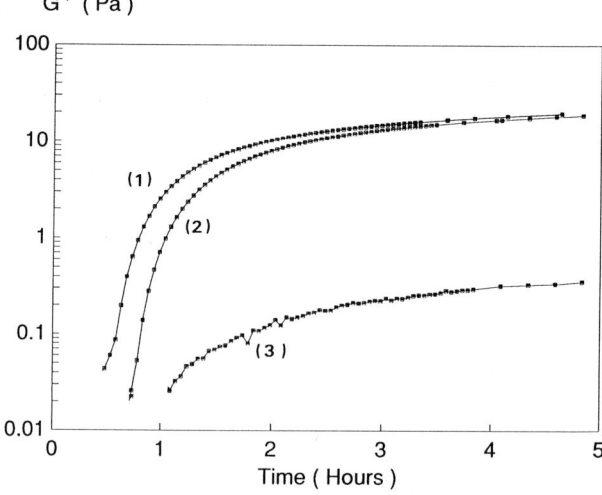

Fig. 4. Variation of G' as a function of time in (in semi-log scales) for T_d = 152°C (curve 1), T_d = 160°C (curve 2), and T_d = 172°C (curve 3)

and the only difference was that the lag period was significantly longer (~42 min). It should be mentioned that such a high G' value means that the gel is rigid and the concentration (1.5%) is much higher than the critical concentration for gelation C_0, as discussed elsewhere [8]. Results obtained for T_d = 172°C were totally different. The final G' value was very low (~0.8 Pa) and the lag time was much longer (~65 min). This was expected, since we characterized a dispersion which cannot be as rigid as a gel.

Conclusion

Contrary to all literature reports, conditions exist which lead to precipitation in concentrated aqueous solution of high molecular weight amylose. The critical parameter is the dissolution temperature T_d. While gels are formed from solutions cooled from T_d < 160°C, amylose crystals of high degree of crystallinity precipitate out of solutions prepared at higher temperatures. The effect of thermal history is completely reversible, since a gel is obtained by redissolving at low T_d a suspension of crystals prepared at higher T_d. The observations suggest a new explanation of the gelation of amylose based on crystallization and a large sensitivity of the kinetics of phase separation to the number of nuclei in the system. From a practical point of view, the present results confirm that the thermally history should be carefully controlled in further studies on amylose gelation. In particular, the 160—170°C range is crucial, and some surprising results and discrepancies reported in the literature may find an explanation in this phenomenon. Finally, thermal history effect can also be of interest when processing starch at elevated temperatures, particularly when dealing with high amylose starches.

References

1. Miles MJ, Morris VJ, Ring SB (1985) Carbohydr Res 135:257-269
2. Ellis HS, Ring SG (1985) Carbohydr Polym 5:201—213
3. Gidley MJ, Bulpin PV (1987) Macromolecules 22:341—346
4. Clark AH, Gidley MJ, Richardson RK, Ross-Murphy SB (1989) Macromolecules 22:346—351
5. Gidley MJ (1989) Macromolecules 22:351—358
6. Doublier JL, Choplin L (1989) Carbohydr Res 193:215—226
7. Kitamura S, Yoneda S, Kuge T (1984) Carbohydr Polym 4:127-136
8. Doublier JL, Llamas G, Choplin L (1990) Makromol Chem Macromol Symp 39:171—177
9. Banks W, Greenwood CT (1975) Starch and its components. Edinburgh University Press, Edinburgh
10. Paschall EF, Foster JF (1953) Makromol Chem 9:73—84
11. Foster JF, Paschall EF (1956) Makromol Chem 21:91—101
12. Huseman E, Pfannemuller B, Burchard W (1984) Makromol Chem 59:1—27
13. Banks W, Greenwood CT (1969) European Polym J 5:649—655
14. Eberstein K, Hopke R, Konieczny-Janda G, Stute R (1980) Stärke 32:397—404

Authors' address:

Dr. Jean-Louis Doublier
INRA-LPCM, rue de la Géraudière
BP 527
44026 Nantes Cedex 03, France

Rheological, potentiometric and ^{23}Na NMR studies on pectin-calcium systems

C. Garnier, M. A. V. Axelos, and J. F. Thibault

Institut National de la Recherche Agronomique, Nantes, France

Abstract: Pectins with degree of methylation lower than 50% are able to gel in presence of divalent ions such as calcium. A phase diagram of a sodium pectinate (DM = 28%) in water, pH 7.4, and 20°C was established, evidencing three phases: sol, gel, and syneresis. From rheological experiments, the calcium concentration at the sol-gel transition was determined, $[Ca^{2+}]_t/[COO^-] \approx 0.168$, for a polymer concentration of 5.8 g/l and it was shown that this transition was accurately described by the scalar percolation theory. Under the same conditions, potentiometric studies showed that the binding of calcium ions seemed to be independent from physical state (sol or gel). Furthermore, this calcium binding decreased only slightly with the amount of added calcium, whereas the sodium one decreased sharply in the same time. By following the evolution of the ^{23}Na NMR longitudinal relaxation time, an abrupt increase was found around $[Ca^{2+}]_t/[COO^-] = 0.182$, which was close to the results of the rheological study.

Key words: Pectin-calcium system; sol-gel transition; rheology; potentiometry; ^{23}Na NMR

Introduction

Pectins are polysaccharides extracted from plant tissues and are mainly composed of α 1-4 D-galacturonic acid units and their methyl esters [1]. Pectins of degree of methylation (DM) (molar ratio of methanol to galacturonic acid) lower than 50% can gel in presence of divalent ions, such as calcium. This ability to form gel is chiefly a consequence of the number and distribution of the charged groups along their backbone [2]. In this study, a phase diagram (sol-gel-syneresis) was established depending on the total amount of calcium and on pectin concentration. Rheological experiments, generally carried out to characterize the physical state of the systems, were performed in order to determine the gel point. An attempt to determine the fraction of calcium really involved in crosslinks was undertaken using potentiometric and NMR methods.

Material and methods

The pectin sample used (C28) was obtained by acid de-esterification from a commercial citrus sample (GENU-X-0907) kindly provided by Copenhagen Pectin Factory (Denmark) as described elsewhere [3]. The DM, determined by titrimetry, is 28%, leading to an average charge density of 1.16, and the anhydrogalacturonic content is 92.7%. The pectin solution was initially adjusted to pH 7.4. Whatever the experimental method used, the pectin-calcium systems were prepared by mixing pectin solution with calcium chloride solution at 70°C during 3 min.

For the establishment of the phase diagram, the test tubes were gently tilted after standing for 48 h at 20°C. When the meniscus could not be seen to deform under its own weight, we said that the system had gelled. Syneresis was detected by the presence of water at the gel surface. Then, by visual investigation, sol-gel and gel-syneresis transition curves could be determined for pectin concentrations higher than 3 g/l.

Investigation of the rheological properties of pectin gels was made with a Carri-Med CS50 dynamic controlled stress rheometer, in oscillatory shear, with a cone plate device (diameter 5 cm, cone angle 4°). The gels were cured in situ and the kinetics of the gel formation were followed at 20°C by measuring G' and G'', the storage and the loss moduli, at a fixed frequency of 0.125 Hz. A mechanical spectrum was recorded on the fully cured gels, which have access to the equilibrium elastic modulus G'_e. A low deformation of 0.04 was maintained whatever the frequency range explored, between 10^{-3} and 5 Hz.

The calcium and sodium activities were determined with ion-selective electrodes (F2112 Ca and G502 Na, Radiometer) on pectin-calcium mixture, for increasing amounts of $CaCl_2$, after waiting for 24 h at 20°C without stirring. A saturated calomel electrode was used as reference. Calibration curves were obtained using NaCl and $CaCl_2$ solutions at 20 ± 0.1°C without stirring and after stabilization. The NaCl and $CaCl_2$ concentrations were measured through conductimetric determinations with silver nitrate. The constance of the electrode response was checked from initial and final calibration.

The sodium NMR measurements were carried out on a Bruker MSL-360 FT spectrometer at 95.27 MHz at 20°C. The pectin-calcium systems were put into a 1-ml cell. Spin-lattice relaxation times T_1 were measured by the inversion recovery method [4].

Results and discussion

From the phase diagram (Fig. 1), established at 20°C, pH 7.4 in a salt-free solution, a polymer concentration C_p = 5.8 g/l (0.022 eq/l) was chosen and the evolution of the equilibrium elastic modulus, the fractional binding of calcium and sodium ions $[Ca^{2+}]_b/[COO^-]$, $[Na^+]_b/[COO^-]$ and sodium longitudinal relaxation time T_1 were followed depending on the amount of calcium added $[Ca^{2+}]_t/[COO^-]$.

Rheological measurements showed that the equilibrium elastic modulus (G'_e) increases with the amount of calcium present in the system (Fig. 2). Above the transition, close to the gel point, G'_e was found to vary with the total calcium concentration $[Ca^{2-}]_t$ as:

$$G'_e \sim |([Ca^{2+}]_t - [Ca^{2+}]_c)/[Ca^{2+}]_c|^2 ,$$

with $[Ca^{2+}]_c$, the value of the total calcium concentration at the gel point, equal to 0.0037 mol/l. The value of the exponent $t = 2$ agrees well with the results already published, $t = 1.93$ [5], which proves that the scalar elasticity percolation theory can accurately describe the properties of the gel transition in such systems. The rheological sol-gel transition was lower than the visual one (0.0037 instead of

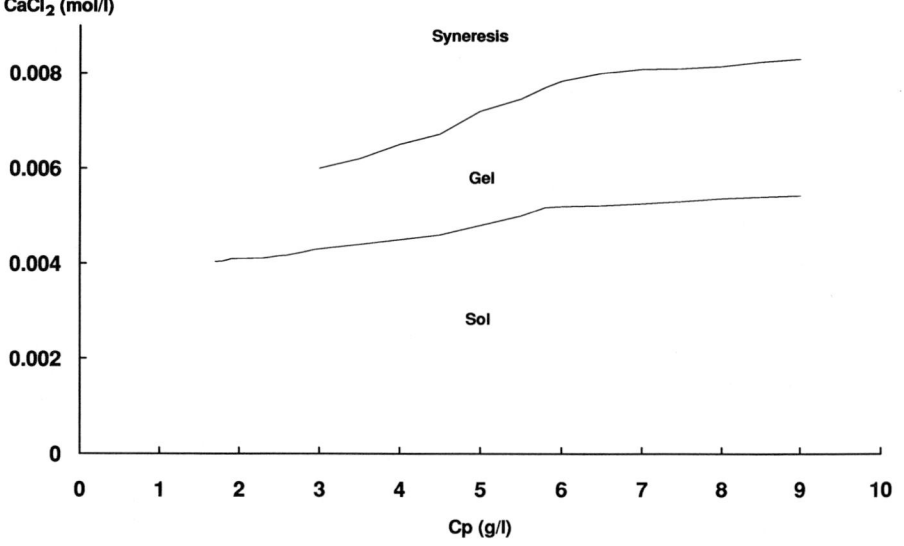

Fig. 1. Phase diagram of the C28 sample in water, pH 7, 20°C as a function of polymer and calcium concentration

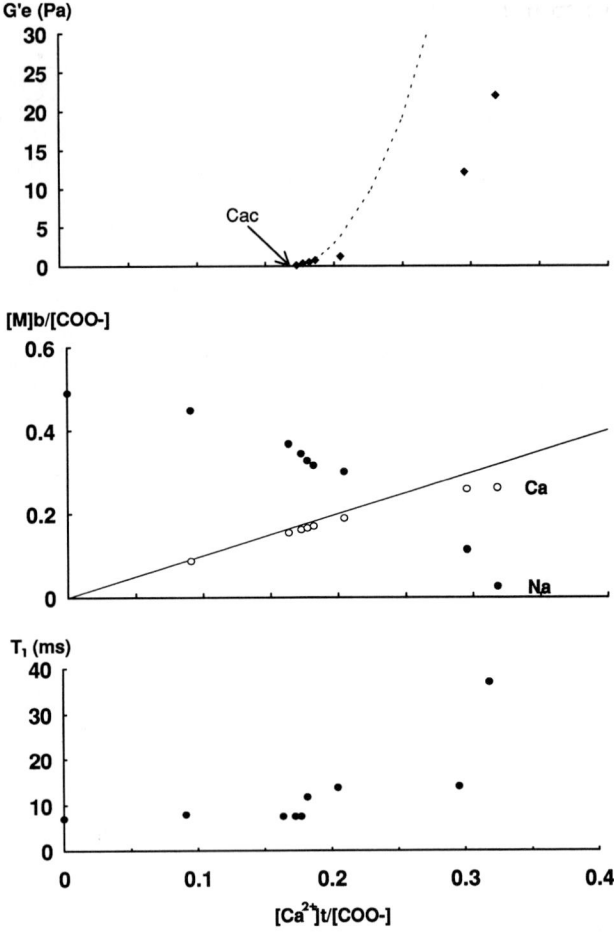

Fig. 2. Evolution of the equilibrium elastic modulus G'_e, fraction of bound ions, $[M]_b/[COO^-]$, and ^{23}Na longitudinal relaxation time T_1 with the amount of calcium chloride added to a sodium pectinate solution

and sodium concentrations were calculated from the measured activities using the Debye-Hückel theory. The values of the selectivity coefficient K_{Ca}^{Na} corresponding to the simple Ca^{2+}—Na^+ exchange and follow the equilibrium:

$$Ca^{2+} + 2\overline{Na^+} \Leftrightarrow 2Na^+ + \overline{Ca^{2+}},$$

where $K_{Ca}^{Na} = (X_{Na^2}/X_{Ca})(\overline{X_{Ca}}/\overline{X_{Na^2}})$; X_i and $\overline{X_i}$ (respectively, the equivalent fractions of counterions i free in solution and thermodynamically bound around the polyion [7]) were found to vary from 41 to 185 for the calcium amount between 0.002 and 0.007 mol/l. This strong dependence of K_{Ca}^{Na} on the total amount of calcium added indicates that no such simple equilibrium may explain the binding of the calcium to the polyion.

A linear dependence of the sodium relaxation time with the amount of sodium chloride added to the sodium pectinate solution (for NaCl concentrations less than 0.1 M), not shown here, was observed in agreement with a rapid exchange between a free state (in solution) and a bound state of the sodium ions associated with the polymer chains [8]. An increase in the amount of free sodium ions was accompanied by an increase in the relaxation time. The addition of calcium chloride to a sodium pectinate solution promoted in the same way an increase in the T_1 values (Fig. 2). Compared to the almost continuously decreasing Na curve obtained from the potentiometric measurements, the NMR results exhibited two well-defined steps. The first one ($[Ca^{2+}]_t/[COO^-] \approx 0.182$) was close to the sol-gel transition obtained from the rheological measurements ($[Ca^{2+}]_t/[COO^-] \approx 0.168$), the second step ($[Ca^{2+}]_{tot}/[COO^-] \approx 0.318$ mol/l) correspond to the gel-syneresis transition. The value of T_1 (37 ms) obtained for this latter calcium concentration was greater than that found for a NaCl concentration of 0.2 M (30 ms), indicating that the sodium concentration in the liquid expelled from the network was very high. However, a linear and continuous increase of the T_1 values, corresponding to the exchange between sodium and calcium and then to the increase of the number of junctions in the system was expected. This discrepancy could be ascribed either to a centrifugation of the sample in the gel state, promoted by the rapid rotational motion and leading to the expulsion of a liquid phase rich in sodium ions, or by the passage from a non-cooperative binding state to a cooperative

0.0051 mol/l), indicating that the visual determination overestimates the transition line.

The potentiometric study showed that the bound calcium first increases linearly with the amount of calcium added and then that the binding affinity decreases progressively but only slightly (Fig. 2). These results are in good agreement with those obtained on the binding of calcium to neutral sodium pectate by Lips et al. [6]. The calcium curve passed through the sol-gel transition, as determined from the rheological measurements, without any noticeable effect, and reaches a value of 0.26 when syneresis occurs. Simultaneously, the amount of bound sodium exhibited a sharp decrease with a downward curvature. Values of the bound calcium one, corresponding to the sol-gel transition and affecting the solvent mobility. In this case, the ^{23}Na NMR response could perhaps no longer be interpreted in terms of rapid exchange between free and bound state.

To conclude, our experiments indicate that the critical exponent of the elastic modulus found in the vicinity of the sol-gel transition is in good agreement with that of the percolation model (1.9). Near the gel point, the calcium effectively bound to the polymer is found to be linearly related to the total amount of calcium which justifies the choice of $[Ca^{2+}]_t$ as a parameter of the cross-linking reaction. The critical regime is quite large, as expected for gelation in semidilute polymer solution. Results of ^{23}Na NMR shows that this method can be used to accurately determine the gel point in such systems, but a more complete interpretation of the data requires further experiments.

Acknowledgement

The authors wish to thank F. Devreux of the "Laboratoire de Physique de la Matière Condensée", Ecole Polytechnique (Palaiseau, France) for providing access to the NMR facilities and for his useful advice.

References

1. Darvill A, Albersheim P, MacNeil M, Lau J, York W, Stevenson T, Thomas J, Doares S, Gollin D, Chelf P, Davis K (1985) J Cell Sci (Suppl) 2:203
2. Axelos MAV, Thibault JF (1991) in: The Chemistry and technology of pectins , Walter RH (Ed) Academic Press, p 109
3. Garnier C, Axelos MAV, Thibault JF (in press) Carbohydr Res
4. Vold RL, Waugh JS, Klein MP, Phillips DE (1968) J Chem Phys 48:3831
5. Axelos MAV, Kolb M (1990) Phys Rev Lett 64:1457
6. Lips A, Clark AH, Cuttler N, Durand D (1991) Food Hydrocoll 5, n° 1/2, 87
7. Rinaudo M, Milas M (1974) J Polym Sci 12:2073
8. Grasdalen H, Kwam BJ (1986) Macromolecules 19:1913

Authors' address:

Catherine Garnier
INRA, B.P. 527
44026 Nantes Cédex 03, France

ID # The formation of semiinterpenetrating polymer networks by photoinduced polymerization

B. Strehmel, D. Anwand, and H.-J. Timpe*)

Institute of Organic Chemistry, Technical University of Merseburg, FRG

Abstract: Semiinterpenetrating polymer networks can be prepared by photopolymerization of epoxy acrylates in thin films of several polymers. The results of the polymerization kinetics give evidence of changes in kinetic relations during the entire polymerization process. Furthermore, the polymers formed at different conversion degrees were investigated using crystal violet as a fluorescence probe. The experimental data obtained by means of the fluorescence technique indicate that the polymerization process most likely takes place in a highly viscous medium. Towards the end, the reaction proceeds in a nearly glassy state until the polymerization is frozen. Results obtained by time-resolved fluorescence spectroscopy and dynamic reaction calorimetry are comparable with the data of the DSC experiments.

Key words: Semiinterpenetrating polymer networks; photoinduced crosslinking; photocalorimetry; rotational mobility; fluorescence probes

Introduction

The photoinduced polymerization, crosslinking of vinyl-based monomers, has become increasingly important in recent years due to the large number of possible applications of these processes. Such applications include microelectronics [1], surface improvement [2], information recording materials [3], and printing plates [4]. In many cases the polymer-forming reactions are carried out in polymeric binders. The presence of such binders is necessary for several of the applications mentioned because the uncured photopolymer must give clear, transparent, and non-tacky films. Furthermore, the physical, chemical, and mechanical properties required of the cured photopolymer can sometimes be achieved only by using polymers in the initial systems. However, so far few data are available on the fundamental aspects of the polymerization reactions which occur in the polymeric binders.

The aim of the present work was to study some physical properties of both the uncured and cured photopolymer and the kinetics of photoinduced crosslinking in polymeric binders. We hoped to establish certain correlations between the physical parameters and the kinetic data of the systems investigated.

All studies were carried out on thin layers (approx. 100 µm), which provide favorable conditions for the photoinduced initiation of polymerization and crosslinking. In addition, it was possible to use fluorescence probes for the description of the polymeric systems.

The problems of free radical crosslinking of the diacrylate of 2,2-bis-[4-(2,3-epoxypropoxy)phenyl]-propan (DDGDA) in several polymeric binders were investigated.

Procedures and methods

Substances

Benzil dimethylketale (BDMK) was taken as a photoinitiator (CIBA-Geigy). The epoxy compound bisphenol-A-diglycidylether (DGEBA) (Shell Chemie GmbH) was recrystallized twice from methanol (m.p. 44°C). Acrylic acid (Merck) was purified in vacuum (b.p. 139°C). The ester (DDGDA) of acrylic acid and DGEBA was then synthesized according to a procedure found in the

---

*) Present address: Polychrome GmbH.

literature on the subject [5]. Crystal violet (Merck) was used as found without further purification. The polymers which served as binders in this study are commercial products which were used without further purification.

Photocalorimetric measurements

The principle of operation of the photocalorimeter and the apparatus used for the kinetic investigations has been described in an earlier paper [6].

Time-correlated single photon counting:

An ILA 120-1 argon ion laser (Carl Zeiss Jena AG) served as the light source. In combination with a modulator (M) and an acousto-optical modulator (AOM) (Friedrich Schiller University Jena), this laser produces periodical pulses with lifetimes of approxiamtely 85 ps and frequencies of 123.2 MHz. The deconvolutions necessary for experimental signals were made using the phase plane method or Marquardt's non-linear calculation algorithm.

The single photon counting (SPC) system was calibrated with 10^{-5} M solutions of pinacyanol (Lambda Physics) in methanol. The lifetime of pinacyanol is approximately 10 ps [7]. The time resolution of the entire equipment is over 50 ps.

Differential scanning calorimetry

Glass transition temperatures (T_g) were determined using DSC-7 equipment (Perkin-Elmer). Before the determination of T_g, the samples were treated at a heating rate of 10°C min^{-1}. The T_g values were determined with second scans at a heating rate of 20°C min^{-1}.

Probe preparation

Acetone solutions (methanol, in the case of the poly(N-vinyl-2-pyrrolidone) system) containing binder (3.5 wt %), photoinitiator (10^{-2} M), crystal violet ($1.2 \cdot 10^{-4}$ M), and monomer (0.07 M) were coated on Ni-plates (diameter: 10 mm) and dried in the dark at room temperature for 24 h. The plates were then irradiated in the calorimeter to determine conversion degrees, and the same probes were used for carrying out fluorescence experiments.

Results and discussion

Kinetics of the photocrosslinking

In all experiments benzil dimethylketale **1** (BDMK) was used as a photoinitiator for the photoinduced formation of radicals. After the excitation of this compound, benzoyl **2** and α-dimethoxybenzyl **3** radicals are formed, both of which are capable of attaching themselves effectively to vinyl monomers [8].

$$PhCO-C(OCH_3)_2Ph$$
$$\mathbf{1}$$
$$\xrightarrow{h\nu} PhCO^\bullet + Ph(OCH_3)_2C^\bullet \quad (1)$$
$$\qquad\qquad \mathbf{2} \qquad\quad \mathbf{3}$$

Substance **4**, an acrylic ester derived from bisphenol-A-diglycidylether and acrylic acid, was used as a monomer in the investigation. Owing to its multifunctionality this monomer is able to form interpenetrating polymer networks.

Poly(methylmethacrylate) **5** (PMMA), poly(N-vinyl-2-pyrrolidone) **6** (PVP) and an alternating copolymer of α-methylstyrene and hydrogen butyl maleate **7** (α-MSMB) served as polymeric binders for the measurements. These polymers possess different glass transition temperatures and oxygen permeabilities [9].

The polymers should therefore influence the photoinduced radical crosslinking, because both properties must affect the network-formation process very strongly.

For quantitative discussions of the experimental data obtained by photocalorimetry, it is important to know the reactions occurring in the systems measured during photoinduced crosslinking. The following reactions were dominant:

Initiation:

$$\text{initiator} \xrightarrow{h\nu} \text{In}^\bullet \quad (2)$$

$$\text{In}^\bullet + \text{monomer} \longrightarrow P_1^\bullet \quad (3)$$

Propagation:

$$P_n^\bullet + \text{monomer} \xrightarrow{k_p} P_{n+1}^\bullet \quad (4)$$

Termination reactions:

first order

$$P_n^\bullet \xrightarrow{k_t^m} \text{polymer} \quad (5)$$

Second-order

$$P_{n\bullet} + P_{m\bullet} \xrightarrow{k_t^b} \text{polymer} \quad (6)$$

Primary radicals:

$$P_{n\bullet} + \text{In}^\bullet \xrightarrow{k_t^{pr}} \text{polymer} \quad (7)$$

The course of polymerization rate R_p can be described by the semi-empirical Eq. (8) [10].

$$R_p = K(x)[M]^a I_0^\beta = K(x)(1-x)^a I_0^\beta = \frac{dx}{dt}, \quad (8)$$

where x: conversion degree, I_0: light intensity; $K(x)$: conversion-dependent value; a: monomer exponent; β: light intensity exponent; t: irradiation time.

A typical R_p-t curve based on calorimetric measurements is given in Fig. 1. The conversion degree x is obtained by integrating R_p. These values are used for further calculations.

Earlier experiments show that exponent a is approximately 1 for many similar photopoylmers [6]. Assuming that this result is also valid for the present systems, Eq. (8) can be very easily integrated. Using this procedure, the relation expressed by Eq. (9) is obtained.

$$-\ln(1-x) = K(x) I_0^\beta t. \quad (9)$$

By means of a double logarithmic plot it is now easy to determine the exponent β. The value of β yields a qualitative information about the termination process. The following conclusions may be drawn relative to the light intensity exponent:

	Termination dominates
β between 0.5 and 1	$\Rightarrow$ by first and second order
$\beta \ll 0.5$	$\Rightarrow$ by primary radicals
$\beta \approx 1$	$\Rightarrow$ only by first order
$\beta \approx 0.5$	$\Rightarrow$ mainly by second order

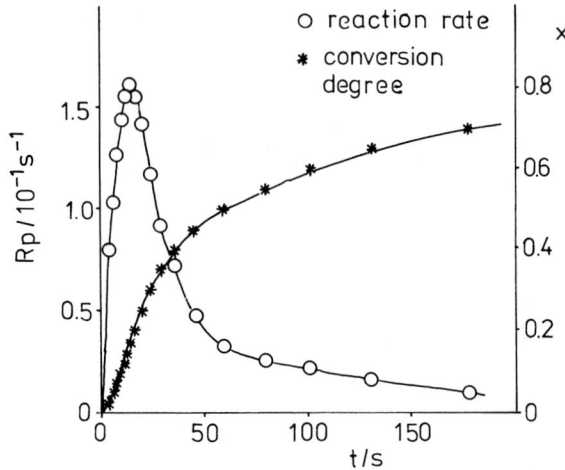

Fig. 1. Dependence of the polymerization rate and conversion degree as a function of irradiation time for the DDGDA/PMMA/BDMK system in air atmosphere ([DDGDA] = 1.84 · 10^{-5} mol · cm^{-2}; [BDMK] = 2.5 · 10^{-6} mol · cm^{-2}; λ = 340—380 nm; m_{binder} = 7 mg, I_0 = 4.15 mW · cm^{-2})

In Fig. 2 the β values are summarized for all the systems investigated. Because reaction conditions vary, the β values change during irradiation. As can be seen from Fig. 2, the polymeric binder also influences the value of β very strongly. Both results indicate that the type of termination also changes with the conversion degree and the chemical nature of the polymer. Furthermore, different β values would have been obtained if the reaction had been carried out in air or in argon atmosphere. As expected, oxygen has significant influence on the network formation process, and the structure of the network formed must also be different. Non-stationary irradiation experiments (postpolymerization) were carried out in order to obtain a better quantitative description of the termination mechanisms. An initial estimate indicates that only propagation steps, and first- and second-order termination reactions are dominant (Eq. (4)—(6)) under such conditions.

The kinetic model need only take into account these three processes to interpret the experimental data. The following three kinds of termination reaction are therefore possible: first-order, second-order, and intermediary mixed-order reactions. Kinetic expressions obtained using reactions (4) to (6) and the procedures outlined in [11] are given in Eqs. (10)—(12), which can be used to describe the postpolymerization results.

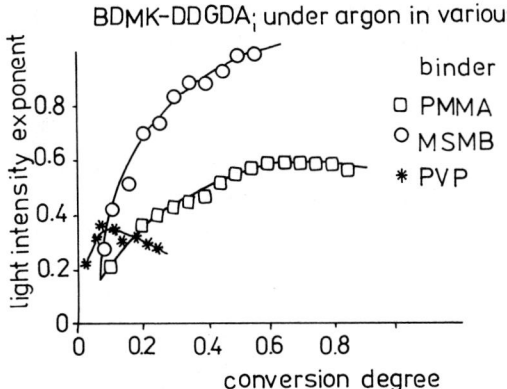

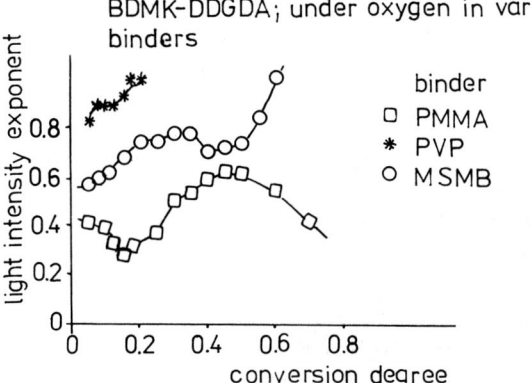

Fig. 2. Dependence of the ligth intensity exponent for the DDGDA-BDMK system in air and inert gas atmosphere as a function of conversion degree and binder composition ([DDGDA] = 1.84 · 10⁻⁵ mol · cm⁻²; [BDMK] = 2.5 · 10⁻⁶ mol · cm⁻²; λ = 340—380 nm; m_{binder} = 7 mg)

First-order termination:

$$-\ln(1 - x_D) = \frac{k_p[P^\bullet]_0}{k_t^m}(1 - \exp(-k_t^m t)) \qquad (10)$$

Second-order termination:

$$-\ln(1 - x_D) = \frac{k_p}{k_t^b}\ln(1 + k_t^b[P^\bullet]_0 t) \qquad (11)$$

Mixed-order termination:

$$-\ln(1 - x_D) = \frac{k_p}{k_t^b}\ln((1 + z)\exp(-k_t^m t)) \qquad (12)$$

$$z = \frac{k_t^b[P^\bullet]_0}{k_t m}, \qquad (13)$$

where (x_D: conversion of polymer formed in the dark period; k_p: rate constant for the propagation reaction; $[P^\bullet]_0$: polymer radical concentration at the beginning of the dark period; k_t^m: rate constant for the first-order termination; k_t^b: rate constant for the second-order termination; z: ratio of the second-order termination to the first order termination.

However, stronger deviations of the experimental data from these expressions can be expected if radicals $P^\bullet$ are kinetically non-equivalent, or if rate constants k_t^m and k_t^b depend on the chain length of $P^\bullet$. Nevertheless, the dependencies corresponding to Eq. (10)—(12) have been measured and quantitatively determined in order to compare them with the results of the stationary experiments.

For the determination of the constants and ratios in Eqs. (10)—(12), the conversion of the double bonds in the dark period x_D were also measured calorimetrically. The experimental parameters were then fitted to the kinetic expressions by means of Marquardt's non-linear algorithm. In each case, only one of the three possible kinetic equations is capable of describing the course of the experimental x_D/t-data.

Postpolymerization results obtained for system BDMK-PMMA-DDGDA are summarized in Table 1. As can be seen, all kinetic parameters depend very strongly on the degree of conversion in irradiation period x_D^0. This is especially evident for the propagation rate and there is also a decrease in the termination rate constants. It can be assumed that changing diffusion conditions of the reaction parameters causes a loss in reactivity. Furthermore, lower values are in many cases found in air atmosphere. This effect was predictable due to the inhibiting and retarding properties of oxygen.

With regard to the termination mechanism, postpolymerization experiments and light intensity exponents determined under stationary irradiation lead to similar results. In argon atmosphere, terminations by radicals (second order or primary radicals; the latter cannot be distinguished by postpolymerization experiments from the former) dominate at small degrees of conversion.

In the presence of oxygen, on the contrary, a first-order termination occurs partially, which correlates with an oxygen-retarding influence.

Polymer characterization with fluorescence probes

The above results demonstrate that kinetic relations are complicated, and that it is therfore

Table 1. Postpolymerization results and light intensity exponents of the DDGDA/PMMA/BDMK system by variation of the conversion degree in the irradiation period x_D^0 and the atmosphere ([DDGDA] = 1.84 · 10^{-5} mol cm^{-2}; [BDMK] = 2.5 · 10^{-6} mol cm^{-2}; λ = 340—380 nm; m_{binder} = 7 mg)

x_D^0	atm.	$k_p[P^\bullet]_0/k_t^m$	k_t^m	$k_p[P^\bullet]_0$	k_p/k_t^b	order*)	β
0.03	argon			0.18	0.032	2	0.20
	air	10	0.03	0.28	0.018	1 + 2	0.40
0.19	argon			0.056	0.093	2	0.35
	air			0.20	0.035	2	0.30
0.45	argon			0.018	0.12	2	0.50
	air	1	0.01	0.012	0.10	1 + 2	0.60

*) Termination order.

necessary to use an additional method to obtain more information about the systems investigated. Most importantly, aspects of mobility must have an important influence on such reaction systems.

The fluorescence probe technique is a very successful method for determining relevant quantities, as has already been shown for other systems [12]. The fluorescence depolarization technique has often been used in the past for viscosity-dependent studies [13]. Because such measurements require special conditions, a dye possessing a very strong viscosity-dependent fluorescence was utilized in the present study. Crystal violet 8 (CV) meets this requirement, and for this reason it was selected as the fluorescence probe. The photophysics of this dye is well known [14].

6

Furthermore, CV possesses the properties necessary for fluorescence probe investigations:
— a sensitive fluorescence signal changing in cases of variable viscosity;
— no significant absorbency in the photoinitiator region.

A model for explaining the viscosity-dependent fluorescence of CV is given in Fig. 3. After excitation, the probe is able to fluorescence normal in the red region. In competition, a rotation process in the excited state can take place, because, in this way, a more stable product with a twisting angle of approximately 90° is formed. This product belongs to the **t**wisted **i**ntramolecular **c**harge **t**ransfer states (TICT) [14]. As is known of such intermediates, it deactivates very fast without radiation or radiates with another frequency of normal fluorescence. The formation of TICT-states is based on an intramolecular charge transfer from the donor part to the acceptor part of the CV molecule. It should be noted that, of all processes, only the formation of the TICT-state is strongly viscosity-dependent. At low viscosities the rotation of molecule parts of CV against the surrounding molecules is sufficiently high. Deactivation via rotation and TICT-formation is therefore very effective, and the resulting red fluorescence from crystal violet is less intensive. At high viscosities the rotation is reduced, and an effective fluorescence can thus be measured.

However, the processes which proceed after the excitation of CV are very fast, hence time-resolved measurements are necessary to monitor the fluorecence lifetime τ_f of the probe. By means of such time-resolved measurements, it is possible to obtain quantitative information about the rotation process. The primary experimental quantity is the fluorescence intensity (see a typical example in Fig. 4; the number of channels is the mean of the delay time). Using Eq. (14), the τ_f value can be calculated.

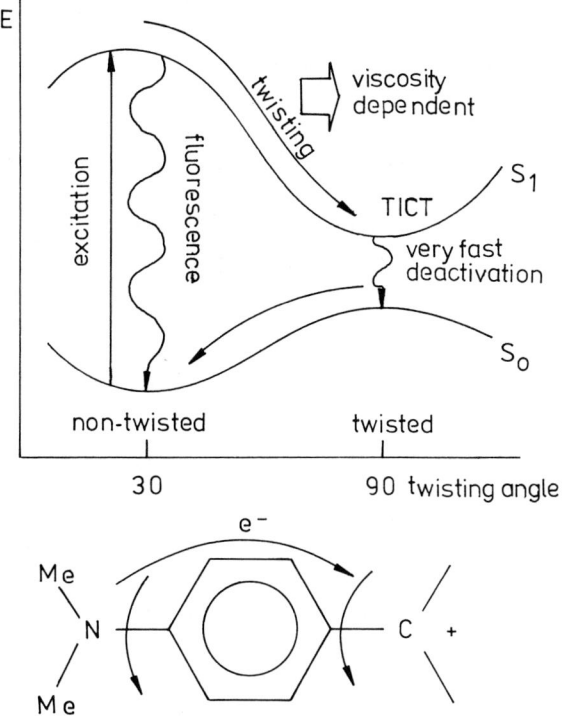

Fig. 3. Photochemical pathways of TICT-fluorescence probe crystal violet

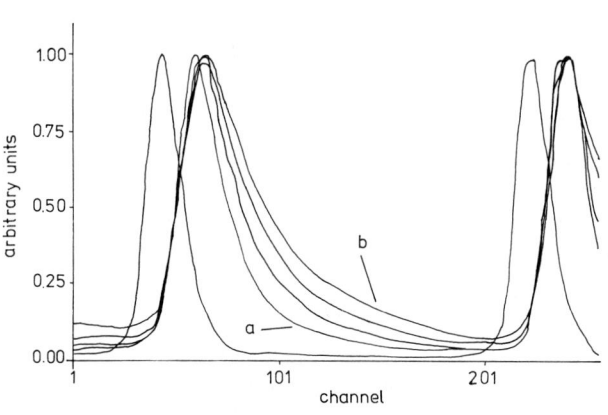

Fig. 4. Fluorescence decay curves of crystal violet in the photopolymer PMMA/DDGDA/BDMK system with respect to irradiation time; a = uncured and b = fully cured system ([DDGDA] = $1.84 \cdot 10^{-5}$ mol $\cdot$ cm^{-2}; [BDMK] = $2.5 \cdot 10^{-6}$ mol $\cdot$ cm^{-2}; λ = 340—380 nm; m_{binder} = 7 mg, I_0 = 4.15 mW $\cdot$ cm^{-2}, air atmosphere, λ_{exc} = 514 nm, λ_{reg} = 630 nm)

$$I(t) \sim \exp(-t/\tau_f) \qquad (14)$$

Between τ_f and the rate constant of rotation k_{rot} the following relation exists:

$$\tau_f = \frac{1}{k_f + k_d + k_{isc} + k_q[Q] + k_{rot}}, \qquad (15)$$

where k_f: constant for the fluorescence; k_d: rate constant for decativation by internal conversion; k_q: rate constant for a bimolecular reaction with a quencher $[Q]$, k_{isc}; rate constant for the isc-process to the excited triplet state.

The rate constant for the rotation k_{rot} can be calculated if the sum $(k_f + k_d + k_{isc} + k_q[Q])$ is known. In liquid nitrogen, where the probe is frozen and rotation cannot occur, this term can be measured. Consequently, under these conditions Eq. (16) holds.

$$\tau_f^0 = \frac{1}{k_f + k_d + k_{isc} + k_q[Q]} \qquad (16)$$

All constants in Eq. (16) are known, and now it is easy to calculate k_{rot} from the lifetime measured.

$$k_{rot} = \frac{1}{\tau_f} - \frac{1}{\tau_f^0} = \frac{1}{\tau_{rot}}. \qquad (17)$$

The quantity of k_{rot} corresponds to the rotational mobility of one part of the dye molecule in the polymeric matrix. The temperature dependence of this value can be described by an Arrhenius plot. Furthermore, the mobility parameters are also connected to the free volume of the matrix. However, for the purposes of this paper, only the mobility parameter k_{rot} is significant because mobility is the controlling factor for kinetics.

Typical fluorescence decay curves are shown in Fig. 4. With prolonged irradiation of the initial system, the decay time increases because the network formed reduces the rotational mobility of the CV probe. A longer fluorescence lifetime must therefore result. This, indeed, has been measured.

The τ_{rot} values obtained using Eqs. (15)—(17) are plotted together with the data of the conversion degree against the irradiation time (see Fig. 5). It is evident that both curves have nearly the same shape. In the first stage of the crosslinking process, a very fast polymer formation was found. But in a second, longer reaction period, the polymerization takes place slowly.

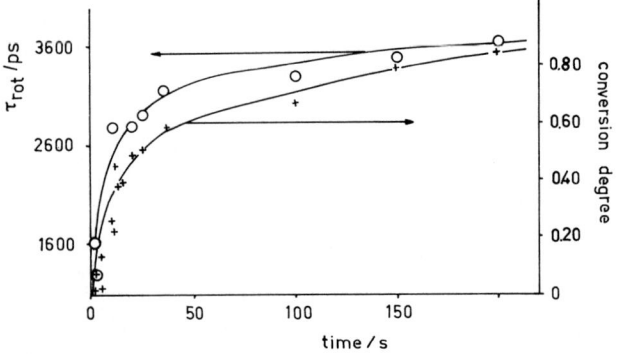

Fig. 5. Rotational lifetime τ_{rot} of CV and conversion degree of the PMMA/DDGDA/BDMK system as a function of irradiation time; for experimental conditions see Fig. 4

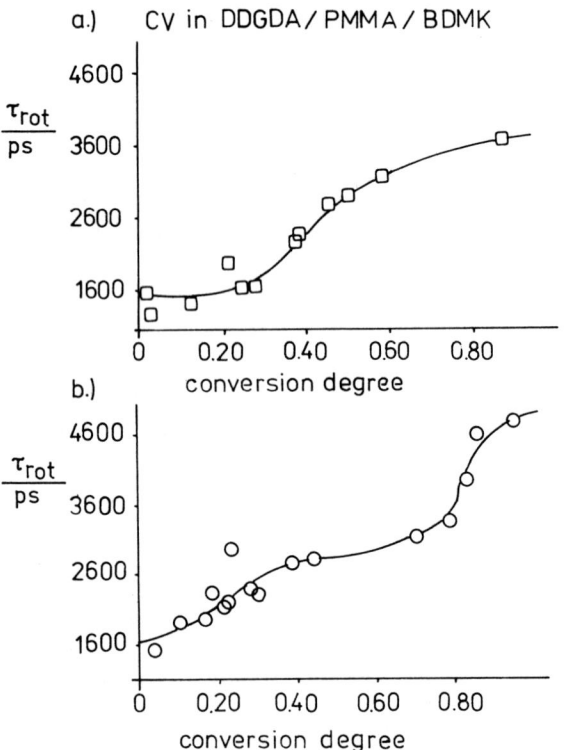

Fig. 6. Rotational lifetime τ_{rot} of CV as a function of conversion degree in the PMMA/DDGDA/BDMK system a) in air and b) in inert gas atmosphere; for experimental conditions see Fig. 4

It is evident that the variation of the R_x values has the same cause as that of the τ_{rot} value. In both cases the increased viscosity due to network formation hinders the processes.

Correlations between rotational lifetimes τ_{rot} and conversion degree x are given in Fig. 6. Again, the very strong influence of oxygen is remarkable. Both the shape and the final τ_{rot} value are different. Higher τ_{rot} values were measured after full exposure in the absence of oxygen. This fact can be explained as the formation of a more dense network in argon atmosphere. Surprisingly, in this case the τ_{rot} values increase very strongly at approximately $x = 0.8$. The intramolecular rotation in the CV molecule is therefore hindered to a greater degree by the polymeric surroundings. The τ_{rot} values measured are in a range similar to those obtained in other organic glasses with the same and similar dyes [14, 15].

However, rotational lifetimes of approximately 1600 ps have also been measured in the initial polymer system. Compared to the results in the viscous solvent glycerol ($\tau_{rot} \approx 300$ ps [15]), these data are much higher. It must therefore be questioned whether the τ_{rot} values alone indicate a glassy state.

DSC measurements

DSC measurements were also carried out in order to obtain information about the T_g value and phase conditions.

With the initial polymeric system and the system at $x = 0.10$, no glass transition temperatures between 30° and 150°C can be measured. It seems that both systems are in the viscoelastic state. In contrast, the fully exposed probe ($x = 0.94$) exhibits an

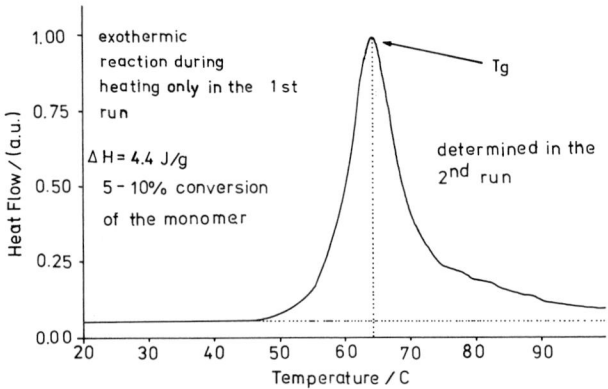

Fig. 7. DSC-measurement for a fully cured PMMA/DDGDA/BDMK-system (composition see Fig. 1) in the first heating rate (heating rate 10°C min^{-1}); T_g was determined in the second heating range

exothermic peak during the first heating run (see Fig. 7). The peak area corresponds to 4.4 J/g. We assume that this peak is caused by occluded radicals in the cured matrix. These radicals are frozen in the glassy state. If the T_g is reached by heating the probe, then the mobility of these radicals is high enough to initiate further polymerization. Thus, the heat value corresponds to an additional double bond conversion of approx. $x = 0.07$. All C=C bonds of the initial monomer can be polymerized together with the photoinduced polymerization. Similar ideas have already been described for analogous systems [16, 17].

In a second heating run of the same probe, the exothermic peak cannot be detected. However, a T_g value of 69°C was obtained in this case. This value is distinctly smaller than that of the pure binder poly(methylmethacrylate) ($T_g = 110$°C). Thus, the network formed during irradiation is miscible with the binder.

References

1. Kloosterboer JG (1988) Adv Polym Sci 84:3—62
2. Baumann H, Timpe H-J (1989) Kunststoffe 79:696—701
3. Böttcher H, Görgens E (1985) Chem Techn 37:137
4. Timpe H-J, Baumann H, Rautschek H, Rautschek M, Müller C (1988) Chem Techn 40:327
5. Mishikubo T, Imaura M, Mizuko T, Takaoka T (1974) J Appl Polym Sci 18:3445—3454
6. Timpe H-J, Strehmel B, Roch F-H, Fritzsche K (1987) Acta Polym 38:238—244
7. Tredwell CJ, Keary CM (1979) Chemical Physics 43:307—316
8. Fleming I "Grenzorbitale und Reaktionen organischer Verbindungen" Weinheim, Verlag Chemie, 1979
9. Müller F-W, Schiller K, Marx J (1985) Angew Makromol Chem 133:75—95
10. Timpe H-J, Strehmel B (1990) Angew Chem 178:131—142
11. Timpe H-J, Strehmel B (1991) Makromol Chem 192:779—791
12. Itagaki H, Horie K, Mita I (1990) Progr Polym Sci 15:361—424
13. Tassin JF, Monnerie L (1983) J Polym Sci: Polym Phys Ed 21:1981—1992
14. Anwand D, Müller F-W, Strehmel B, Schiller K (1991) Makromol Chem 192:1981—1991
15. Strehmel B, Strehmel V, Timpe H-J, Urban K (1992) European Polymer Journal 25:525—533
16. Timpe H-J, Strehmel B, Schiller K, Stevens S (1988) Makromol Chem, Rapid Commun 9:779—791
17. Klosterboer JG, van de Hai GMM, Boots HMJ (1984) Polym Comm 25:354—357

Authors' address:

Dr. Bernd Strehmel
GNF e.V.
AG "Laserdiagnostik"
Rudower Chaussee
D-O-1199 Berlin, FRG

Observation of the curing process of epoxy resins by inverse gas chromatography

H. Wetzel, K.-G. Häusler, and M. Fedtke

Department of Chemistry, Institute for Macromolecular Chemistry, Technical University Merseburg, FRG

Abstract: Inverse gas chromatography (IGC) was used to observe the curing of non-modified and diol modified systems of epoxy resins on the basis of diglycidyl ether of bisphenol A (DGEBA). This method can be used for determining the gel point, the times of vitrification and curing of the respective system. The activation energy of the crosslinking reaction and the diffusion coefficients of the injected solvent in the curing and cured epoxy resins can be obtained by varying the conditions of gas chromatography.

Key words: Epoxy resins; curing process; gel point; glass transition; inverse gas chromatography

Introduction

Epoxy resins have excellent thermal, mechanical, and chemical properties. Networks with widely differing properties can be synthesized by chemical modification, and variation of the physical conditions during the curing process, respectively. The knowledge of the relationship between chemical composition and curing conditions, and the properties of the cured network provide a possibility of deliberately producing materials with optimal properties. Therefore, the development and improvement of methods for analytical examination of the overall curing process are of great practical importance. In addition to viscosity, acoustic, electrical and thermoanalytical measurements, inverse gas chromatography is also a suitable method for monitoring the crosslinking reaction of epoxy resins up to the completion of the crosslinking structure, and hence the transition from the fluid to the solid.

In an experiment using IGC the species of interest is not the injected sample, as in usual gas chromatography, but the stationary phase, which is usually made up of a polymer-coated support. A GC column can be packed with this support. The epoxy resin under investigation is characterized by means of retention data of injected volatile test substances.

Experimental

Preparation of the epoxy resins

DGEBA was purified by recrystallization of Epilox A 17-00 (Leuna-Werke) from acetone-methanol (F. = 42 °C). DGEBA was well mixed with diamino diphenyl sulphone (DDS) in a molar ratio of 1:1. These mixtures can be stored for some time and used for measurements without further pretreatment [1].

For a second system of reaction, DGEBA was reacted with butan-1,4-diol (BU) in a molar ratio of 1:0.25 and 3 Mol-% magnesium perchlorate as curing agent ($Mg(ClO_4)_2$, Merck) in a thermostated double-jacketed glass vessel at 100 °C. After conversion of about 15% the reaction mixture becomes homogeneous and clear. After cooling to room temperature, these preproducts can be stored for some days without measurable reaction.

An inert support was coated with the systems of epoxy resins using the "Fritting Method" proposed by Struppe [2]. The stationary phase obtained was aged at room temperature and filled into a GC-column.

Measurements

The measurements were carried out using a high temperature gas chromatograph GCHF 18.3 (Fa. Chromatron).

The test substance used was methylethylketone. Methylethylketone (0.01...0.1 µl) was dosed by means of a sample injected valve for better reproducibility of the dosed amount. The pure carrier gas was divided for reaching the necessary dilution of the solvent-carrier gas mixture. The essentially smaller part was conducted through the solvent, which resulted in the solvent being enriched in the carrier gas. The split parts were then reunited and flow through the sample loop. This mixture enters the GC-column when the injected valve is turned. The chromatogram obtained was evaluated using Integrator IC 100 A of the firm Laboratorni Pristroje Praha.

The experimental equipment is represented in Fig. 1.

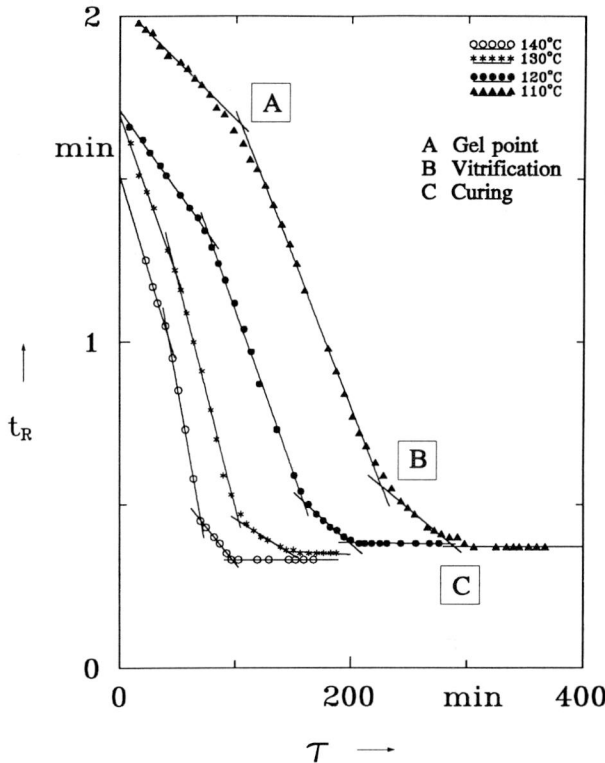

Fig. 2. Dependence of retention time on curing time for DGEBA/DDS

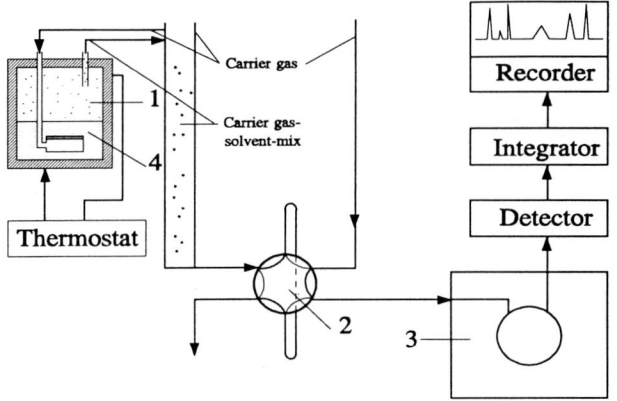

Fig. 1. Experimental equipment of the IGC (1 = wash bottle; 2 = injection valve; 3 = GC with GC-column; 4 = solvent)

Results

The DGEBA/DDS system described by Grozdov et al. [3] was first re-examined in order to test the applicability of the method and to utilize the knowledge obtained for diol modified epoxy resins. It could be confirmed that the curing process can be observed by the decrease in retention time and change in peak height, in dependence on curing time (Figs. 2 and 3).

For the sake of more exact determination of the characteristic transitions, however, we only used the retention time/curing time curve which can be divided into four sections of different slopes.

The first section up to point A can be assigned to the pregel state. Retention proceeds by absorption in the polymer matrix. Point A is the gel points. It is associated with a change in the rate of decrease of retention time. The strong increase of viscosity beyond the gel point essentially reduces the capability of the solvent to penetrate into the polymer phase.

The contribution from the retention volume, which is based on adsorption on the polymer surface, to the total retention volume becomes greater. As a consequence, the peak height stops decreasing and starts increasing again shortly before the time of glass transition is reached. According to Guillet and Lipson [4], the system here is in the region of non-equilibrium absorption. That means, due to the small diffusion coefficient the diffusion rate becomes so slow that the solute does not have time to reach an equilibrium partition between polymer and carrier gas.

The second change in the slope of the retention time/curing time curve (Point B, Fig. 2) is interpreted as the time of glass transition. Now the polymer is presented as a solid in a glassy state.

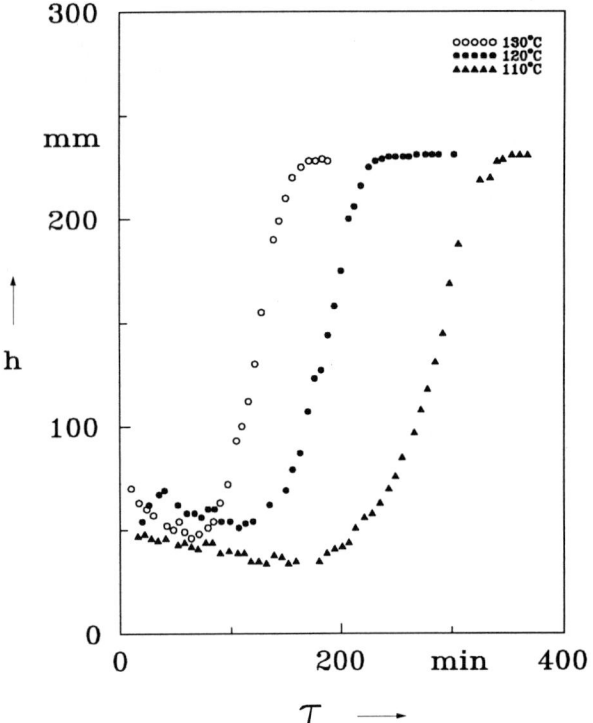

Fig. 3. Dependence of peak height on curing time for DGEBA/DDS

In most cases, however, the reaction is not yet terminated. The crosslinking reactions occur as before but proceed at a much slower reaction rate. Due to further solidification of the structure, it is now hardly possible for the solvent to penetrate into the polymer. The contribution of surface adsorption to the sorption increases strongly, which becomes evident in a strong increase of peak heights.

The time at which peak height and retention time assume a constant value is defined as the time of the apparent end of curing (Point C, Fig. 2). Retention now almost exclusively occurs by condensation and adsorption on the polymer surface.

In order to investigate the effects of temperature (T) on the curing time, curing was examined at differing temperatures within the range of 110 ... 140°C (Fig. 2). A rise in temperature caused an accelearation of the reaction.

The optimal parameter for the crosslinking of the resins investigated can be selected from a plot of the gel points, glass transitions and end points of curing in the form of an Arrhenius- or a TTT-diagram according to Gillham [5], in dependence on the reaction time and temperature, considering the density of crosslinking achieved.

An activation energy (E_A) of the curing process of 55 kJ/mole was found from a plot of gel points (t_g) in the Arrhenius-diagram (Fig. 4, Eq. (1)).

$$\ln t_g = \ln t_{g0} + \frac{E_A}{R} \cdot \frac{1}{T} \quad (1)$$

(R gas constant).

The activation energy determined from plotting the times of glass transition over the temperature was 51 kJ/mole.

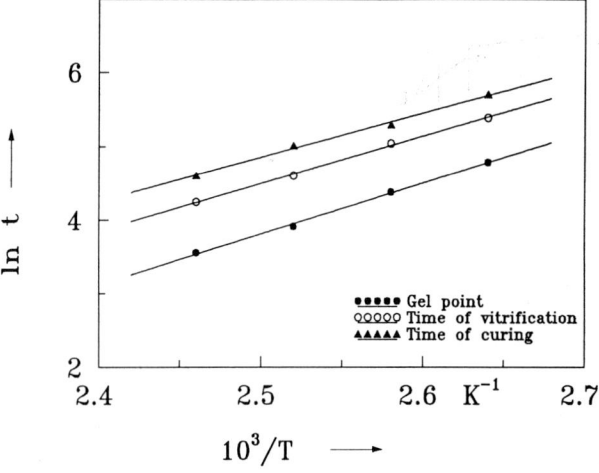

Fig. 4. Arrhenius-diagram for the curing reaction for DGEBA/DDS

The knowledge obtained from the non-modified epoxy resin was transferred to diol modified resins. The DGEBA/BU system was hardened using $Mg(ClO_4)_2$ as a catalyst. The forms of the curves obtained (Figs. 5 and 6) differ from those of the DGEBA/DDS system mentioned above. Peak heights do not show the strong increase after glass transition. This is due to the decrease of crosslinking density and internal softening caused by insertion of diole. Low crosslinking density and internal softening result in the fact that the diffusion coefficient of the solvent in the polymer is still relatively high; hence, the contribution of absorption of solvent in the polymer phase to the total retention volume still causes peak broadening.

The glass temperature of the DGEBA/BU/$Mg(ClO_4)_2$ system is at 125°C [6], which is rather low as compared with the curing temperature. This

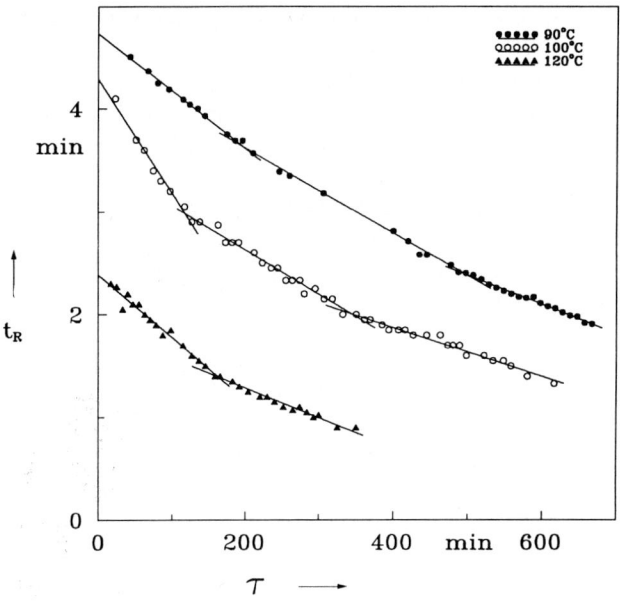

Fig. 5. Dependence of retention time on curing time for DGEBA/BU/Mg(ClO$_4$)$_2$

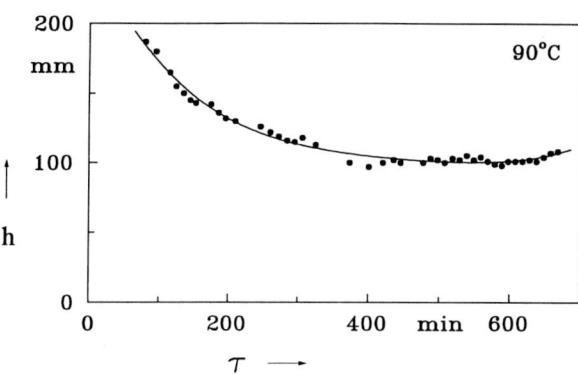

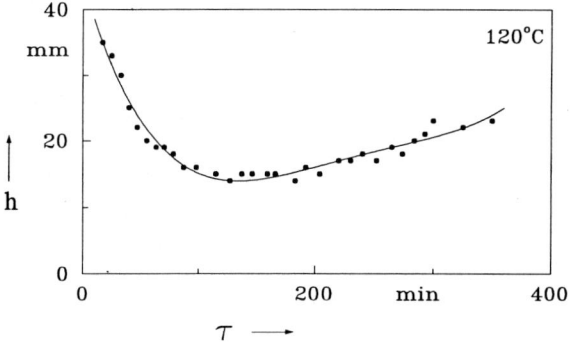

Fig. 6. Dependence of peak height on curing time for DGEBA/BU/Mg(ClO$_4$)$_2$

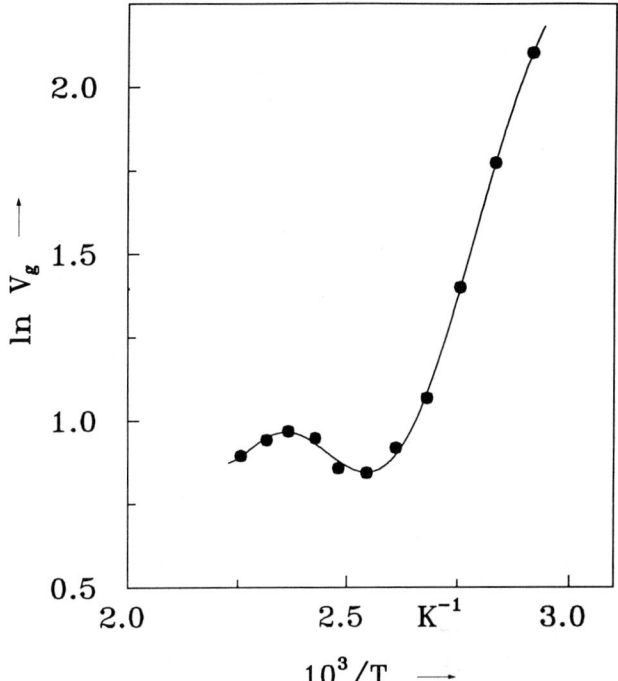

Fig. 7. Retention diagram for DGEBA/BU/Mg(ClO$_{42}$)

small difference between hardening and glass temperature may well be another reason for the small increase in peak heights.

The glass temperature of the cured DGEBA/BU/Mg(ClO$_4$)$_2$ network was determined by IGC according to the method described by Guillet and Lipson [4] (Fig. 7, see Eq. (2)). It was found that glass transition occurs in a temperature range of 105 ... 145 °C. 125 °C (minimum) was found to be the mean glass temperature.

$$\log V_g = \frac{H_L}{2{,}3\,RT} + \text{const.} \qquad (2)$$

(H_L solution enthalpy of the polymer substance).

This value is in good agreement with the glass temperature determined by means of thermomechanical analysis [6].

A fact of great important for polymer applications is the capability of small molecules of diffusing into and through the polymer. At 180 °C, the diffusion coefficient (D_L) of the methylethylketone used for IGC was determined in the DGEBA/BU/Mg(ClO$_4$)$_2$ system cured at 120 °C (Fig. 8).

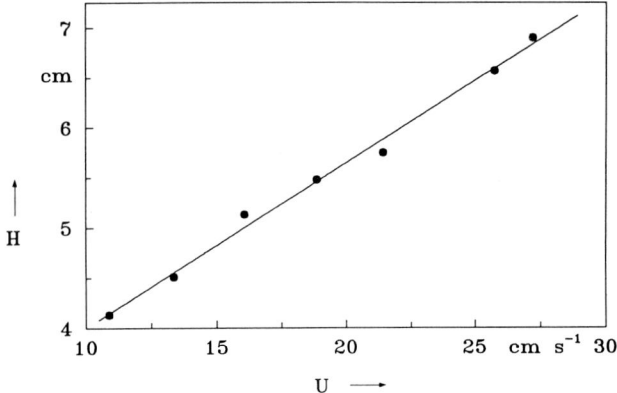

Fig. 8. Van-Deemter-diagram for DGEBA/BU/Mg(ClO$_4$)$_2$

By varying the carrier gas velocity (u), a diffusion coefficient of $9.8 \cdot 10^{-10}$ cm^2 s^{-1} was obtained from the retention parameters by means of the mass transfer coefficient (C, see Eq. (3)) of the Van-Deemter-equation (Eq. (4)):

$$C = -\frac{8 d^2 k'}{\pi^2 D_L (1 + k')} \quad (3)$$

$$H = A + \frac{B}{u} + C \cdot u, \quad (4)$$

where H is the plate height of the column, A, B are constants, k' is the capacity or distribution coefficient, and d is the thickness of stationary phase).

It could be shown that the IGC is suitable to observe the whole curing process of nonmodified and diol modified systems of epoxy resins on the basis of diglycidyl ether of bisphenol A. IGC is a simple and cheap method for determining the gel point, time of vitrification and end point of curing of the respective system. On the other hand, it is also possible, after curing, to determine glass transition temperature, diffusion coefficient and solubilities for probes in the polymer, as well as numerous thermodynamical parameters.

References

1. Wetzel H, Häusler K-G (1990) Plaste und Kautschuk 37:219—224
2. Leibnitz E, Struppe HG (1984) Handbuch der Gaschromatographie, Leipzig, Akad Verlagsges Geest & Portig K.-G., pp 198
3. Grozdov AG, Stepanov BN (1980) Polym Sci U.S.S.R. 22:11—15
4. Lipson JEG, Guillet JE (1982) In: Developments in Polymercharacterisations-3, Dawkins JV (ed) London, Appl Science, pp 33—74
5. Wisanrkkit G, Gillham JK (1990) J of Coatings Technology 62:35—50
6. Häusler K-G, Tänzer W, Kunze A (1991) Acta Polymerica 42:565—570

Authors' address:

Prof. Dr. Karl-Georg Häusler
Dipl.-Chem. Hendrik Wetzel
Technical University Merseburg, Department of Chemistry
Institute of Macromolecular Chemistry
Geusaer Str. 88
D-O-4200 Merseburg, FRG

Fluorescence probe studies during the curing of epoxy systems

B. Strehmel[1], M. Younes[2], V. Strehmel[1], and S. Wartewig[2]

Departments of Chemistry[1] and Physics[2], Technical University Merseburg, FRG

Abstract: The curing of bisphenol-A-diglycidylether with diaminodiphenylmethane has been investigated by monitoring the fluorescence decay of relaxing stilbazolium dyes. Because these compounds show a very strong viscosity-dependent fluorescence, it is possible to obtain information about the rotational mobility of the probe within the curing system investigated. The conversion degree of epoxy groups was determined by Raman spectroscopy. Further, the time of gelation was measured by a cycloviscograph (torque measurement). Results obtained by Raman spectroscopy and torque measurements are compared with data obtained by time-resolved fluorescence measurements. The fluorescence lifetimes measured during the curing process increase very fast near the gelation point. Investigations of the uncured system lead to the conclusion that two kinds of diffusion can be studied by using such probes: Stokes diffusion and diffusion into free volume.

Key words: Bisphenol-A-diglycidyl ether; amine curing; fluorescence probes; stryryl compounds

Introduction

We report on the curing of bisphenol-A-diglycidylether (DGEBA) I in the presence of diaminodiphenylmethane (DDM) II.

Earlier, this reaction was already investigated by application of different analytical methods [1, 2]. The results obtained yield information about macroscopic properties like viscosity changes, network density, and others. Fluorescence probes provide information about microscopic parameters like mobility and microviscosity. Microscopic together with macroscopic parameters give more details about the processes occurring in such complicated systems. In other studies, the curing of epoxides has already been investigated with the fluorescence technique [3]. But, very often, the fluorescence depolarization method has been used to obtain information about microviscosity parameters [4]. The disadvantage of this method lies in the small time region and is caused by the ratio of rotational time and fluorescence lifetime. However, with stationary polarization experiments the network-forming process of vinyl compounds has been successfully analyzed [4]. In other studies the self fluorescence of curing epoxides has been used for indication [5]. The structure of the emitting species has only been postulated because the emitting product was not isolated.

Quantitative information about microscopic parameters can be obtained from dyes possessing a strong viscosity-dependent fluorescence. In this case, the pathways leading to viscosity-dependent effects are known.

Experimental part

Substances

The epoxy compound DGEBA (Leuna-Werke AG) was recrystallized from acetone-methanol-mixture and the diaminodiphenylmethane DDM (Merck)

was recrystallized twice from methanol. The purity of the compounds was characterized by ^{1}H-NMR, ^{13}C-NMR, IR and HPLC. The stilbazolium compounds were synthesized as described elsewhere [6] and the dye quinaldin blue was used as a commercial product from Aldrich without further purification.

For monitoring the curing process by time-resolved fluorescence the following substituted styryl- and cyanine dyes were used:

[Chemical structures: DASPMI, DASPI, JSMPI, QB]

The advantages of these dyes are that they are stable and do not bleach (determined by absorption spectroscopy). Furthermore, they do not accelerate the curing reaction, which was proved by means of torque measurements in the presence and absence of fluorescence probes in the system.

Measurements

The viscosities of DGEBA/DDM mixtures between 30° and 70°C were determined with a Höppler viscometer. Glycerol, for which the temperature dependence of viscosity is known [7], was used as a standard. Torque measurements were carried out with a cycloviscograph from Brabender.

Fluorescence studies based on time-correlated single photon counting measurements were performed with the argon ion laser ILA 120-1 (Carl Zeiss Jena AG) as light source. The set-up has been described elsewhere [8]. Deconvolutions necessary for the processing of the experimental signals were done by using the phase plane method or Marquardt's non-linear least square algorithm.

Conversion of epoxy groups was derived from the Raman intensity ratio of the band at 916 cm^{-1} (epoxy group) with respect to the band at 640 cm^{-1} (aromatics). The computer-aided Raman spectrometer used was equipped with the argon ion laser ILA 120 and a double grating monochromator GDM 1000 (both from the Carl Zeiss Jena AG).

Sample preparations

The fluorescence experiments with mixtures of DGEBA and DDM (molar ratio = 2:1) were directly carried out in 5-mm glass tubes at 70°C in the single photon counting set-up. The dye was dissolved before the curing reaction started.

Results and discussion

The curing of DGEBA with DDM was started by addition of the primary amino group to DGEBA, which leads in the first reaction step to secondary amino groups and secondary OH-groups. In the following steps the epoxy groups attach to these secondary OH-groups and secondary amino groups and insoluble polymer networks result.

The Raman band intensity of the epoxy group decreases during network formation while the intensity of the aromatics is nearly constant. The conversion degree x of epoxy groups was calculated by the following formula:

$$x = 1 - \frac{[I_{(epoxy)}(t)] \cdot [I_{(aromat)}(t=0)]}{[I_{(aromat)}(t)] \cdot [I_{(epoxy)}(t=0)]}. \quad (1)$$

The Raman spectra at the reaction time $t = 45$ and $t = 300$ min are plotted in Fig. 1a and b. The conversion degree according to Eq. (1) is shown in Fig. 1c.

The dyes used possess a strong viscosity-dependent fluorescence. This phenomenon can be explained by a rotation of a molecular part in the excited state because more stable products are formed in that way. Those processes can be explained by the formation of twisted intramolecular charge transfer states (TICT) whereby an electron transfer occurs within the twisted molecule in the excited state [9]. Quantumchemical studies (CNDO calculations) are useful to describe such reactions [9].

From time-dependent fluorescence experiments these rotational processes can be determined quantitatively. The fluorescence intensity decay can be described by Eq. (2).

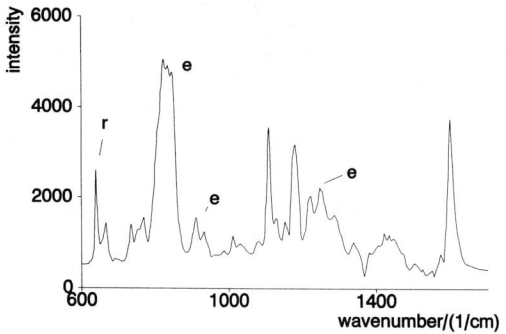

Fig. 1a. Raman spectrum of the mixture [DGEBA]/[DDM] = 2:1 (molar ratio) at 70°C; cure time 45 min, excitation wavelength 488 nm, r = reference (aromat), e = epoxy

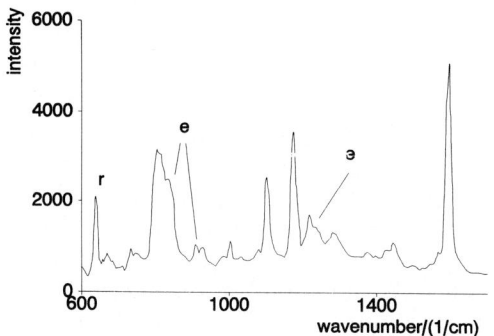

Fig. 1b. Raman spectrum of the mixture [DGEBA]/[DDM] = 2:1 (molar ratio) at 70°C after 300 min curing; excitation wavelength 488 nm, r = reference, e = epoxy

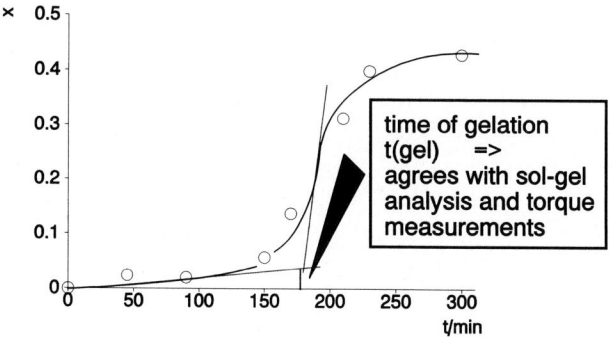

Fig. 1c. Conversion degree of epoxy groups (x), calculated with Eq. (1), in the system [DGEBA]/[DDM] = 2:1 (molar ratio) at 70°C as a function of the reaction time (t)

$$I_f(t) \sim \exp(-t/\tau_f) . \qquad (2)$$

According to Scheme 1, τ_f can be expressed by Eq. (3)

$$\tau_f = \frac{1}{k_f + k_{nr} + k_{rot}} , \qquad (3)$$

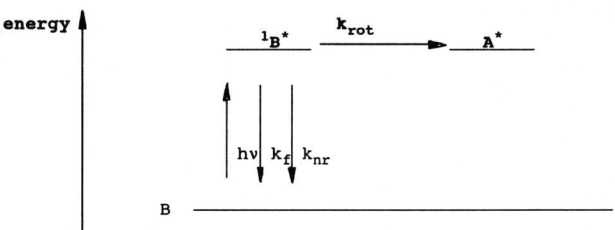

Scheme 1: Photochemical pathways of a fluorescence probe B occurring a deactivation by rotation of one molecule part in the excited state of B (B^*) to a energetically more stable twisted product in the excited state (A^*), rate constant k_{rot}; abbreviations of the other rate constants in this scheme are described in the text

where k_f-rate constant for fluorescence, k_{nr}-rate constant for viscosity independent radiationless processes, and k_{rot}, the most important constant, is the rate constant for the rotational process of a molecular group in the TICT-state. Only this process is strongly viscosity- and temperature-dependent and can be related to the rotational mobility of the probe in the matrix. The temperature dependence of k_f and k_{nr} can be neglected because these values are very small compared with those of k_{rot}. It is also possible that the first excited singlet state is quenched by other molecules. This process is also viscosity- and temperature-dependent [10], but, in the system investigated, such reaction does not occur.

For the calculation of k_{rot} the value of the sum ($k_f + k_{nr}$) has to be known. The related parameter τ_f^0 is given by Eq. (4)

$$\tau_f^0 = \frac{1}{k_f + k_{nr}} \qquad (4)$$

and can be measured by experiments at liquid nitrogen temperatures where the rotation is frozen in. For the different dyes in the system DGEBA/DDM = 2:1 (molar ratio) the following τ_f^0 values were determined: DASPMI: 4800 ps; DASPI: 4000 ps; JSMPI: 4550 ps; QB: 2700 ps.

From Eqs. (3) and (4) follows:

$$k_{rot} = \frac{1}{\tau_f} - \frac{1}{\tau_f^0} = \frac{1}{\tau_{rot}}. \quad (5)$$

Because the rotational process is based on first-order kinetics, the rotational lifetime (τ_{rot}) of the probe can be used to quantitatively describe that process.

In fluid systems, the rotational process is the fastest of all the deactivation processes of the excited state. If this process is restricted, then the τ_f or τ_{rot} values increase. Thus, changes of τ_f indicate changes of the rotational mobility of the probe molecule in the polymeric matrix. An increase of τ_f is caused by a stronger hindrance of the probe rotation due to the surroundings. In this case, k_{rot} decreases and, therefore, τ_f approaches the value of τ_f^0.

For an uncured system at different temperatures the influence of viscosity changes on τ_{rot} is plotted in Fig. 2. From the experimental values τ_f and τ_f^0 the τ_{rot}-value was calculated with the expression of Eq. (5). It has been found that the following relation is fulfilled for an uncured system:

$$\tau_{rot} = C\eta^a. \quad (6)$$

The exponents a, determined by a double logarithmic plot, are listed in Table 1.

The a-values give the following information:

$a = 1 \Rightarrow$ the dye nearly quantitatively describes the Stokes diffusion

Table 1. Viscosity exponent a for different probes measured in the system [DGEBA]/[DDM] = 2:1 (molar ratio) in the uncured state (error limit ±0.07)

Probe	DASPMI	DASPI	QB	JSMPI
a	0.30	0.45	0.55	0.53

$a < 1 : 1 - a \Rightarrow$ part of the activation energy describing the diffusion into free volume,

$a \Rightarrow$ part of the activation energy describing Stokes diffusion.

Both activation energies (Stokes diffusion and diffusion into free volume) show the two kinds of motion monitored by the probes used. This fact can also be explained by the microviscosity theory by Gierer and Wirtz [11], which was also successfully applied to fluorescence probes in other viscous systems [8].

For the curing process of the system studied the τ_{rot} values are calculated at different reaction times. The curve obtained is plotted in Fig. 3. It is evident from this picture that the point of inflection can be related to the gelation point (determined by torque measurements [1]). At this time the viscosity increases very fast, leading to the reduced rotational mobility of the probe molecule. Furthermore, in the first reaction part before the gelation time is approached a change of the mobility can already be registrated (Fig. 3).

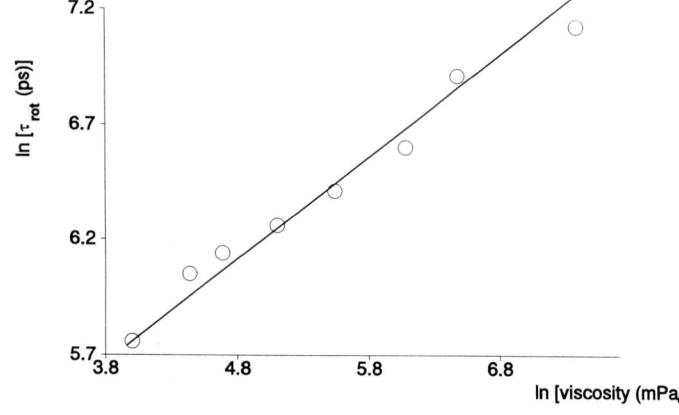

Fig. 2. DASPI in [DGEBA]/[DDM] = 2:1 (molar ratio). Logarithm of the rotational time τ_{rot} versus the logarithm of the viscosity at different temperatures for the uncured system

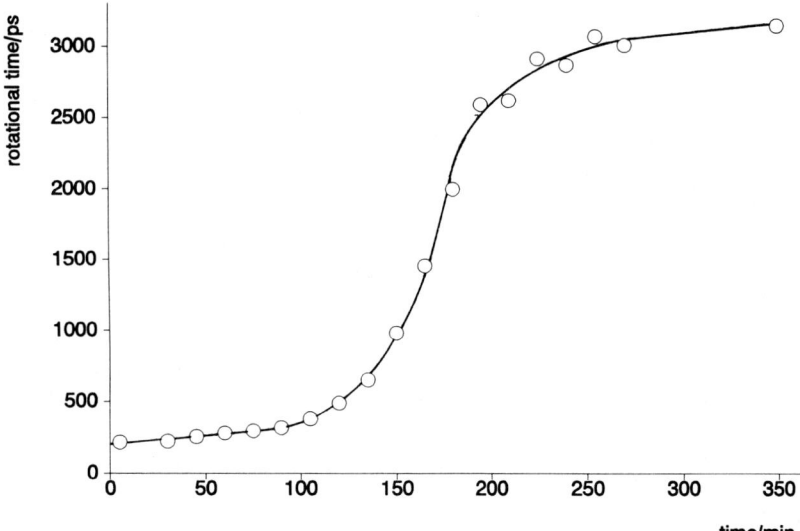

Fig. 3. QB in [DGEBA]/[DDM] = 2:1 (molar ratio). Dependence of the rotational time τ_{rot} on the rection time (t)

This small effect is also shown in the plot of the rotational lifetimes against the degree of conversion (see Fig. 4). The reaction of the epoxide compound with the aromatic amine is an autocatalytic process [12] which can also be observed in the course of the conversion-time-curve in Fig. 1c. The fastet acceleration of this reaction occurs in the region of the gelation point. Therefore, the τ_{rot}-values obtained for the different fluorescence probes used are plotted as a dependence of the conversion.

It is obvious that all probes possess the point of inflection at about the same conversion. This means that all probes describe the variation in the course of the curing process in the same manner. Furthermore, it is evident that the mobility of the fluorescence probes changes with prolonged curing reaction ($x > 0.2$).

References

1. Strehmel V, Fryauf K, Sommer C, Arndt K-F, Fedtke M (1992) Angew Makromol Chem 196:195—206
2. Dusek K (1987) Adv Polym Sci 78:1—60
3. Shmorhun M, Jamieson AM, Simha R (1990) Polymer 31:812—817
4. Fuhrmann J, Leicht R (1981) Polymer Bulletin 4:141—148; Fuhrmann J, Leicht R (1980) Colloid Polym Sci 258:631—637
5. Levy RL, Schwab SD (1988) ACS Symp Ser 367:113—121
6. Vogel M (1991) Dissertation Freiburg
7. Landolt-Börnstein: Zahlenwerte und Funktionen aus Physik, Chemie, Astronomie, Geophysik und Technik, 6th Edition, Vol. 5
8. Strehmel B, Strehmel V, Timpe H-J, Urban K (1992) Eur Pol j 28:525—533
9. Rettig W (1986) Angew Chem 98:969—986
10. Chu DY, Thomas JK (1989) J Phys Chem 93:6250—6257
11. Gierer A, Wirtz K (1953) Z Naturforsch 8a:532—546
12. Mijovic J, Fishbain A, Wijaya J (1992) Macromolecules 25:979—985

Authors' address:

Dr. Bernd Strehmel
GNF e.V.
AG "Laserdiagnostik"
Rudower Chaussee
D-O-1199 Berlin, FRG

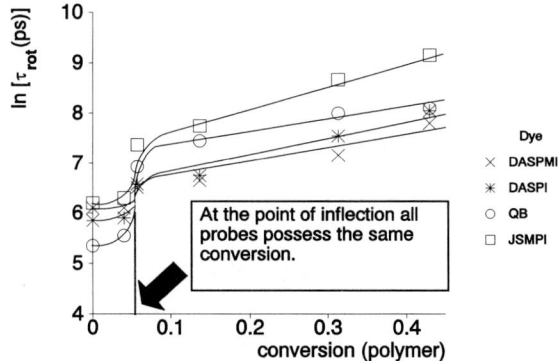

Fig. 4. Logarithm of the rotational time τ_{rot} as a function of conversion degree of epoxy groups for different dyes in the system [DGEBA/DDM] = 2:1 (molar ratio)

Conditions of formation and equilibrium swelling of polymer networks formed by protein macromolecules

Yu. Mel'nichenko and V. Klepko

The Institute of Macromolecular Chemistry, Ukrainian Academy of Sciences, Kiev, Ukraine

Abstract: Kinetics and extent of equilibrium swelling of gelatin hydrogels formed under different conditions were studied as a function of polymer concentration ϕ, temperature of sol-gel transition T_g and aging time t_a. It is shown that swelling kinetics can be described by the Peters and Candau approach. The extent of equilibrium swelling Q_e is related to the conditions of gel formation by equation $Q_e \sim \phi^{1-\nu} t_a^{-0.15} T_g^{0.1}$ where $\nu = 0.4$. The last result is indicative of the applicability of Flory's classical theory to describe swelling of polymer gels formed by complex protein macromolecules.

Key words: Gelatin; gel; sol-gel transition; swelling

Gelatin gels are widely used in photo- and holography, medicine and food industries, etc. Their application usually includes stages of gelatin gel swelling in water or mixed solvent. In the present work, we investigated the interrelation between conditions of formation and features of gelatin gels swelling.

The samples were prepared from photographic grade gelatin (with the parameters given in [1]) free of any additives. The hot gelatin solution was poured on thin glass substrates which then were rapidly cooled to provide sol-gel transition. To prepare gels we used solutions with gelatin volume fractions of ϕ_0 = 0.021, 0.0308, 0.0625, 0.1288; temperatures of gelation T_{gel} = 4°, 10°, 22°C and aging times t_a = 1, 3, 5 h. The extent of gel swelling $Q(t)$ at a moment of time t was measured by sample weighing after immersion in water and subsequent determination of the ratio of $V(t)/V_0 = Q(t)$, where V and V_0 are the current and initial gel volume, respectively. As is seen from Fig. 1, after approximately 15 h the flat gel samples formed, with heights of about 1 mm, experience a nearly equilibrium swelling with an extent of $Q_e = V_e/V_0$.

To describe the kinetics of gelatin gel swelling, we used the solution of Tanaka, Hocker and Benedek (Eq. (2)) for polymer network motion under swelling in the form of Peters and Candau [3] for infinite flat gel samples:

$$\Delta a(t) \approx \frac{8}{\pi^2} \Delta a_0 \exp(-t/\tau),$$

$$\tau = \frac{a_e^2}{\pi^2 D(1 + \mu/M_{os})}, \quad (1)$$

where $\Delta a(t) = a_e - a(t)$; $\Delta a_0 = a_e - a_0$; a_0, a_e and $a(t)$ are the thickness of a gel at initial, final (equilibrium value) and current moment of time, respectively; D = collective diffusion coefficient; μ and M_{os} = shear and osmotic longitudinal moduli of a gel, respectively. It should be stressed that, in general, Eq. (1) is predicted to be valid for the long time swelling behavior. Kinetic curves of Fig. 1 in theoretical coordinates of $\ln[\Delta Q(t)/\Delta Q_0] = \ln[\Delta a(t)/\Delta a_0]$ vs. time are shown in Fig. 2, confirming the expected logarithmic dependence of relative extent of swelling on t in quite a broad range of the last variable values.

Using the equation [4]:

$$M_{os}/\mu = 2.25 + [1 + 1/2(\phi_e/\phi_0)^{2/3}]/$$
$$[1 - 1/2(\phi_e/\phi_0)^{2/3}], \quad (2)$$

for the gels above, we obtain μ/M_{os} = 0.26 and 0.32, respectively. Processing of the data by Eq. (1) with the obtained values of μ/M_{os} gives $D = 1.4 \times 10^{-7}$ cm^2 s^{-1} for the first and $D = 1.8 \times 10^{-7}$ cm^2 s^{-1}

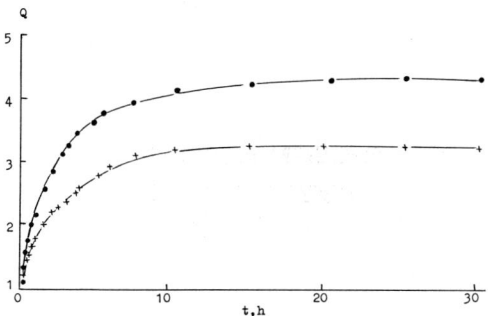

Fig. 1. Kinetics of gelatin gels swelling Q with $\phi_0 = 0.0308$ (×) and $\phi_0 = 0.0625$ (●). $T_{gel} = 22\,°C$, $t_a = 1$ h

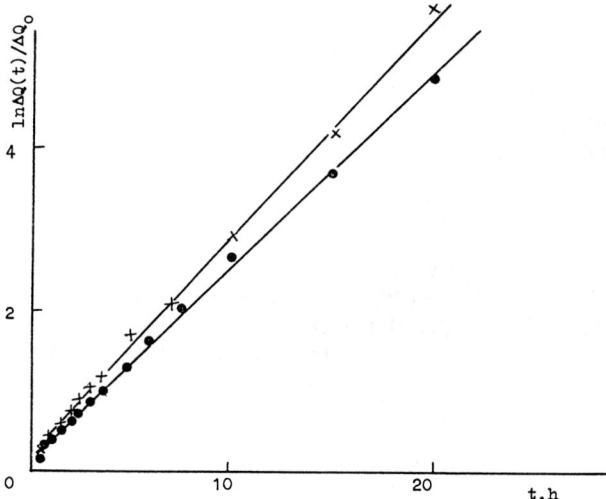

Fig. 2. Kinetics of gelatin gels swelling in logarithmic plot corresponding to Eq. (1). Symbols as in Fig. 1. $\Delta Q(t) = Q_e - Q(t)$; $\Delta Q_0 = Q_e - Q_0 \equiv Q_e - 1$

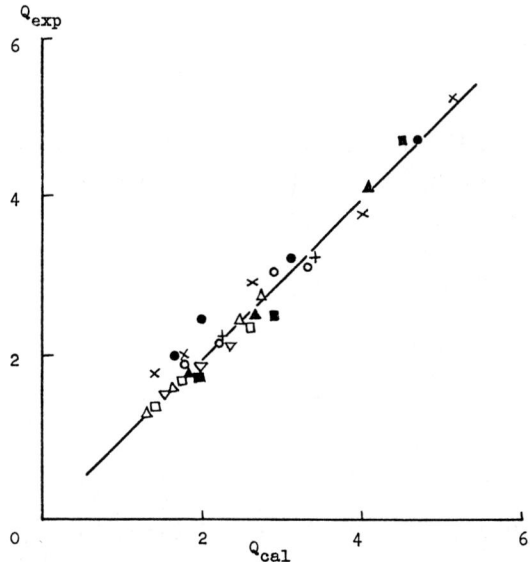

Fig. 3. Correlation between experimental and calculated by Eq. (3) values of equilibrium swelling for gelatin gels formed under different conditions: t_a, h: 1 (●), 3 (×) and 5 (△) for $T_g = 22\,°C$; t_a, h: 1 (○), 3 (▽) and 5 (□) for $T_g = 10\,°C$; t_a, h: 1 (+), 3 (■) and 5 (▲) for $T_g = 4\,°C$

for the second sample. These values of collective diffusion coefficients are in agreement with those of Amis et al. [5] determined by quasi-elastic light scattering.

Computer processing of all the data obtained for different samples had shown that the influence of initial conditions of gelatin gel formation on the extent of their equilibrium swelling Q_e can be described by the general equation:

$$Q_e \sim \phi_0^{1-\nu} t_a^{-0.15} T_g^{0.1}, \qquad (3)$$

where $\nu = 0.4$ (see Fig. 3). The last result shows that Flory's theory is applicable for the description of swelling of polymer gels formed by complex protein macromolecules.

References

1. Mel'nichenko Yu, Gomza Y, Shilov V, Kuzilin Yu (1991) J Photgr Sci 39:133—138
2. Tanaka T, Hocker L, Benedek G (1973) J Chem Phys 59:5151—5156
3. Peters A, Candau S (1988) Macromolecules 21:2278—2282
4. Hecht A, Geissler E (1983) Polym Commun 24:98—100
5. Amis E, Jamney P, Ferry J, Yu H (1983) Macromolecules 16:441—446

Authors' address:

Dr. Y. Mel'nichenko
The Institute of Macromolecular Chemistry
Kharkov Rd 48
253660 Kiev, Ukraine

Cross-link fluctuations: NMR properties and rubber elasticity

M. G. Brereton

IRC in Polymer Science and Technology, University of Leeds, U.K.

Abstract: A scale invariant model for the residual tensorial spin-spin couplings between nuclei has been used to investigate the NMR properties of deformed elastic networks. A general expression is derived which includes the effects of cross-link fluctuation dynamics and a non-isotropic deformation dependence of the fluctuation volume. The theoretical results are used to re-interpret the experimental results of Sotta and Deloche on the doublet structure observed in the ^{2}H NMR spectra of deformed PDMS networks in terms of strain-dependent cross-link fluctuations. Previously, this feature has been used as evidence for an orientation-induced weak nematic network ordering. From the experimental results the magnitude of the junction fluctuations in the undeformed state is estimated to be approximately 1/3 of the phantom network result. Using a theory developed by Flory, these restricted junction fluctuations can be related to the elastic properties of the network. In particular, they were related to the deviations of the elastic modulus from the phantom chain result. When expressed in terms of the Mooney-Rivilin constants C_1 and C_2, the Flory theory together with the decuced value of the junction fluctuations obtained from the NMR results gives $C_2/C_1 = 0.18$. Using literature data, this value is in close agreement with the value $C_2/C_1 = 0.21$ obtained from mechanical measurements on almost identical networks.

Key words: NMR; cross links; rubber elasticity

Introduction

The purpose of this paper is to show how the fluctuations of cross links in strained networks can be related to specific features of NMR relaxations. A general expression will be derived for the transverse relaxation of magnetisation of nuclei attached to polymer chains between cross-link points in a deformed network for abitrary dynamics and deformation dependence of the cross-link points. The theoretical results will be used to analyse the experimental work of Deloche and Sotta [1] on uniaxially deformed poly(dimethylsiloxane) networks using deuterium nuclear magnetic resonance (^{2}H NMR). Particular attention will be given to the differeing contributions to the NMR relaxation from the static network structure as described by the mean distance between cross links and the fluctuations of the cross-link points themselves. The cross-link fluctuations will be shown to give rise to an imposed oscillation on the relaxation of nuclear magnetisation (doublet splitting). The frequency of the oscillations is determined by the magnitude of the junction fluctuations.

The extent of the network junction fluctuations, as revealed by this NMR experiment, can be well correlated with the mechanical properties of the network by utilising a theory due to Flory [2]. Flory argued that the presence of real inpenetrable chains (entanglements) would both restrict and impede the fluctuations of the cross-link points and make the fluctuations strain dependent. This is in contrast to the original phantom chain model for rubber elasticity proposed by James and Guth [3], where the junction points fluctuate in a volume which is independent of the state of strain and of the same order as the size of the chains between the cross links. In Flory's model the fluctuations increased in the direction of the strain, leading to a transition from an affine to non-affine chain deformation as the strain is increased. The subsequent deviations from the phantom chain model where shown by Flory to be consistent with the phenomenological Mooney Rivilin [4] description.

The basic approach to deriving the NMR properties of networks is described in the next section and will follow that of Cohen Addad [5—7] and the further developments proposed by the present author [8, 9]. The residual energy of the tensorial spin-spin couplings between the nuclei depends on the dynamics, state of strain and configurational properties of the hindered motions of the monomer units between the cross links. If the individual monomer motions are fast, then the NMR properties are transferred from the local scale associated with the monomers to a semi local scale associated with the the end-to-end vector between the cross-link points. At this level the residual interaction energy becomes scale invariant and enables exact analytic expressions to be derived [10]. The dynamics of the monomer units play no further role other than to re-scale the coupling constant v (dipolar or quadrupolar) and to transfer attention to the relatively slower dynamics and configurational properties of the network cross-link points.

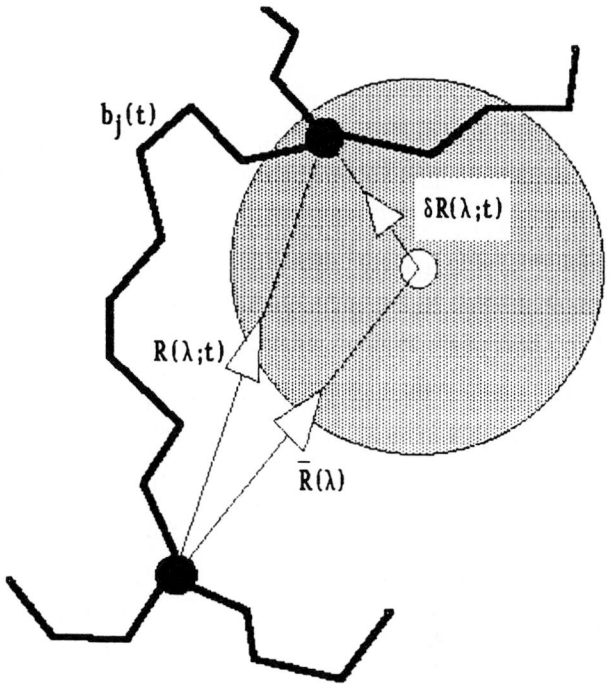

Fig. 1. The cross-link fluctuation volume is shown as the shaded area. $\bar{R}(\lambda)$ is the mean vector distance between crosslinks and $\delta R(\lambda;t)$ the instantaneous fluctuation

NMR properties of a deformed network

The model is shown in Fig. 1. The chain between the cross-link points is considered to be composed of N statistical segments $\{b_j\}$, each of average length b. A state of deformation λ exists in the network and the instantaneous vector $R(\lambda;t)$ joining two particular cross-link points can be expressed as

$$R(\lambda;t) = \bar{R}(\lambda) + \delta R(\lambda;t) ,$$

where $\bar{R}(\lambda)$ is the mean vector distance between the two particular junction points and $\delta R(\lambda;t)$ describes the fluctuations of the junction points about the mean positions which, following Flory, are considered to be strain dependent.

The simplest theoretical model with which to study the transverse NMR relaxation in polymer molecules consists of two spin 1/2 nuclei (a proton pair) a vector distance d apart, fixed to a bond in a chain of identical bonds. The dipolar interaction energy between these spins is

$$\varepsilon(t) = v\{3\cos^2 a(t) - 1\} , \qquad (1)$$

where v is a coupling constant and $a(t')$ is the angle d makes with the applied field at a time t'. The same form of result holds for quadrupolar interactions when the asymmetry parameter can be neglected. A full account of the use of ^{2}H NMR in oriented systems has been given by Samulski [11]. This interaction leads to a dephasing of the transverse components and subsequent relaxation of the magnetisation is described by

$$G(t) = \left\langle \cos v \int_0^t (3\cos^2 a(t') - 1)dt' \right\rangle . \qquad (2)$$

When details about the dynamics of the monomeric units can be completely ignored, the NMR properties can be transferred to a coarse scale described in terms of statistical bond vectors $\{b_j\}$. This approach was pioneered by Cohen Addad [5—7]. If the dynamics of the semi-local, statistical vectors $\{b_j\}$ are also fast, then the process can be continued [9] and the NMR properties are further re-scaled to a space scale determined by the end-to-end vector $R(\lambda)$. At this scale the dipolar interactions lead to a relaxation of transverse magnetisation $M_+(t)$ given by

$$M_+(t) = M_0 \left\langle \cos \frac{3v}{2N} \cdot \int_0^t \frac{\{2Z^2(\lambda;t') - X^2(\lambda;t') - Y^2(\lambda;t')\}dt'}{Nb^2} \right\rangle . \quad (3)$$

The $X(\lambda;t)$, $Y(\lambda;t)$, $Z(\lambda;t)$ are the coordinates of the chain end-to-end vector $R(\lambda;t)$ with the other end set at the origin. N is the number of statistical vectors $\{b_j\}$ comprising the chain between the crosslink points.

The advantage of the repeated re-scaling is that the coordinates (X, Y, Z) may be treated as independent Gaussian random variables and, hence, the mathematical problem is completely posed by the term

$$g(v,t) = \left\langle \exp \frac{3\omega i}{2N} \int_0^t \frac{X^2(t')}{Nb^2} dt' \right\rangle$$

$$= \langle g\{v,t;X(\lambda;t)\} \rangle . \quad (4)$$

Then the normalised relaxation function $G(t) = M_+(t)/M_0$ is given by

$$G(t) = \text{real part: } [g(2v,t)g(-v,t)g(-v,t)] . \quad (5)$$

For a netowrk with fluctuating crosslinks, set $X(\lambda;t) = \bar{X}(\lambda) + \delta x(\lambda;t)$ in (4) for $g\{v,t;X(\lambda;t)\}$ to give

$$g\{v,t;X(\lambda;t)\}$$
$$= \left\langle \exp \frac{3vi}{2N} \frac{1}{Nb^2} \left[\bar{X}^2(\lambda)t + 2\bar{X}(\lambda) \right. \right.$$
$$\left. \left. \cdot \int_0^t \delta x(\lambda;t')dt' + \int_0^t \delta X^2(\lambda;t')dt' \right] \right\rangle . \quad (6)$$

The averaging in (6) now requires a dynamical averaging over the crosslink fluctuations $\delta X(\lambda;t')$ and a static averaging over the mean distance $\bar{X}(\lambda)$ between neighbouring pairs of cross-links. The details are mathematical and draw on previous work [8—10]. An outline of the evaluation of the auxillary function $g\{v,t;X(\lambda;t)\}$ for abitrary junction fluctuation dynamics is presented in the Appendix.

For the purpose of simplifying the discussion given here, the particular case of junction fluctuations described by a single relaxation time τ will be discussed for times $t \gg \tau$ and fast on the NMR time scale i.e. $v\tau/N \ll 1$. Under these conditions it is shown in the Appendix that the relaxation function (6) can be evaluated to give

$$g\{v,t;X(\lambda;t)\}$$
$$= \exp \frac{3vi}{2N} t \left[\frac{\bar{X}^2(\lambda) + \langle \delta X(\lambda)^2 \rangle}{Nb^2} \right]$$
$$- \left\{ \frac{3v}{N} \right\}^2 \frac{\tau t}{2} \frac{\langle \delta X^2(\lambda) \rangle}{Nb^2}$$
$$\cdot \left[\frac{\bar{X}^2(\lambda) + \langle \delta X(\lambda)^2 \rangle}{Nb^2} \right] . \quad (7)$$

If the distribution of mean chain end-to-end vectors $\bar{R}$ is assumed to be Gaussian, so that for each component,

$$h(\bar{X}) \approx \exp \frac{-3}{2} \frac{\bar{X}^2}{\langle \bar{X}^2(\lambda) \rangle} , \quad (8)$$

then $g\{v,t;X(\lambda;t)\}$ can be averaged over the mean chain end-to-end vectors to give

$$\left[1 - \frac{vi}{N} t \frac{\langle \bar{X}^2(\lambda) \rangle}{Nb^2} \right]^{-1/2}$$
$$\cdot \exp \left[\frac{3vi}{2N} t \frac{\langle \delta X(\lambda)^2 \rangle}{Nb^2} - \left\{ \frac{3v}{N} \right\}^2 \right.$$
$$\left. \cdot \frac{\tau t}{2} \left\{ \frac{\langle \delta X(\lambda)^2 \rangle}{Nb^2} \right\}^2 \right] \quad (9)$$

for $v\tau/N \ll 1$.

In constructing the full relaxation function from (9) the following combinations of Cartesian coordinate terms occur, which can be better written as:

$$\{2\bar{Z}^2(\lambda) - \bar{X}^2(\lambda) - \bar{Y}^2(\lambda)\} = 2\bar{R}^2(\lambda)P_2(\vartheta) , \quad (10)$$

where ϑ is the angle $R(\lambda)$ makes with the applied magnetic field B and $P_2(\vartheta)$ is a Legendre polynomial. For each pair of neighbouring crosslink points the angle ϑ will be different, however for the averaged junction fluctuations which are non-isotropic around the deformation direction, we have

$$\{2\langle \delta Z(\lambda)^2 \rangle - \langle \delta X(\lambda)^2 \rangle - \langle \delta Y(\lambda)^2 \rangle\}$$
$$= 2\langle \delta R(\lambda)^2 \rangle P_2(\Omega) , \quad (11)$$

where Ω is the angle between λ, and B is independent of the bond angle ϑ. The result for the full relaxation function can be written as

$$G(t;\lambda) = \int d\bar{R} h(\bar{R}) e^{-t/T_2(\bar{R})} \cos \Delta(\bar{R})t , \quad (12)$$

where $h(\bar{R})$ is the distribution function for the mean chain end-to-end vectors between the cross links, $\Delta(\bar{R})$ and $T_2(\bar{R})$ both depend on the deformation λ through $\bar{R}$ and are given by

$$\Delta(\bar{R}) = \frac{3\nu}{N}\frac{1}{Nb^2}$$
$$\cdot [\bar{R}^2(\lambda)P_2(\vartheta) + \langle\delta R(\lambda)^2\rangle P_2(\Omega)] \quad (13)$$

$$T_2^{-1}(\bar{R}) = \frac{9\nu^2}{2N}\frac{\tau}{N}$$
$$+ \left[2\frac{\langle\delta Z^2(\lambda)\rangle}{Nb^2}\frac{\bar{Z}^2(\lambda)}{Nb^2} - \frac{\langle\delta X^2(\lambda)\rangle}{Nb^2}\right.$$
$$\left.\cdot\frac{\bar{X}^2(\lambda)}{Nb^2} - \frac{\langle\delta Y^2(\lambda)\rangle}{Nb^2}\frac{\bar{Y}^2(\lambda)}{Nb^2}\right]. \quad (14)$$

Comparison with the experimental work of Sotta and Deloche

Deuterium NMR has been extensively used to study the orientational order induced in deformed rubber networks [12—15]. Sotta and Deloche [1] performed experiments on poly(dimethylsiloxane) networks using deuterium NMR. When a uniaxial force was applied a doublet splitting was immediately observed which varied with the angle Ω between the force direction and the applied field as $P_2(\Omega)$. Sotta and Deloche interpreted this uniaxiality in terms of an orientational nematic field and wrote the relaxation function similarly to (12) as

$$G(t) = \int dR\, h(R)e^{-t/T_2}\cos\Delta_R t. \quad (12)$$

T was arbitrarily assumed to be independent of R and identified with the usual NMR relaxation time related to the molecular motion. Δ_R is the residual quadrapolar interaction term, given by Sotta and Deloche as

$$\Delta_R = \frac{\nu_Q}{5N}\left[3\frac{R^2}{Nb^2}P_2(\vartheta) + \frac{2}{3}\frac{U}{(5-U)}\right.$$
$$\left.\cdot(\lambda^2 - 1/\lambda)P_2(\Omega)\right]. \quad (16)$$

ω_Q is the quadrapolar coupling constant and U characterises the strength of the orientational field postulated by Sotta and Deloche. In this work junction fluctuations are ignored and R represents the end-to-end distance between the cross links.

The result (12) together with (13, 14), derived in this paper on the bais of strain-dependent junction fluctuations, is similar in form to that given by Sotta and Deloche. The first term in both expressions is derived from the static network structure. The second terms are formally similar but are derived from quite different physical considerations. The second term of (13) can be compared with the corresponding one (16) of Sotta and Deloche by assuming that the junction fluctuations deform affinely and identifying the magnitude $\langle\delta R_0^2\rangle$ of the unperturbed fluctuations with the orientational field parameter U of Sotta and Deloche by

$$\frac{\langle\delta R_0^2\rangle}{Nb^2} \equiv \frac{2}{3}\frac{U}{(5-U)}. \quad (17)$$

Sotta and Deloche report that the qualitative features of the doublet structure seen in the NMR spectra of end-linked PDMS chains is fitted by adjusting $U \sim 0.5$. By adopting the alternative explanation proposed in this paper, this implies that the junction fluctuations $\langle\delta R_0^2\rangle$ in the undeformed network PDMS are $\sim 1/15$ of the mean square chain length.

In the next section the effect of the restricted junction fluctuations on the elastic properties of the PDMS network is discussed on the basis of the Flory's theory.

Correlation with the elastic network properties

In a phantom network of functionality φ the junction fluctuations are given by [16]

$$\delta R_{\text{ph}}^2 = \frac{(\varphi - 1)}{\varphi(\varphi - 2)}Nb^2. \quad (18)$$

The PDMS networks used by Sotta and Deloche were hexafunctional ($\varphi = 6$) and hence $\delta R_{\text{ph}}^2 \sim Nb^2/5$. From the re-interpretation of the NMR data given in this paper the actual junction fluctuations $\langle\delta R_0^2\rangle$ are $Nb^2/15$, i.e. they are smaller by a factor of 3. In the theory of Flory this reduction of junction fluctuations is assigned to entanglements and the subsequent effect on the mechanical properties of the network is written in terms of the tensile force f for a simple elongation as

$$f = f_{\text{ph}} + f_e. \quad (19)$$

From Flory's work [2], we have

$$f_e/f_{ph} = \frac{2}{\varphi - 2} \left\{ \frac{\lambda K(\lambda^2) - \lambda^{-2} K(\lambda^{-1})}{\lambda - \lambda^{-2}} \right\}, \quad (20)$$

where the functional form of $K(\ )$ is given in Flory's paper. It is highly non-linear in the deformation λ and depends on a parameter $\varkappa$ given by the ratio of the phantom network crosslink fluctuations to the entangled network fluctuations:

$$\varkappa = \frac{\Delta R_{ph}^2}{\Delta s_0^2} = \frac{\varphi - 1}{\varphi(\varphi - 2)} \frac{Nb^2}{\Delta s_0^2}. \quad (21)$$

In the limit $\lambda \to 1$ of small deformations it can be shown that

$$f_e/f_{ph} \to \frac{2}{\varphi - 2} \left\{ \frac{\varkappa^2(1 + \varkappa^2)}{(1 + \varkappa)^4} \right\}. \quad (22)$$

The term $\Delta s(\lambda)$ (Flory's notation) is related to the full junction fluctuations $\delta X(\lambda)$ used in this paper by

$$\frac{1}{\delta X(\lambda)} = \frac{1}{\Delta X_{ph}} + \frac{1}{\Delta s(\lambda)}. \quad (23)$$

If $\delta X(\lambda) \ll \Delta X_{ph}$, then $\delta X(\lambda) \sim \Delta s(\lambda)$.

The expression developed by Flory for the entanglement contribution f_e and hence f is highly non-linear in the deformation ratio λ but is qualitatively similar to the phenomenological Mooney-Rivlin form:

$$f = (2C_1/L_0)(\lambda - \lambda^{-2}) + (2C_2/L_0)(1 - \lambda^{-3}), \quad (24)$$

where C_1 and C_2 are constants and L_0 is the undeformed length of the sample. The ratio C_2/C_1 of the Mooney-Rivlin constants can be identified as:

$$\frac{C_2}{C_1} = \frac{f_e}{f_{ph}}\bigg|_{\lambda=1}, \quad (25)$$

where f_{ph} is the first term and f_e the second term of (18). Hence the Mooney-Rivilin coefficients can be identified with the $\varkappa$ parameter of Flory's theory by

$$\frac{f_e}{f_{ph}}\bigg|_{\lambda=1} = \frac{2}{\varphi - 2} \frac{\varkappa^2(1 + \varkappa^2)}{(1 + \varkappa)^4} = \frac{C_2}{C_1}, \quad (26)$$

which gives $\varkappa = 25/8 = 3.125$. Consequently, from the Flory model, i.e. expression (20) we have

$$\frac{f_e}{f_{ph}}\bigg|_{\lambda=1} = 0.18 = \frac{C_2}{C_1}. \quad (27)$$

In the literature Mark et al. [17] have measured this ratio for hexafunctional networks of $M_n = 11\,300$ and find the value

$$\frac{C_2}{C_1} = 0.21. \quad (28)$$

For the shorter chains used by Sotta and Deloche ($M_n = 9700$) this ratio is expected to be slightly lower. In any case, the agreement is excellent and, together with the Flory theory, clearly supports the interpretation of the doublet structure in the NMR spectra as arising from network junction flutuations.

Conclusions

In this paper a scale invariant model for the transverse relaxation of nuclear magnetisation has been used to investigate the NMR properties of deformed elastic networks. The cross-links have been assumed to be both deformation-dependent and dynamically fluctuating. A general result has been presented showing the influence these features have on the NMR relaxation. An outstanding result is the effect of a non isotropic deformation dependence of the junction fluctuations. This was shown to lead directly to oscillations in the relaxation function or to a doublet splitting in the resonance spectrum. The frequency of the oscillations is directly governed by the magnitude and deformation dependence of the cross-link fluctuations. These result were used to interpret the experimental work of Sotta and Deloche of the ^{2}H NMR spectra of deformed PDMS networks. These workers interpreted the doublet structure in terms of a weak nematic ordering described by an orientational mean field parameter U. Agreement with the results presented in this paper is achieved if the junction fluctuations are assumed to deform affinely and U is identified with the junction fluctuations $\langle \delta R_0^2 \rangle$ through

$$U/5 \sim 3 \langle \delta R_0^2 \rangle / 2Nb.$$

Using the value $U = 0.5$ chosen by Sotta and Deloche to fit their results gives the size of the junction fluctuations in the undeformed state as $\langle \Delta R_0^2 \rangle \approx Nb^2/15$. These fluctuations are more restricted, by approximately a factor of 3, than are the phantom chain junctions. Using a theory developed by Flory, these restricted junction fluctuations can be related to the elastic properties of the network. In particular, they relate to the deviations of the modulus from the phantom chain result. When expressed in terms of the Mooney-Rivilin constants C_1 and C_2, the Flory theory gives $C_2/C_1 = 0.18$, using the deduced value of $\langle \Delta R_0^2 \rangle$ from the NMR results. This value is in close agreement with the value $C_2/C_1 = 0.21$ taken from the literature on almost identical networks.

Appendix

The calculation of

$$I(t) = \left\langle \exp\left[a \int_0^t x(t')dt' + \frac{b}{2}\int_0^t x(t')^2 dt' \right] \right\rangle \quad (A1)$$

where

$$a = \frac{3vi}{N}\frac{\bar{X}(\lambda)}{Nb^2} \quad b = \frac{3vi}{N}\frac{1}{Nb^2}$$

and

$$x(t') = \delta X(\lambda; t') .$$

It is assumed that the cross-linked fluctuations $\delta X(\lambda; t')$ form a Gaussian random process and that the brackets in (A1) are evaluated using the generalised Gaussian distribution function

$$\exp -\frac{1}{2} \iint \delta X(t_1) C(t_1 - t_2) \delta X(t_2) dt_1 dt_2 . \quad (A2)$$

The function $C(t_1 - t_2)$ is related to the fluctuation function

$$\langle \delta X(\lambda; t') \delta X(\lambda; 0) \rangle = \gamma(t) \quad (A3)$$

in a manner to be described shortly.

The method to be employed has been described in detail in [8, 9] and utilises a Fourier transform δX_a of $\delta X(t)$ on the finite time interval $0 - t$:

$$\delta X_a = \frac{1}{t}\int_0^t dt' \delta X(t') \exp\{2\pi i a t'/t\} \quad (A4)$$

and the reverse transformation

$$\delta X(t) = \sum_{a=-\infty}^{\infty} \delta X_a \exp\{-2\pi i a t'/t\} . \quad (A5)$$

Using these transformations, together with (A2), the term (A1) for $I(t)$ can be written as

$$I(t) = \prod_a \int dx_a \exp\left[-\frac{t^2}{2} \sum_{a=-\infty}^{\infty} x_a C_{a\beta} x_\beta + a t x_0 + b t^2 \sum_{a=-\infty}^{\infty} |x_a|^2 \right], \quad (A6)$$

where

$$C_{a\beta} = \int_0^t \int_0^t dt_1 dt_2 C(t_1 - t_2) \exp 2\pi i a(t_1 - t_2)/t$$

$$= 2t \int_0^t dt' C(t') \exp\{a - \beta\}t'/t\}$$

$$\cdot \frac{\sin\{\pi(a - \beta)(t - t'/2)/t\}}{\pi(a - \beta)} . \quad (A7)$$

The largest term of $C_{a\beta}$ occurs when $a + \beta = 0$ and diagonalises the integrand in (A6). The integrals are standard and give

$$I(t) = \prod_a \left[\frac{1}{1 - \frac{b}{tC_a}} \right]^{1/2} \exp\left\{ \frac{a^2 t^2}{2(t^2 C_0 - bt)} \right\}. \quad (A8)$$

$C_a \equiv C_{aa}$ can be related to the fluctuation correlation function (A3). From the distribution function for the x_a, i.e. the first term in (A6), we have

$$\langle |x_a|^2 \rangle = [t^2 C_a]^{-1} . \quad (A9)$$

Using (A7) for C_a,

$$\langle |x_a|^2 \rangle = \frac{1}{t^2} \int_0^t \int_0^t dt_1 dt_2 \langle x(t_1) x(t_2) \rangle$$

$$\cdot \exp\{2\pi i a(t_1 - t_2)/t\} = \gamma_a . \quad (A10)$$

Hence,

$$I(t) = \prod_a \left[\frac{1}{1 - bt\gamma_a} \right]^{1/2} \exp\left\{ \frac{a^2 t^2 \gamma_0}{2(1 - bt\gamma_0)} \right\} . \quad (A11)$$

For the simple case where the cross-link fluctuations can be described by a single relaxation time τ, i.e.

$$\langle \delta X(\lambda;t)\delta X(\lambda;0)\rangle = \gamma(t) = \delta X^2(\lambda)\exp(-t/\tau),\quad (A12)$$

γ_a can be readily evaluated. Details have already been given in [8, 9], and for $t > \tau$ the result can be written as

$$\gamma_a = \frac{2\tau}{t}\frac{1}{1 + (2\pi a\tau/\tau)^2}\delta X^2(\lambda),\quad (A13)$$

and

$$\gamma_0 = \frac{2\tau}{t}\delta X^2(\lambda).$$

The infinite product in (A11) can be evaluated [8] to give

$$\prod_a\left[\frac{1}{1 - bt\gamma_a}\right]^{1/2} = \exp -\beta t,\quad (A14)$$

where, in terms of the original variables (A1), β is given by

$$\beta = \frac{1}{2\tau}\left\{\sqrt{1 - 2\frac{iv\tau}{N}\frac{3\delta X^2(\lambda)}{Nb^2}} - 1\right\}.\quad (A15)$$

In the limit of fast fluctuations on the NMR time scale, i.e. $v\tau/N \ll 1$

$$\beta = \frac{3i}{2}\frac{v}{N}\frac{\delta X^2(\lambda)}{Nb^2}$$
$$+ \frac{1}{2\tau}\left\{\frac{v\tau}{N}\frac{3\delta X^2(\lambda)}{Nb^2}\right\}^2,\quad (A16)$$

and

$$bt\gamma_0 = \frac{2v\tau}{N}i3\frac{\delta X^2(\lambda)}{Nb^2} \ll 1.$$

Hence the full result for $I(t)$ can be written

$$I(t) = \exp -\left[3i\frac{\delta X^2(\lambda)}{Nb^2} + \frac{v\tau}{N}\frac{3\delta X^2(\lambda)}{Nb^2}\right.$$
$$\left.\cdot\left\{\frac{3\bar{X}^2(\lambda)}{Nb^2} + \frac{3\delta X^2(\lambda)}{Nb^2}\right\}\right]2N.\quad (A17)$$

This is the result used in the paper.

References

1. Sotta P, Deloche B (1990) Macromolecules 23:1999
2. Flory PJ (1977) J Chem Phys 66:5720
3. James H, Guth E (1947) J Chem Phys 15:651; (1947) J Chem Phys 15:669
4. Rivlin R (1948) Phil Trans R Soc A241:379
5. Cohen-Addad JP, Dupeyre R (1983) Polymer 24:400
6. Cohen-Addad JP (1983) Polymer 24:1128
7. Cohen-Addad JP, Guillermo J (1984) J Polym Sci 22:931
8. Brereton MG (1989) Macromolecules 22:3667—3674
9. Brereton MG (1990) Macromolecules 23:1119—1131
10. Brereton MG (1990) J Chem Phys 94:2136—2142
11. Samulski ET (1985) Polymer 26:177
12. Deloche B, Samulski ET (1981) Macromolecules 14:575
13. Deloche B, Beltzung M, Herz J (1982) J Phys Lett 43:L763
14. Sotta P, Deloche B, Herz J, Lapp A, Durand D, Rabadeaux J (1987) Macromolecules 20:2769
15. Sotta P, Deloche B, Herz J (1988) Polymer 29:1171
16. Flory P (1976) J Proc R Soc (London) A351:351
17. Mark JE (1982) Adv Poly Sci 44:1

Author's address:

M. G. Brereton
IRC in Polymer Science
and Technology
University of Leeds
Leeds LS2 9JT, U.K.

Segmental orientation in filled networks

K. Baumann and W. Gronski

Materials Science Center, University of Freiburg, FRG

Abstract: The segmental orientation of uniaxially strained cis-polybutadiene networks containing deuterated polybutadiene and filled with carbon black was probed by measuring the quadrupolar splitting in ^{2}H NMR experiments. The quadrupolar splitting and the line shape are not influenced by the filler content. Possible reasons of this unexpected behavior are discussed.

Key words: <u>c</u>is-1,4-polybutadiene networks; <u>c</u>arbon black; <u>o</u>rientation; $^2\underline{H}$ <u>N</u>MR

Introduction

The role of fillers in reinforcement of rubbers has been studied in numerous mechanical and dynamic mechanical experiments [1, 2]. Since the stress distribution in a filled system is directly connected to the distribution of segmental orientation, one can expect to learn more about the origins of reinforcement by appropriate orientation measurements. The simple strain birefringence method cannot be used in nontransparent systems such as rubbers filled with carbon black. It is the purpose of the present paper to introduce ^{2}H NMR as a method for the measurement of orientation in filled elastomers. The method has been applied first to unfilled, chemically crosslinked networks swollen in a deuterated solvent [3, 4] and networks containing deuterated labeled chains such as polydimethylsiloxane [5] and polybutadiene networks [6]. Brief reports have also been made about thermoplastic polyurethanes [7] and poly(styrene-b-isoprene-styrene) block copolymers [8]. The first report on ^{2}H NMR orientation studies on filled systems was on natural rubber vulcanizates filled with carbon black and swollen in deuterated butadiene oligomers as probe molecules [9]. No influence of the filler content on orientation was found in these studies. In contrast to this finding, a distinct effect of the filler was reported in carbon-black-filled SBR vulcanizates using deuterated benzene as a probe of orientation [10]. Since the orientation of low molecular weight probe molecules might not be a reliable measure of segmental orientation in heterogeneous systems, we used filled networks containing deuterated polymer chains in the present study.

Theoretical background

The potential of the deuteron NMR method for orientation analysis rests on the anisotropy of the interaction of the quadrupolar moment of the deuterium nucleus with the electric field gradient of the CD bond. The interaction leads to a quadrupolar splitting of the ^{2}H NMR resonance. The splitting is proportional to the temporal average of the second Legendre polynomial $P_2(\theta)$

$$\Delta\nu = \nu_Q \overline{P_2(\cos\theta)}, \quad (1)$$

where ν_Q is the static quadrupolar interaction constant of the order of 200 kHz, and θ is the instantaneous angle between the CD bond and the static magnetic field B_0. If a preaverage is taken over the reorientation of the CD bonds faster than ν_Q^{-1} ($10^{-6} - 10^{-5}$ s) the quadrupolar splitting of the uniaxially strained network can be written

$$\Delta\nu = \bar{\nu}_Q P_2(\cos\Omega)\langle P_2(\cos\delta)\rangle; \quad (2)$$

$\bar{v}_Q$ is the preaveraged interaction constant for the interaction of the quadrupole moment with the averaged electric field gradient. The averaged efg is directed along the direction of a chain segment about which fast conformational rearrangements are taking place. δ is the angle between this segment and the direction of strain, and Ω is the fixed angle between the direction of strain and the static magnetic field. The segment can be considered as quasistatic on the ^{2}H NMR scale having correlation times $\tau > v_Q^{-1}$. The residual quadrupolar interaction constant scales as $\bar{v}_Q \propto N^{-1}$, N being the number of monomer units of the segment [11, 12]. For 1,4-polybutadiene the length of the segment on the dynamical scale of the ^{2}H NMR experiment has been estimated to consist of 10 monomer units [6]. From Eq. (2), one infers that the quadrupolar splitting is proportional to the orientation function $\langle P_2(\cos\delta)\rangle$ of the segment with respect to the axis of uniaxial strain. Therefore, apart from a different definition of the segment, similar information on segmental orientation can be obtained from a ^{2}H NMR experiment as from birefringence measurements. The results of ^{2}H NMR experiments on strained rubbers have been interpreted on the basis of expressions for segmental orientation derived from theories of rubber elasticity [5, 6]. According to the classical theory of affine deformation of junction displacement vectors, $\langle P_2(\cos\delta)\rangle$ changes proportional to the deformation function $\lambda^2 - \lambda^{-1}$ and scales with n_s^{-1} where n_s is the number of statistical segments per network chain between chemical crosslinks [13]. Since the true stress also varies in the same way, the ratio of segmental orientation and true stress is constant. This is the statement of the stress optical law.

Complications may arise if either magnetically nonequivalent deuterons are present in the polymer chains, or the network is heterogeneous with spatially varying cross-link densities. According to Eq. (2), both effects give rise to a distribution of quadrupolar splittings. Complex line shapes may also arise if the segmental distribution is nonuniaxial because of a nonuniform distribution of stresses in the sample. A deviation from uniaxiality can be tested by rotating the axis of macroscopic strain with respect to the direction of the static magnetic field [5]. In the uniaxial case the quadrupolar splitting disappears at the magic angle $\Omega = 55°$ and the line shape of the unstrained sample is obtained.

Experimental

Deuterted $d_{1,4}$-butadiene was obtained from thermal decomposition of deuterated 3-sulfolene-$d_{2,5}$. Partially deuterated poly(cis-$d_{1,4}$-butadiene) was synthesized with the Ziegler Natta catalyst diisobutylaluminiumhydride, ethylaluminiumsesquichloride, neodymversatate (molar ratio 40:1:1) in n-hexane. The weight average molecular weight was between M_w = 150 000 and 300 000 g/mol and the heterogeneity was 4.0. The polymer contained 97 mol% cis-1,4 units. The deuterated polymer was mixed with undeuterated cis-polybutadiene (CB 10, Hüls AG) in the ratio 1:9. Networks filled with carbon black N 330 were prepared by Degussa AG containing the following ingredients in phr: cis-PB 100, ZnO 4.0, stearic acid 2.0, tetramethylthiuramdisulfid 2.0, N 330 0/25/50. The ^{2}H NMR experiments were carried out with a Bruker CXP 300 spectrometer at 46.07 MHz in a saddle coil with a $\pi/2$ pulse of 6—7 ms. The stretching experiments were carried out on rubber pieces of 10 × 0.5 × 1.0 mm^3 in a hand-driven stretching device with the stretching direction parallel to the magnetic field.

Results

In Fig. 1 the nominal stress of the unfilled and filled samples is shown in dependence of the extension ratio for the first deformation cycle. The stress increases strongly with increasing filler content. Therefore, if the stress optical law is also valid for

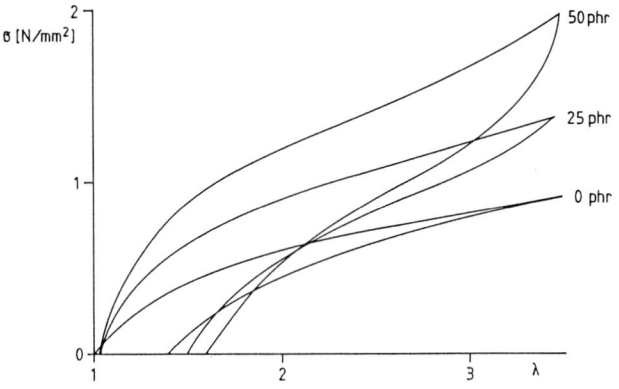

Fig. 1. Nominal stress in dependence of extension ratio I of cis-1,4-polybutadiene networks with different content of carbon black N330 at constant crosslink density (2.0 phr thiuram)

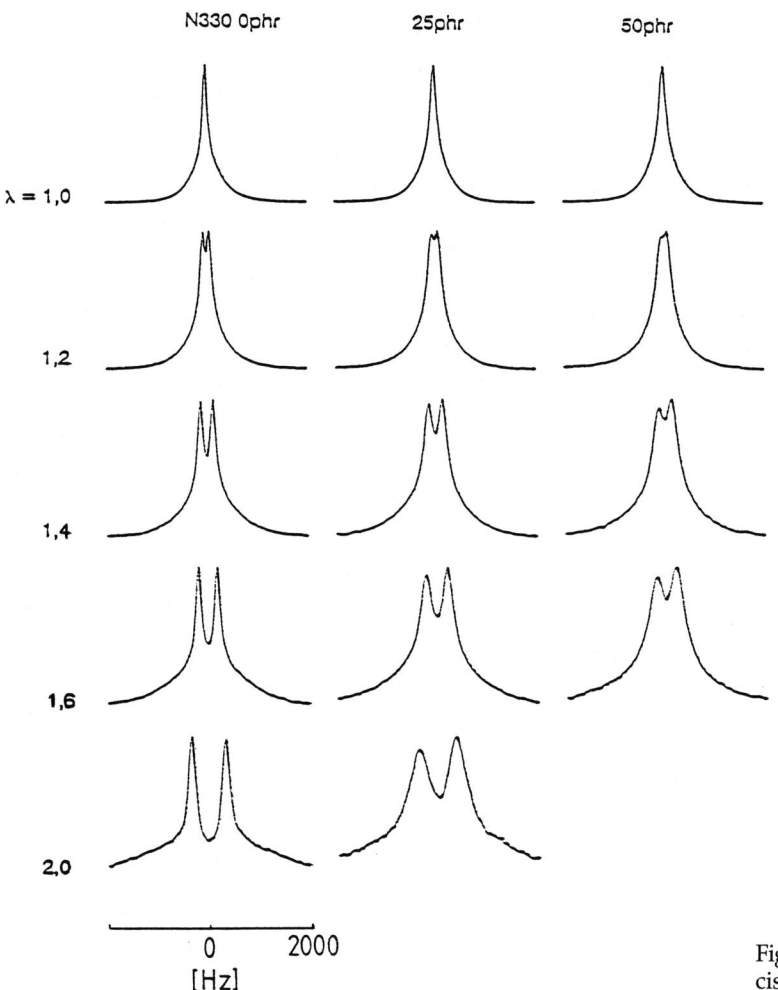

Fig. 2. ^{2}H NMR spectra of cis-1,4-polybutadiene networks of Fig. 1 as a function of the extension ratio

filled systems, it is expected that the quadrupolar splitting as a measure of orientation should exhibit a similar strong dependence on the filler content as does the stress. In Fig. 2 the ^{2}H NMR spectra of the networks are shown as a function of strain and filler content for the first stretch. Qualitatively, one can already see that the expected behavior does not occur. With increasing filler content at constant deformation the main effect appears to be simply an overall broadening of the doublet line shape at nearly constant splitting and asymmetric intensity increase in the outer signal wings. This is shown in Fig. 3, where the quadrupolar splitting is plotted versus the classical deformation function. Inasmuch as the quadrupolar splitting is taken as a measure of the average orientation it behaves as a normal unfilled rubber following the classical $\lambda^2 - \lambda^{-1}$ dependence on the extension ratio.

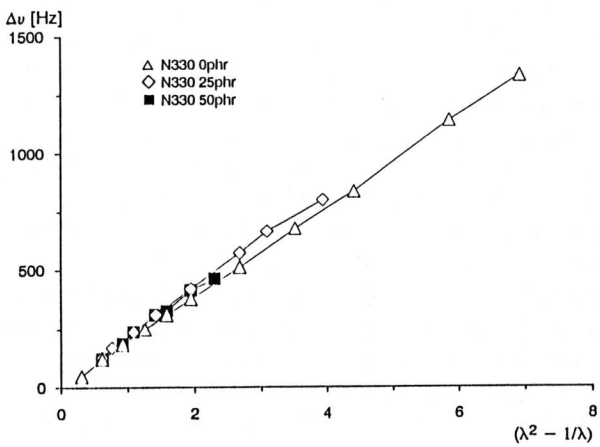

Fig. 3. Quadrupolar splittings of the ^{2}H NMR spectra of Fig. 2

Discussion

The large hysteresis of the stress strain curves in Fig. 1 demonstrates the well known fact that, in the presence of the filler, irreversible structural changes of the network filler system are enhanced with increasing filler content. These changes are generally thought to be a consequence of an inhomogeneous deformation of the rubber enhanced in the filled systems by overstrained interfiller network chains partly detaching from the filler surface during deformation [14]. The hysteresis of the samples is accompanied by a strong set after retraction even in the unfilled rubber (Fig. 1). The latter fact points to flow which is a consequence of a relatively large sol fraction produced by the mixing process. This has been shown by ^{1}H and ^{2}H NMR transversal relaxation measurements [15]. In Fig. 1 the usual reinforcement effect in rubbers filled with active fillers is observed as the enhancement of stress with increasing filler content at constant strain in the extension curve. In a continuum model description the reinforcement is considered as a consequence of an inhomogeneous stress distribution in the rubber matrix changing with the deformation dependent distribution of carbon black agglomerates. The effect of stress is usually rationalized in terms of a strain enhancement factor describing the enhancement of the average strain in the matrix relative to the macroscopic strain [16].

The main effect to be discussed is that the reinforcement effect observed in the stress-strain curve is not observed in the orientation function, as demonstrated by the behavior of the quadrupolar splitting in Figs. 2 and 3. Is the stress optical law violated in filled networks? To examine this question, one has to consider that the validity of the stress optical law assumes uniaxial orientation of the rubber. Only in this case is the second moment of the orientation distribution proportional to the true stress. This is usually true in an uniaxially strained chemically crosslinked rubber [3]. In a filled rubber there is a spatially varying distribution of stresses with a high concentration in the neighborhood of the particles in the direction of strain. The macroscopic stress is a volume average of the components in the direction of macroscopic strain. The distribution of stresses is associated with a corresponding distribution of orientations. If the stress or strain distribution in the rubber matrix in known the complete line shape of the ^{2}H NMR signal could, in principle, be calculated by the weighted superposition of resonance doublets of the orientation distribution. The quadrupolar splitting or, equivalently, the second moment of the orientation distribution as measured at maximum height of the line shapes in Fig. 2 will, in general, not be proportional to the average macroscopic stress as it was in the case of an uniaxial system.

The extent of deviation from uniaxiality could, in principle, be measured by rotating the stress axis with respect to the magnetic field. If the doublet collapses to a single line at the magic angle and if the line shape is the same as the line shape of the isotropic sample this would prove the uniaxiality of the system [5]. The relative difference to the spectrum of the unstrained rubber would be a direct measure of the volume fraction of the rubber matrix in which a nonuniaxial stress distribution is present. Due to experimental difficulties, this experiment could not yet be performed. However, the observation that the quadrupolar splitting in Fig. 3 follows the expected dependence on deformation of an unfilled system is a strong evidence that at least the major part of the rubber matrix is deformed in an uniaxial way, though obviously at the much lower stress level of the unfilled system.

This finding could be explained in terms of a "two network" deformation model [17] of filled systems in which the mechanical response is described by a parallel coupling model where one part A is formed by the pure rubber phase, whereas a second part B contains the filler and a fraction of the rubber coupled in series. Since the macroscopic strain must be equal in both parts, the rubber phase in part B is overstrained since the filler can be considered as undeformable. The normal behavior of the pure rubber in A is then associated with a major part of the ^{2}H NMR line shape and its orientation should approximately be given by the quadrupolar splitting measured at maximum peak height. The overstrained part of the rubber in part B is expected to give rise to a second doublet with larger quadrupolar splitting in the wings of the signal. The latter would explain the asymmetric appearance of the line shapes. It is the quadrupolar splitting of this second part which is expected to depend on the filler content.

To test the model, the free induction decay of the strained sample was approximated by the superposition of two exponentials modulated by the frequencies of the quadrupolar splittings of the two rubber phases in the state of strain according to

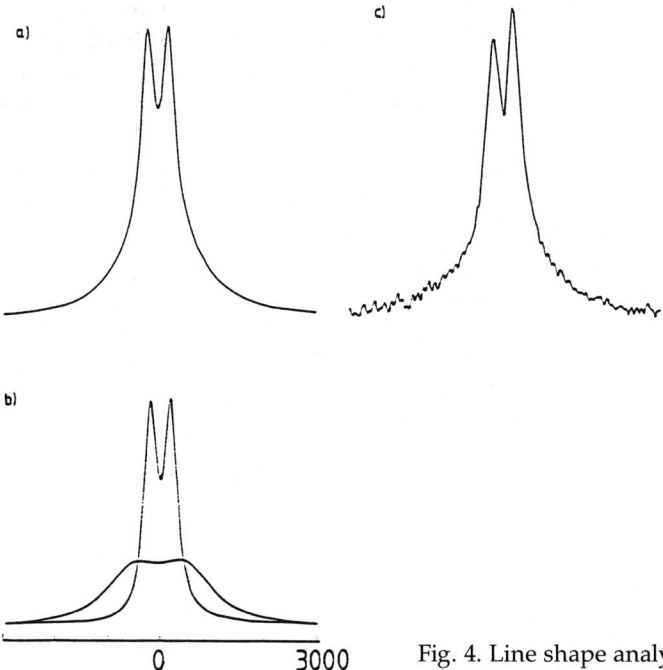

Fig. 4. Line shape analysis of a cis-1,4-polybutadiene network containing 50 phr carbon black at an extension ratio of $\lambda = 1.6$. a) total simulated spectrum, b) spectral components, c) experimental line shape

$$M(t) = A \exp(-t/T_{2A}) \cos(2\pi \Delta v_A t)$$
$$+ B \exp(-t/T_{2A}) \cos(2\pi \Delta v_B t), \quad (3)$$

where $T_{2A,B}$ are the transversal relaxation times of the rubber in A and B. The individual components and their superposition in the frequency domain are compared in Fig. 4 with the experimental line shape for a particular example. In all cases the experimental line shapes can be fitted extremely well. The quadrupolar splitting of the narrow and broad doublets supposed to correspond to the normal and overstrained rubber components are plotted in Fig. 5 versus the classical deformation function $\lambda^2 - \lambda^{-1}$. At about 30 vol% (50 phr) carbon black the quadrupolar splitting of the broad component is much greater than the splitting of the narrow component which is in accordance with the above "two network" interpretation. However, Fig. 2 shows that also the line shapes of the unfilled samples are not homogeneous. A similar decomposition into two components was made for the unfilled sample in order to fit the experimental line shapes. The quadrupolar splittings of the unfilled sample obtained from the line shape analysis are also given in Fig. 5. Both splittings are of similar magnitude as those of the filled sample. Therefore, one can conclude that both components observed in the filled samples are components characteristic for the unfilled rubber matrix. They cannot be explained by the presence of a parallel coupling model of a rubber component under normal strain and a component in an overstrained state.

An explanation for the absence of any filler effect can then only be sought in the very small portion of overstrained chains contributing to the transversal relaxation function. By solid echo experiments, it was checked whether there is a solid-like contribution resulting from oriented chains of hindered mobility, without any positive result. This finding is compatible with Bueches model of reinforcement [14]. According to this model, the load is essentially held by overstrained chains attached to neighboring carbon black particles acting in parallel across a given cross-section of the sample. From the parameters of this model, which were estimated in [11], a very small fraction of ca. 10^{-3} of all chains is in this highly strained state. Such a small number is beyond the sensitivity of detection in an NMR experiment.

One could argue that this unexpected behavior is a consequence of phase separation of deuterated

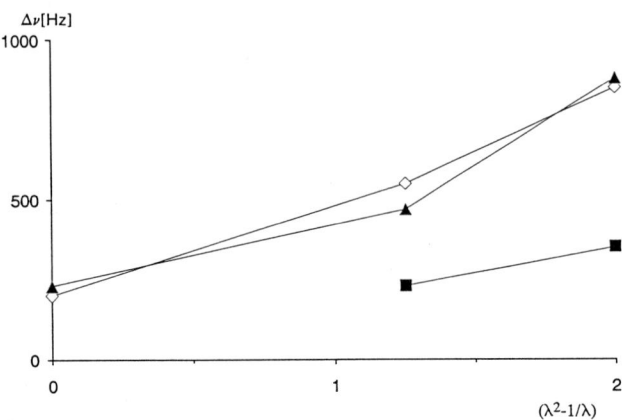

Fig. 5. Quadrupolar splitting of the spectral components as a function of deformation. Broad component of the unfilled (▲) and filled (◇) vulcanizate containing 50 phr carbon black. Narrow component (■) of filled and unfilled samples

and nondeuterated polybutadiene. In fact, an LCST behavior has been observed in mixtures of fully deuterated and non-deuterated 1,4-polybutadiene [18]. At the vulcanization temperature of 140°C a phase-separated structure may be frozen in by the curing process. During mixing, due to unfavorable anisotropic interactions, deuterated droplets could be excluded from highly sheared regions near filler particles and then frozen by crosslinking. In the strained state the deuterated droplets would then be situated in the less strained regions of the "two network" system discussed above. Hence, they would be oriented essentially as the pure rubber matrix.

Finally, we present the most probabable explanation by considering in more detail the factors which contribute to stress and strain in a filled rubber. According to current understanding the main mechanism contributing to an increased Youngs modulus in a filled rubber is given by a hydrodynamic viscosity increase and the adsorption of polymer chains on the filler surfaces. Replacing viscosities by moduli, the first effect is described by [19]

$$E = E_0(1 + 2.5c + 14.1c^2) \qquad (4)$$

where c is the effective filler concentration including the effect of occluded rubber and E_0 is the modulus of the unfilled rubber. Equation (4), valid at low strain, has been proposed to also describe the strain amplification at large macroscopic strain discussed above in connection with the "two network" model. Strain amplification is considered to be a major reinforcement mechanism at large strain. The second factor, adsorption, can be interpreted on a molecular scale. Adsorptive polymer-filler contacts will act in a similar way as chemical crosslinks giving an additional contribution to the modulus according to rubber elasticity theory. This contribution will be proportional to the total filler surface in the system.

In principle, strain amplification and increased crosslink density by adsorptive bonds should enhance the segmental orientation in filled systems as compared to the unfilled rubber. As far as the effect of additional adsorptive bonds is concerned, this effect is dependent on the total surface area and not primarily on the volume fraction of the filler. In cases where the dispersion of the filler is low, the effect will also be small. In the case of an inferior dispersion giving rise to a heterogeneous distribution of large agglomerates there will be a correspondingly small orientation effect induced by adsorptive bonds. In such a heterogeneous system the agglomerates will undergo large irreversible rearrangements and desagglomeration processes which will relax local strains. Strain relaxation will be enhanced if also flow processes take place in the rubber matrix. The presence of viscous flow is seen in the large permanent set of the samples investigated, also present in the unfilled rubber (Fig. 1). Both effects will lead to diminished orientation. Notwithstanding these effects, the reinforcement in a heterogeneous partly viscous system is still present due to the large frictional resistance of the motion of the filler partices and sliding of polymer chains at the filler surfaces. The reinforcement in these systems will primarily be operative by virtue of the hydrodynamic effect which has no direct influence on segmental orientation.

It is suggested that the observed apparent discrepancies are caused by the heterogeneous distribution of the filler in the rubber, i.e., insufficient dispersion of carbon black. In similar experiments on mixtures of undeuterated and deuterated polydimethylsiloxane filled with hydrophobized or nonhydrophobized silica of different specific surface area the quadrupolar splittings of the filled systems were found to be even smaller than in the unfilled networks [20]. Contrary to expectation, the lowest values were found for the system with unhydrophobized silica with high specific surface area. As in the vulcanizates investigated here, also in these systems unusually large hysteresis and permanent set were observed. This indicates that, in filled vulcanizates with insufficient dispersion of the filler, the decrease of orientation due to heterogeneity overrides the increase of orientation associated with enhanced polymer surface contacts. The reason for imperfect dispersion in the samples is probably a consequence of the inefficient mixing in a laboratory mixer of the small amount of polymer given by the limited amount of deuterated material available.

The question remains as to the significance of the components derived from the line shape analysis. In a more detailed analysis of transversal relaxation [15] it is shown that the relaxation function of networks generally contains three components, a Gaussian-like component of intercrosslink chains decaying on a time scale of a few ms, an exponential component with medium decay time of ca 10 ms, and a third exponential component of longer decay times. These components correspond to the relaxation of intercrosslink chains, of dangling chain ends, and to the relaxation of a sol fraction usually present only in small amount. Since in a single pulse experiment the nonexponential behavior of the intercrosslink part cannot be distinguished from an exponential decay because of magnetic field inhomogeneities, this part was approximated by an exponential function in the present analysis. The second slower decaying exponential then corresponds to dangling free ends and free chains without differentiating between these two parts. In a previous ^{2}H NMR analysis of uniaxially strained cis-1,4-polybutadiene networks over a large range of extension ratios it was shown that the quadrupolar splitting of the broad component, i.e., the orientation of intercrosslink chains, is always greater than that of the narrow component of free dangling chains [21]. At large extension ratios $\lambda > 2.0$ a different dependence on λ can be derived for the different structural parts of the network. While the orientation of network has the same $\lambda^2 - \lambda^{-1}$ dependence as the true stress, the orientation of dangling chains varies proportional to λ. This is in accordance with the fact that entropic elastic re-energy is stored only in network chains joining network junctions.

Conclusion

The experiments have shown that deuterium NMR on strained filled elastomers containing deuterated polymer yields information on segmental orientation. The expected correlation of orientation and filler content, as in the case of stress, has not been found. This can be explained most probably by a heterogeneous distribution of the filler particles. In future experiments it will be very important to control the dispersion and morphology of the samples in order to establish a meaningful relationship between segmental orientation and the degree of filling and filler properties.

Acknowledgements

The authors are indebted to Dr. B. Freund, Degussa AG for preparation of the vulcanizates and to Prof. Dr. H. G. Kilian for valuable discussions. Financial support by the Deutsche Kautschukgesellschaft is gratefully acknowledged.

References

1. Kraus G (1978) In: Eirich F (ed) Science and Technology of Rubber. Academic Press, New York, p 339
2. Kilian HG, Schenk H, Wolff S (1987) Colloid Polym Sci 265:410
3. Deloche B, Samulski ET (1981) Macromolecules 14:575
4. Sotta P, Deloche B, Herz J (1988) Polymer 29:1171
5. Sotta P, Deloche B (1990) Macromolecules 23:1999
6. Gronski W, Stadler R, Maldaner Jacobi M (1984) Macromolecules 17:741
7. Gronski W, Emeis D, Brüderlin A, Maldaner Jacobi M, Stadler R, Eisenbach CD (1985) British Polymer J 17:103
8. Gronski W, Stöppelmann G (1986) Progr in Polymer Spectroscopy, Proc of the 7th Europ Symp on Polymer Spectroscopy, Dresden 1985, Teubner Texte zur Physik 9:148
9. Baumann K, Gronski W (1989) Kautschuk, Gummi, Kunstst 42:383
10. Simon G (1991) Polymer Bulletin 25:365
11. Brereton MG (1990) Macromolecules 23:1119
12. Cohen Addad JP (1983) J Chem Phys 43:1509
13. Erman B, Monnerie L (1985) Macromolecules 18:1985
14. Bueche F (1960) J Appl Polym Sci 10:10714
15. Simon G, Baumann K, Gronski W (1992) Macromolecules 25:3624
16. Mullins L, Tobin R (1965) J Appl Polym Sci 9:2993
17. Duschel J, Göritz D (1992) Polymer (in press)
18. Bates FS, Dierker SB, Wignall GD (1986) Macromolecules 19:1938
19. Smallwood HM (1944) J Appl Phys 15:758
20. Litwinov V, unpublished results
21. Gronski W, Forster F, Pyckhout-Hintzen W, Springer T (1990) Makromol Chem Macromol Symp 40:121

Authors' address:

Prof. W. Gronski
Institut für Makromolekulare Chemie
Universität Freiburg
Stefan-Meier-Str. 31
D-W-7800 Freiburg, FRG

Field-cycling NMR relaxation spectroscopy of molten linear and cross-linked polymers. Observation of a $T_1 \propto \nu^{0.25}$ law for semi-global chain fluctuations

H.-W. Weber[+]), R. Kimmich[+]), M. Köpf[+]), T. Ramik[+]), and R. Oeser[*])

[+]) Universität Ulm, Sektion Kernresonanzspektroskopie, Ulm, FRG
[*]) Institut Laue Langevin, Grenoble, France

Abstract: Melts of polydimethylsiloxane, polyisoprene, polyisobutylene, and polyethylene were studied by the aid of proton relaxation spectroscopy predominantly using the field-cycling technique. Spin-lattice relaxation in the rotating frame ($T_{1\rho}$) and transverse relaxation were investigated in addition. With spin-lattice relaxation the total frequency range was 10^3 to $3 \cdot 10^8$ Hz. Linear as well as cross-linked polymers with varying molecular weights and mesh lengths, respectively, were studied. Close to or above the critical molecular weight, peculiar frequency dependencies of the spin-lattice relaxation times were found which can be described by power laws $\omega^{1/4}$ or $\omega^{1/2}$. Depending on the molecular weight and the temperature, a crossover between these frequency dependences takes place. For low molecular weights and high temperatures the indication of a low-frequency plateau appears. Empirically the dispersion can be described well by the aid of the intensity function

$$I(\omega) = \frac{I(0)}{1 + c_1 \omega^{1/4} + c_2 \omega^{1/2}} \ .$$

The molecular weight dependences of the low-frequency spin-lattice relaxation times and of the transverse relaxation time suggest that the corresponding fluctuations have partly local and partly global features. The experimental findings are discussed in terms of the three-component scheme of chain fluctuations previously reported.

Key words: Polymer melts; networks; NMR relaxation; field-cycling; contour-length fluctuation

Introduction

The reorientational fluctuations of polymer melts and networks are composed of local or global, and — with respect to the solid-angle range — restricted or unrestricted components. The correlation and intensity functions of the nuclear spin couplings are correspondingly complicated. One of the basic questions is whether distinct components can be defined and to what extent they can be distinguished and studied separately in experiments. Well-known schemes of global chain modes are the Rouse model [1, 2], the reptation model [2, 3], and the contour-length fluctuation model [4], for instance.

In our previous work [5], we developed a concept based on three components of chain fluctuations (Fig. 1). That is i) the restricted local segment reorientation about the chain axis within the so-called "tight tube" (component A), ii) the reorientation by longitudinal (or locally reptational) curviliniear chain diffusion along the coiled conformation of the tight tube (component B), and iii) global reorientation modes due to contour-

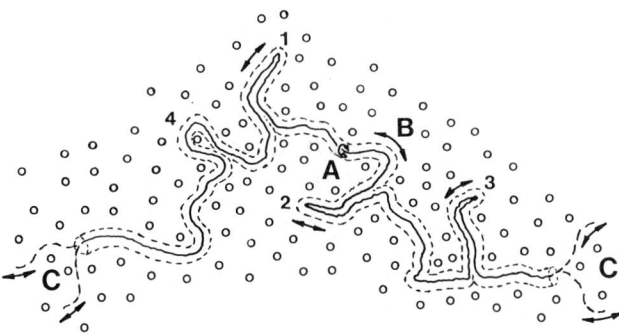

Fig. 1. Schematic representation of a polymer chain in a polymer matrix. The letters A, B, and C refer to the three dynamic components described to in the text. The dashed lines represent the "tight tube". The numbers 1, 2, and 3 indicate chain folds or excursions the fluctuation range of which corresponds to the Doi/Edwards "tubelike regions" [2, 7]

length fluctuation or even by reptational disengagement of the chains from their original tubes (component C).

The idea of "tube" concepts is to introduce a geometric structure schematically representing the toplogical restraints which are expected in a condensed and entangled polymer system. While the diameter of the tight tube essentially is given by the nearest neighbor distance of the chain axes (which can be measured by the aid of wide-angle x-ray scattering, e.g. [6]), the Doi-Edwards tubelike regions are much more extended. The diameters estimated from fits to experimental plateau modulus values are about six times larger than those of the tight tubes [2]. This, at first sight, contradictory situation can be rationalized by considering "folds" or "lateral excursions" of the tight tube fluctuating in a range corresponding to the Doi-Edwards tubelike region. A treatment using statistical physics procedures led to a mean fold length in the order of the segment length of the primitive path [7], i.e., to the same dimension of the tube as predicted by Doi and Edwards on the basis of a quite different concept [2].

The tube diameter in any case depends on the degree of dilution by solvents. This could be demonstrated by an NMR study where the whole range from the melt to dilute solution was investigated [8].

Contour-length fluctuation by fluctuation of the ensemble of folds or lateral excursions of a chain implies the distinction of three chain blocks with respect to dynamics. Doi suggested first that chain end blocks disentangle by contour-length fluctuation much faster than by reptational disengagement [4]. In [9] it was shown that a three-block concept permits the closed description of chain dynamics in the whole range of molecular weights (M): Above the critical molecular weight (M_c) the central chain block is responsible for the features of dynamics often declared as "entangled behavior". The length of this block decreases with decreasing molecular weight. The critical molecular weight reflects the situation when the central block does not exist anymore so that below M_c the chain dynamics is solely governed by the characteristics of the chain-end blocks. The molecular weight of the chain end blocks (M_e) turned out to be constant above M_c [10]. This can be understood by the anomalous displacement behavior of segments in the vicinity of the terminating chain folds.

The three-component/three-block chain dynamics concept correctly predicts a series of peculiar findings in NMR experiments with polymer melts and solutions. This particularly refers to parameter-independent features such as characteristic molecular weight (Table 1) or special values of exponents in limiting power laws of dynamic quantities. A list of dependencies explained in this way can be found in [7], Table 1. Comprehensive reports are given in [5, 11, 12].

The purpose of the present study is to clarify two questions arising in context with chain dynamics in concentrated polymer liquids. Chemically cross-linked polymers are expected to show the same local components of the chain dynamics as linear chains. The global processes, however, should be changed strongly. Information of particular interest refers to the effective molecular weight of networks in comparison to linear chains.

The definition of independent components B and C in principle is only possible if their time scales are very different [13]. Reorientation by contour-length fluctuation and by reptational displacements along the tight tube are related to one another. The latter process simultaneously is the prerequisite for the fluctuation of chain folds. Hence, there will be ranges of the molecular weight, of the frequency and of the temperature where the time scales overlap. The second question thus refers to the characterization of chain dynamics in this overlap region. It will be shown that the chain dynamics in this range leads to a hitherto not recognized frequency dependence of the spin-lattice relaxation time (T_1).

Table 1. Characteristic molecular weights evaluated from transverse relaxation data ([6] and this work)

		Temp. °C	M_c	M_{BC}
PE	(protons)	150	—	52 000
		200	5 600	70 000
PS	(quatern ^{13}C)	200	31 000	71 000
PIB	(protons)	100	15 000	80 000
		140	15 000	133 000
		180	15 000	215 000
PI	(protons)	20	19 000	70 000
		40	19 000	80 000
		80	19 000	119 000
PI	(quatern. ^{13}C)	80	20 000	140 000
PTHF	(protons)	50	10 000	100 000
PDMS	(protons)	80	20 000	$\geq$ 200 000

The conclusions of this particular study are based on experiments with polyethylene (PE), polyisoprene (PI), polyisobutylene (PIB) and polydimethylsiloxane (PDMS). Most of the data were, however, recorded with linear and cross-linked PDMS. This substance was investigated many times in the past. The literature comprises reports on studies based on NMR relaxation [14—21], mechanical relaxation [22—24], neutron scattering [20, 25—27], dielectric relaxation [28, 29], and measurements of the self diffusion coefficient using an NMR field-gradient technique [30]. The present investigation concentrates on field-cycling NMR relaxation spectroscopy (e.g., [31]) supplemented by conventional spin-lattice relaxation spectroscopy (e.g., [32]), nuclear spin-lattice relaxation in the rotating frame (e.g., [33]), and transverse relaxation (e.g., [32]). Taken together, a range of fluctuation rates of about 10^2 to 10^9 s^{-1} was accessible.

Experimental

The field-cycling NMR relaxation measurements were carried out with a home-built apparatus using a liquid nitrogen cooled copper magnet. The proton frequency range was $2 \cdot 10^3$ to $2 \cdot 10^7$ Hz. Spin-lattice relaxation times in the laboratory or rotating frame and transverse relaxation times were recorded with Bruker SXP 4-100 and MSL 300 spectrometers operating at 90 and 300 MHz, respectively.

Linear PDMS samples in a range of weight-averaged molecular weights M_w = 340 to 250 000 were purchased from Polysciences, Warrington, Pennsylvania and Polymer Standard Service, Mainz, FRG. The ratio of weight average and number average molecular weights, M_w/M_n, of the samples originating from the latter source was specified as being better than 1.2.

PDMS networks with different mesh molecular weights ranging from M_n = 700 to 20 000 were prepared using tetrafunctional crosslinks. The tetrafunctional crosslinks were cross- (tetraallyloxyethane) or ring-shaped (tetravinylcyclotetrasiloxane). The polydispersity of the mesh molecular weights varied in the range M_w/M_n = 1.6 to 1.8.

Polyisoprene (M_w = 280 000; M_w/M_n = 1.05; 95% 1,4 isomers) and polyisobutylene (M_w = 610 000; M_w/M_n = 1.25) were purchased from Polymer Standard Service, Mainz, FRG.

Frequency dependences

Figure 2 to 5 show the frequency dependence of the proton spin-lattice relaxation time in the laboratory and rotating frame, T_1 anbd $T_{1\rho}$, respectively. The total frequency range is $1 \cdot 10^3$ to $3 \cdot 10^8$ Hz. The data refer to linear PDMS, PI, PE, and PIB at various temperatures and for various molecular weights above M_c.

Apart from indications of a low-frequency plateau found with low molecular weights and high temperatures, the frequency dependencies can be analyzed in regimes of different power laws. Depending on the molecular weight and the temperature, the influence of two distinct power laws is evident:

$$T_1 \propto \nu_0^{0.25}$$
$$T_{1\rho} \propto \nu_1^{0.25}, \quad (1)$$

and

$$T_1 \propto \nu_0^{0.5}$$
$$T_\rho \propto \nu_1^{0.5}, \quad (2)$$

where $2\pi\nu_0 = \omega_0 = \gamma B_0$ and $2\pi\nu_1 = \omega_1 = \gamma B_1$. γ is the gyromagnetic ratio, B_0 is the external magnetic field, and B_1 the amplitude of the radio-frequency field in the resonantly rotating frame.

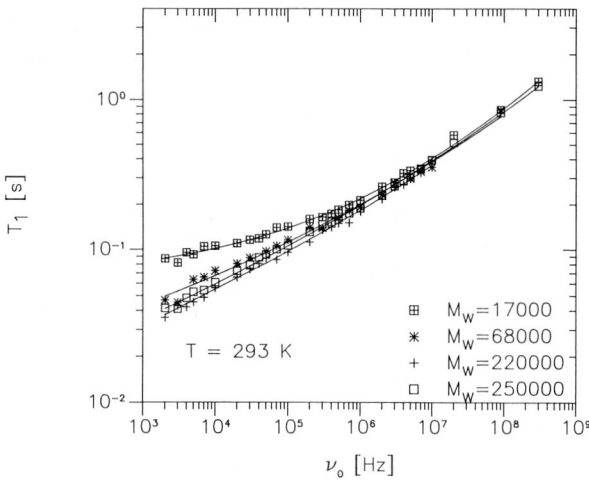

Fig. 2. Frequency dependence of the proton spin-lattice relaxation time of melts of linear PDMS at 293 K. The molecular weights were chosen close or above M_c. The solid lines represent fits on the basis of Eq. (3)

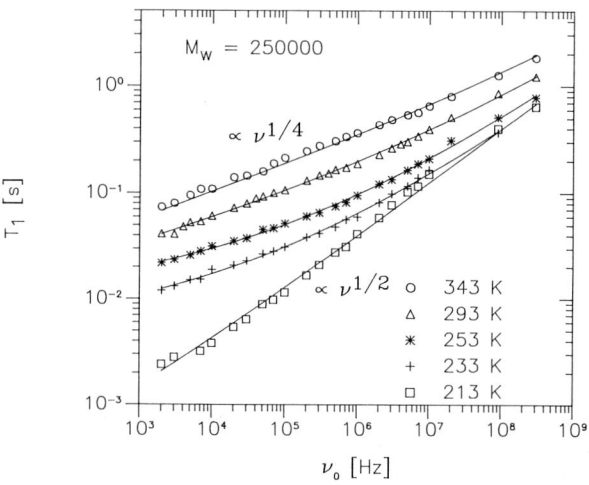

Fig. 3. Frequency dependence of the proton spin-lattice relaxation time of melts of linear PDMS with $M_w = 250\,000$

In some cases, a crossover from the slope 0.5 (tending to appear at high frequencies) to the slope 0.25 (tending to appear at low frequencies), which finally terminates in a low-frequency plateau, was observed. It is shifted to lower frequencies with decreasing temperatures. In other cases only one of the slopes characterized the dispersion in the whole frequency range.

The empirical intensity function

$$I(\omega) = \frac{I(0)}{1 + c_1 \omega^{1/4} + c_2 \omega^{1/2}} \quad (3)$$

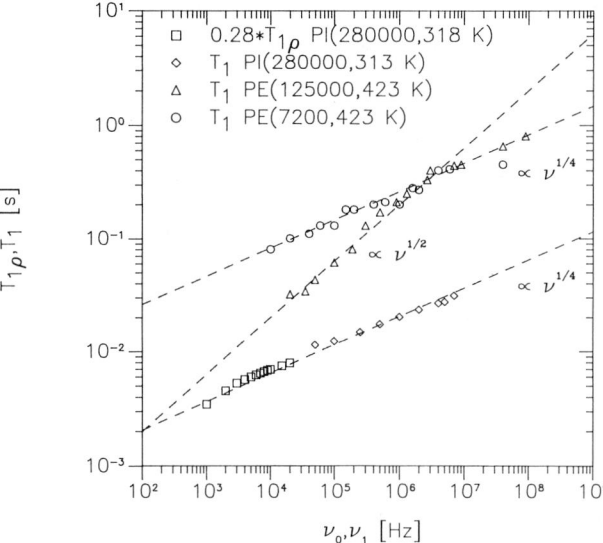

Fig. 4. Frequency dependence of the proton spin-lattice relaxation times T_1 and $T_{1\rho}$ of melts of linear polymers with different molecular weight M_w. The PE data are from [37]. Depending on the molecular weight, the data can be represented in a wide range by straight lines with slopes 1/4 or 1/2

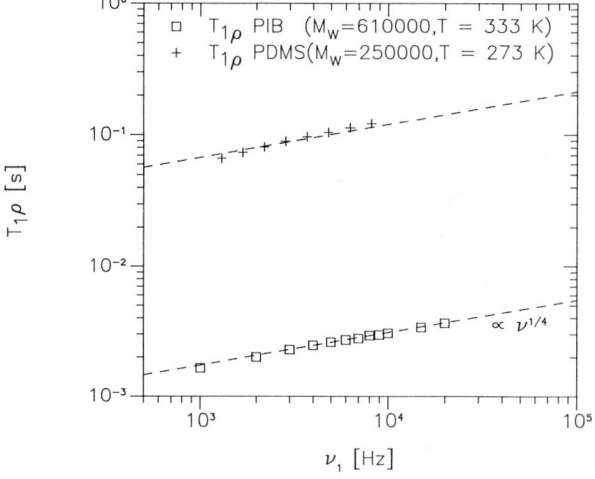

Fig. 5. Frequency dependence of T_ρ of melts of linear PIB ($M_w = 610\,000$) and linear PDMS ($M_w = 250\,000$). In both cases, the data can be represented by straight lines with a slope 1/4

provides a good description of the data. The solid lines in Figs. 2 and 3 were fitted to the experimental data on this basis. The parameters were $I(0)$ and the time constants c_1 and c_2.

Figures 2 and 3 show the frequency dependence of T_1 in PDMS melts with the temperature and the molecular weight as parameters. The temperature

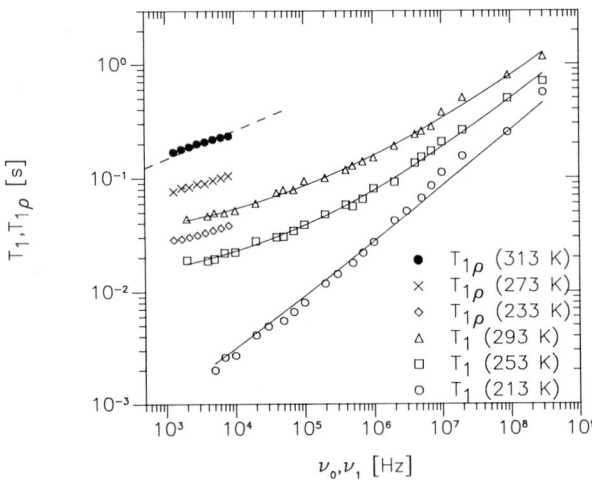

Fig. 6. Frequency dependence of the proton spin-lattice relaxation times T_1 and T_ρ of melts of cross-linked PDMS. The mesh molecular weight (number average) was $M_n = 10\,000$. The solid lines represent fits on the basis of Eq. (3)

range was 213 to 343 K. The material was liquid with the exception of 213 K where partial crystallization already took place. At this temperature, the sample therefore was annealed overnight in order to have stable measuring conditions. The corresponding T_1 dispersion is then expected to be governed by the amorphous regions of the sample.

The solid lines represent fits on the basis of Eq. (3). The parameters vary in ranges $10^{-2} < c_1 < 10^0$ and $10^{-4} < c_2 < 10^{-2}$. At the edge temperatures the data are described by the limiting power laws $T_1 \propto \omega^{1/2}$ and $T_1 \propto \nu^{1/4}$, respectively, in the whole frequency range.

Fgirue 6 shows the frequency dependence of the spin-lattice relaxation times measured with cross-linked PDMS. The curves are characterized by the same features as with linear polymers. The cross-links become perceptible only with mesh molecular weight less than about 3000.

Molecular weight dependencies

Above M_c where the molecular-weight dependence of free volume becomes negligible, variations of the relaxation times with the chain length are indicative for global processes. In our previous work [5, 9, 34] strong effects of that kind were reported for T_2 (defined as the decay time to $1/e$ in the experiments as well as in the theoretical model

calculations). The so-called double-bend behavior following from the existence of two characteristic molecular weights, M_c and M_{BC}, was found for polyethylene, polyisoprene, polyisobutylene, polytetrahydrofurane, and for polystyrene melts (Tab. 1).

An equivalent behavior was observed with linear PDMS melts. Figure 7 shows data of linear PDMS representing the molecular weight dependence of T_1 measured at $\nu_0 = 2$ kHz, T_2 (Hahn spin-echo decay time to $1/e$), T_{2L} (slowly decaying component as defined in Ref. [10]), and $T_{1\rho}$ measured at $\nu_1 = 1.3$ kHz. The double-bend shape of the T_2 molecular-weight dependence is only partly visible because the molecular weights obviously do not exceed the characteristic molecular weight M_{BC} very

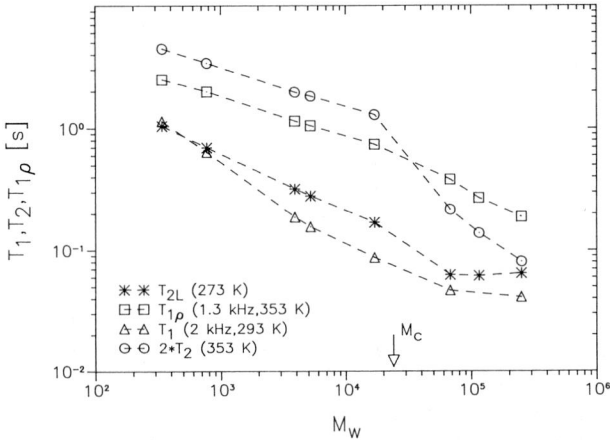

Fig. 7. Molecular weight dependence of different relaxation times of linear PDMS. T_2 is defined as the decay time to $1/e$. T_{2L} characterizes the slowly (and exponentially) decaying component of the spin-echo curve evaluated as described in [10]. The non-vanishing molecular weight dependencies of T_1 (2 kHz), T_ρ (1.3 kHz) and T_2 above $M_c \approx 20\,000$ indicate the influence of global processes. Below M_c the molecular weight dependence of the free volume is relevant in addition. In contrast to T_2, T_{2L} tends to be constant for $M_w > M_c$ in agreement with its assignment to dynamic chain-end blocks of constant length [10]

much. The influence of global processes is, however, evident for the molecular weights above M_c. As expected T_{2L} does not show this effect. The explanation is that it can be assigned to chain-end blocks governed by chain-length independent dynamics [10].

With T_1 and $T_{1\rho}$ the double-bend behavior should appear again in principle. The frequencies in these experiments were, however, relatively large compared with the frequency corresponding to the local field which finally determines T_2. Therefore, the molecular-weight dependence above M_c is visible only by way of suggestion. This finding is analogous to the fact that with polystyrene the T_2 double-bend behavior is observable with ^{13}C resonance, but not with protons [34].

Discussion

The frequency dependence of the spin-lattice relaxation times of polymer melts in the laboratory and in the rotating frame has been studied in detail with a series of polymer melts. More than five decades of the frequency were covered. Four different polymers (PDMS, PI, PIB, PE) were investigated. For $M > M_c$, two peculiar power laws, namely $v^{0.25}$ and $v^{0.5}$ became evident. These power laws appeared also in relaxation experiments with PDMS networks. The fact that an equivalent behavior was found in all cases suggests that the phenomena are universal.

The underlying fluctuations are in the neighborhood of global processes: Although the observed molecular weight dependencies of the spin-lattice relaxation times in the kHz regime are not very strong, that of the transverse relaxation time is. This means that the lower end of the frequency window of the spin-lattice relaxation experiments is only weakly overlapping with the range of effective frequencies which are relevant for T_2. In terms of the three-component scheme, the conclusion is that we are dealing with the combined effect of components B and C in a time scale where both processes contribute to the intensity function value at the frequency effective in the experiment. With molecular weights approaching M_c, a terminal low-frequency plateau finally is indicated as expected due to the influence of component C.

The $v^{1/2}$ behavior was already observed in previous studies [9, 11, 12, 35, 36] referring to polyethylene and polyethylene oxide. The interpretation is conclusively possible by the aid of component B, i.e., reorientations mediated by the quasi one-dimensional displacement of segments along the coiled tight tube. The crucial prerequisite of this model behavior is that the conformation of the tight tube is stationary in the whole time scale relevant for the experiment.

Component C, on the other hand, is a process tending to change the tight-tube conformation mainly as a consequence of the contour-length fluctuation due to the fluctuating fold ensemble. One might argue that the corresponding global modes are quite slow compared with the correlation time of component B so that the chain "sees" a stationary tube in the time scale of component B. The truth is, however, that contour-length fluctuation also implies higher modes representing the fluctuation of a local fold and that the time scale of component B is much longer than expected from the time constant appearing in the corresponding correlation function. The one-dimensional character of the diffusion process leads to an extremely slowly decaying correlation function.

Here, the notion of a chain subdivided into subchains by folds which lead to a partial decoupling effect may be illustrative: Component B refers to the curvilinear diffusion of subchains and has therefore a predominantly local character. Displacements of subchains simultaneously cause fluctuations of the folds and, hence, of the contour-length. Thus, there must be intermediate time or frequency regimes in which the two components cannot be treated independently. They rather form a common component which must be treated in an integrated way. The regime of mutual dependence is considered to be reflected by the observed $v^{0.25}$ frequency dependencies. Only with the lowest molecular weights can a terminal plateau, due to the independent action of component C, be expected and is indicated in the experiments.

The frequency dependencies reported in this article can be described by aid of the empirical intensity function, Eq. (3). An analytical formalism linking the frequency dependencies with the time dependencies of the segmental mean-square displacement is in progress. On the basis of this theory the mean-square displacement behaviour predicted by the Doi/Edwards model can be translated into the corresponding T_1 dispersion and thus be compared with experimental data.

Acknowledgement

We thank J. Wiringer and for the cooperation in the course of this work. Financial support by the Deutsche Forschungsgemeinschaft is gratefully acknowledged.

References

1. Rouse PE (1953) J Chem Phys 21:1272—1280
2. Doi M, Edwards SF (1986) The Theory of Polymer Dynamics. Clarendon Press, Oxford
3. de Gennes PG (1971) J Chem Phys 55:572—579

4. Doi M (1981) J Polymer Sci, Polym Lett 19:265—273
5. Kimmich R, Schnur G, Köpf M (1988) Progr NMR Spectr 20:385—421
6. Zalwert S (1970) Makromol Chemie 131:205—216
7. Kimmich R, Köpf M (1989) Progr Colloid Polym Sci 80:8—20
8. Köpf M, Schnur G, Kimmich R (1988) J Polym Sci: Polym Lett 26:319—323
9. Kimmich R, Bachus R (1982) Colloid Polym Sci 260:911—936
10. Kimmich R, Köpf M, Callaghan P (1991) J Polym Sci: Polym Physics Ed 29:1025—1030
11. Callaghan PT (1988) Polymer 29:1951—1959
12. Huirua TM, Wang R, Callaghan PT (1990) Macromolecules 23:1658—1664
13. Kimmich R (1977) Polymer 18:233—238
14. Andrianov KA, Zhadnov AA, Litvinov VM, Larrukhin BD, Isitsishvili VG (1975) Vysokomol Soyed A17:1323—1326
15. Litvinov VM, Lavrukhin BD, Zhdanov AA, Andrianov KA (1977) Polymer Sci USSR 19:2330—2340
16. Simon G, Birnstiel A, Schimmel K-H (1989) Polym Bull 21:235—241
17. Folland R, Steven JH, Charlesby A (1978) J Polym Sci: Polym Physics Ed 16:1041—1057
18. Litvinov VM, Lavrukhin BD, Zhdanov AA (1985) Polymer Sci USSR 27:2786—2794
19. Grapengeter H-H, Kosfeld R, Offergeld H-W (1980) Colloid Polym Sci 258:564—569
20. Grapengeter H-H, Alefeld A, Kosfeld R (1987) Colloid Polym Sci 265:226—233
21. Burnett LI, Rottler CL, Laughon DH (1978) J Polym Sci: Polym Physics Ed 16:341—347
22. Andrady AL, Llorente MA, Mark JE (1991) Polym Bull 26:357—362
23. Kosfeld R, Heß M (1981) Angew Makromol Chemie 95:149—154
24. Oppermann W, Rehage G (1981) Colloid Polym Sci 259:1177—1189
25. Ewen B (1984) Pure Appl Chem 56:1407—1422
26. Oeser R, Ewen B, Richter D, Farago B (1988) Phys Rev Letters 60:1041—1044
27. Richter D, Ewen B, Farago B, Wagner T (1989) Phys Rev Letters 62:2140—2143
28. Adachi H, Adachi K, Ishida Y, Kotaka T (1979) J Polym Sci: Polym Physics Ed 17:851—857
29. Naoki M, Kondo S (1983) Polymer 24:1139—1144
30. Kimmich R, Unrath W, Schnur G, Rommel E (1991) J Magn Reson 91:136—140
31. Noack F (1986) Progr NMR Spectr 18:171—276
32. Abragam A (1961) The Principles of Nuclear Magnetism. Clarendon Press, Oxford
33. Look DC, Lowe IJ (1966) J Chem Phys 44:2995—3000
34. Kimmich R, Rosskopf E, Schnur G, Spohn K-H (1985) Macromolecules 18:810—812
35. Kimmich R (1975) Polymer 16:851—852
36. Schneider H, Hiller W (1990) J Polym Sci: Polym Physics Ed 28:1001—1014
37. Koch H, Bachus R, Kimmich R (1980) Polymer 21:1009—1016

Authors' address:

H.-W. Weber
Sektion Kernresonanzspektroskopie
Universität Ulm
D-W-7900 Ulm, FRG

Segmental orientation of "long" and "short" chains in strained bimodal PDMS networks: A ^{2}H-NMR study

B. Chapellier[1]), B. Deloche[1]), and R. Oeser[2])

[1]) Laboratoire de Physique des Solides, Universite Paris-Sud, Orsay, France
[2]) Institut Laue Langevin (ILL), Grenoble, France

Abstract: The segmental orientational order generated in bimodal PDMS networks under uniaxial stress is probed, using deuterium-NMR (^{2}H-NMR). The method consists in observing partially time-averaged nuclear (quadrupolar) interactions. The induced segmental order is measured on both kinds of network chains: "short" chains (M_n = 3000) and "long" chains (M_n = 25 000). The order depends on the "long"-chain mass percentage ω (15% $\leqslant$ ω $\leqslant$ 80%). Moreover, for a given ω and a given applied stress the segmental orientation is very similar for both kinds of chains. This result is inconsistent with independent chain descriptions of networks.

Key Words: Bimodal network; deuterium NMR; orientational order; PDMS network

Introduction

Recently, great effort has been devoted to the development of spectroscopic techniques (i.e., infrared dichroism, polarized fluorescence, nuclear magnetic resonance (NMR)) sensitive to the microscopic chains behavior in dense amorphous polymer systems. In particular, deuterium-NMR (^{2}H-NMR) has already been used to monitor the orientational behavior of chain segments of polymeric networks under uniaxial stress [1, 2]. The present work extends those previous NMR studies to the case of bimodal networks made of mixtures of "long" and "short" chains [3]. Specifically, we report herein direct ^{2}H-NMR measurements of stress-induced segmental ordering carried out either on "short" chains or on "long" chains of a given bimodal distribution. Then, the variation in orientational order rate is studied as a function of the "long"-chain mass percentage. The interpretation of our results requires to consider local orientational interchains effects.

Experimental

Samples

^{2}H-NMR experiments have been performed on end-linked tetrafunctional polydimethylsiloxane (PDMS) bimodal networks. The network synthesis and labeling closely follow the procedure mentioned in [4], except that the crosslinkers have been substituted by tetravinylsiloxane[1]). The "short" chains had a number-average molecular weight M_n of 3000 g/mol, and the "long" chains one of 25 000 g/mol. In each case the molecular weight distribution is about 1.8. As indicated in Table 1, each bimodal network is characterized by the *mass* percentage ω of "long" chains among "short" chains. Moreover, for a given fraction ω, two complementary networks were investigated: One with only "long" chains perdeuterated (samples S, M, D, in Table 1) and the other one with only "short" chains perdeuterated (samples s, m, d in Table 1). This specific labeling of the bimodal networks is the source of contrast in the ^{2}H-NMR experiments.

Stretching device

Sample elongation was performed as described earlier [1]. Both ends of the sample were gripped in jaws. One of these jaws was fixed while a calibrated screw moved the other one along the NMR tube.

---

[1]) It should be noted that siloxane derivatives have greater elastic moduli than ethane derivatives.

Table 1. Characteristics of the bimodal networks composed of "long" chains (M_n = 25 000) and "short" chains (M_n = 3000). p = slope of the different straight lines in Fig. 1

Sample	Long-chain mass % (ω)	Deuterated chains	Slope $p = \Delta\nu/\lambda^2 - \lambda^{-1}$
3000	0	short	86.5
D	15	long	80
d	15	short	65
M	50	long	62
m	50	short	68.5
S	80	long	43.5
s	80	short	42
25 000	100	long	34

The sample elongation was monitored before and after each NMR experiment with a micrometer on a microscope stage. In this fashion, the elongation ratio $\lambda = L/L_0$ (where L and L_0 were the lengths of the network elongated and relaxed, respectively) was measured within an accuracy of a few percent.

NMR conditions

A CXP-90 Bruker spectrometer was used, operating at 12.8 MHz with a conventional electromagnet (1.9 Tesla). The magnetic field was normal to the uniaxially applied constraint. The spectra were obtained by Fourier transform averaged free induction decays.

^{2}H-NMR background

Fast anisotropic reorientations of a C-D bond lead to a quadrupolar interaction which is no longer time-averaged to zero [5, 6]. When its motion is uniaxial around a macroscopic symmetry axis, such a residual interaction splits the liquid-like NMR line into a doublet whose spacing is in frequency units:

$$\Delta\nu = 3/2\,\nu_q P_2(\cos\Omega)\,\langle P_2(\cos\theta(t))\rangle\,, \quad (1)$$

where ν_q denotes the static quadrupolar coupling constant ($\nu_q \approx 175$ kHz). The angles in the second Legendre polynomials depend on the experimental geometry and the molecular dynamics: Ω is the angle between the spectrometer magnetic field and the sample symmetry axis, namely, the axis of the applied constraint ($\Omega = 90°$ in Eq. (1), so that $P_2(\cos\Omega) = 1/2$), $\theta(t)$ is the instantaneous angle between the C-D bond and the constraint axis. The brackets indicate an average over the motion faster than the characteristic ^{2}H-NMR time ν_q^{-1}. Then, $\langle P_2(\cos\theta(t))\rangle$ is the orientational order parameter of the C-D bond with respect to the symmetry axis of the sample.

Results

The ^{2}H-NMR spectra of a bimodal network change from a single line into a resolved quadrupolar doublet ($\Delta\nu \neq 0$) as the sample is uniaxially deformed ($\lambda > 1$) (Fig. 1). The finite value of the splitting $\Delta\nu$ is indicative of uniaxial chain segments ordering along the applied force direction [1, 8].

Figure 2 shows the variation of the splitting $\Delta\nu$ with the elongation ratio λ. In the low deformation limit, the data fit with a $\lambda^2 - \lambda^{-1}$ law, as was observed previously in unimodal networks [1, 8]. The corresponding slope $p = \Delta\nu/(\lambda^2 - \lambda^{-1})$, which describes the efficiency of the stress-induced orientational process, appears very similar for two complementary bimodal networks (samples M and m).

As shown in Fig. 3, this striking feature remains valid for each of the bimodal networks that has been investigated (see Table 1). Moreover, the value of the slopes of the bimodal networks decreases roughly linearly as the mass fraction ω of the "long" chain increases from 0% (unimodal network, Mn = 3000) to 100% (unimodal network, Mn = 25 000).

Discussion

A NMR study of the order generated in strained unimodal PDMS networks as a function of the size of the chains between crosslinking junctions has already been performed [8, 9]. The present data concerning the two unimodal networks (Fig. 2) emphasize the result of this previous study: The slope p decreases as the mesh size of the unimodal network increases. It has been shown that the variation of p is closely related to the elastic properties

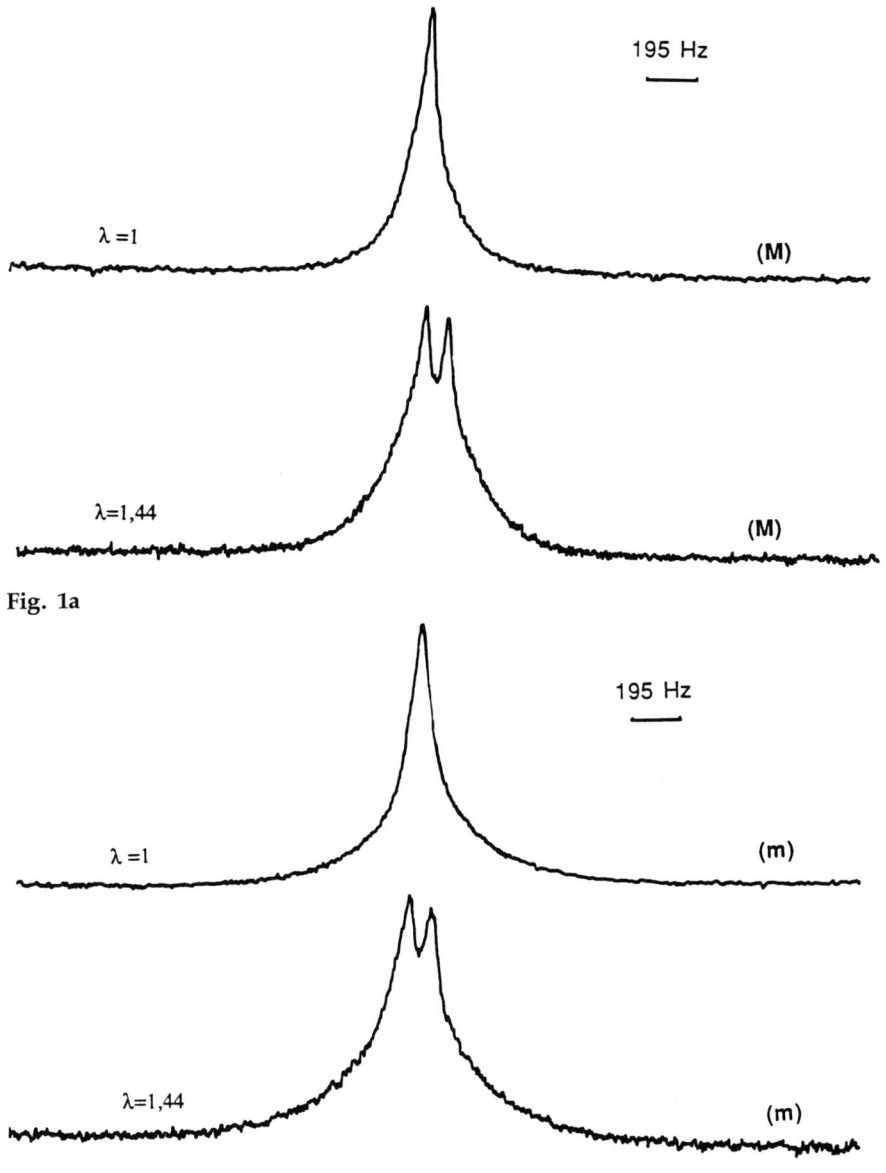

Fig. 1. a) ²H-NMR spectra of sample M in the relaxed state ($\lambda = 1$) and in an alongated state ($\lambda \approx 1.44$). b) ²H-NMR spectra of sample m in the relaxed state ($\lambda = 1$) and in an elongated state ($\lambda \approx 1.44$)

of the network [8]. Accordingly, when bimodal networks are considered the observed value of the slope p is intermediate between the unimodal ones. Moreover, as shown in Fig. 3, when ω increases from 15% to 80%, the slope decreases regularly. This decrease corresponds to a change of elasticity of the rubber as "short" chains are progressively substituted by "long" chains [3, 10].

The main result that we would like to point out here is the following. A bimodal network is a very interesting system since it allows one to compare the behavior of "short" and "long" chains within the same elastic medium. Figure 3 shows clearly that the degree of orientational order induced for both kinds of network chains is very similar, independent of the composition ω of the bimodal network[2]). This similarity is very surprising and contrary to the classical description of rubber elasticity [12]. Indeed, according to those independent chain descriptions, the stress-induced orienta-

---

[2]) This experimental result has been recently observed on bimodal polytetrahydrofuran networks using infrared spectroscopy (see reference [11]).

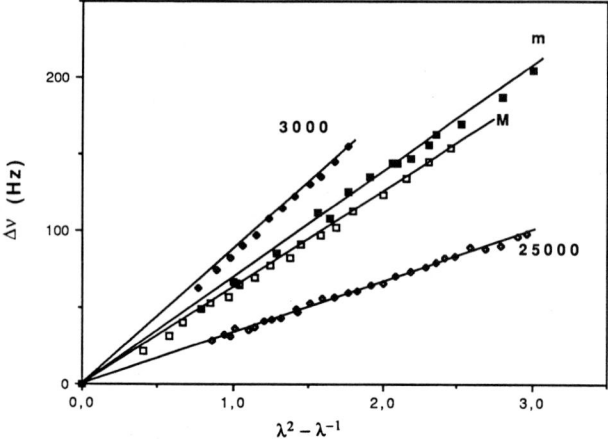

Fig. 2. Quadrupolar splitting Δv of the two unimodal networks ($M_n = 3000$ and $M_n = 25\,000$) and of the two bimodal networks M and m ($\omega = 50\%$) versus $\lambda^2 - \lambda^{-1}$. For reasons of clarity, all the samples studied (see Table 1) have not been reported in this figure

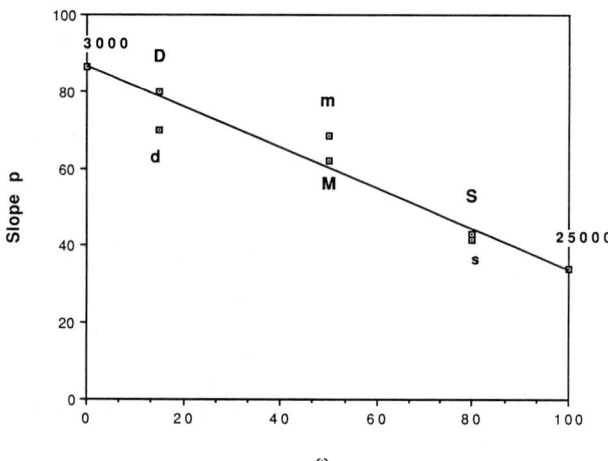

Fig. 3. Slopes $p = \Delta v/(\lambda^2 - \lambda^{-1})$ plotted against the mass percentage ω of "long" chains

tion depends on the number of statistical segments between two crosslinking junctions, thus, in the case of bimodal networks the segmental orientation of "short" chains should be greater than the one of "long" chains [13].

The interpretation of our results considers that (cooperative) orientational correlations between chain segments take place in such dense, liquid-like polymers, so that both kinds of chains are locally coupled in their orientational behavior [7, 14]. Under these conditions, our present observations appear very complementary to those obtained a few years ago on free PDMS polymer chains dissolved inside a PDMS network [15].

Acknowledgements

We acknowledge the financial support provided by the Direction des Recherches Etudes et Techniques (DRET).

References

1. Deloche B, Beltzung M, Herz J (1982) J Phys Lett 43:L763
2. Gronski W, Stadler R, Jacobi MM (1984) Macromolecules 17:741
3. Andrady AL, Llorente MA, Mark JE (1980) J Chem Phys 72(4):2282
4. Beltzung M, Picot C, Rempp P, Herz J (1982) Macromolecules 15:1594
5. Cohen and Reiff (1975) Solid State Phys 5:321
6. Samulski ET (1985) Polymer 26:177
7. Sotta P, Deloche B (1990) Macromolecules 23:1999
8. Dubault A, Deloche B, Herz J (1984) Polymer 25:1405
9. Dubault A, Deloche B, Herz J (1987) Prog Colloid Polym Sci 75:45
10. Llorente MA, Andrady AL, Mark JE (1981) J Polym Science 19:621
11. Hanyu A, Stein RS (1991) Makromol Chem, Macromol Symp 45:189
12. Roe RJ, Krigbaum WR (1964) J Appl Phys 35:2215
13. Higgs PG, Ball RC (1988) J Phys France 49:1785
14. Sotta P, Deloche B, Herz J, Lapp A, Durand D, Rabadeux JC (1987) Macromolecules 20:2769
15. Deloche B, Dubault A, Herz J, Lapp A (1986) Europhys Lett 1(12):629

Authors' address:

Bernard Chapellier
Laboratoire de physique des solides
Université Paris-Sud
91405 Orsay cedex, France

Small chains in a deformed network.
A probe of heterogeneous deformation?

F. Zielinski, M. Buzier, C. Lartigue, J. Bastide*), and F. Boué

Laboratoire Léon Brillouin, CEA Saclay, Gif sur Yvette, France
*) Institut Charles Sadron, Strasbourg, France

Abstract: We describe and discuss the scattering $I(q)$ of neutrons at small angle by polymer (polystyrene) networks containing either labeled (deuterated) solvent (i) or deuterated chains (ii) incorporated in the network before crosslinking, but not linked to the network ("trapped"). With labeling (i), the swelling of the gels above the volume of preparation always leads to an increase of the scattering. This is observed for gels synthesized in solution or close to the bulk state. With labeling (ii) (in undeformed state), the mixing of free chains is at random for light crosslinking. For increasing crosslinking, additional fluctuations appear and increase progressively up to complete phase separation. In deformed state (with labeling (ii)), the anisotropy is classical (elliptical, $S(q_\parallel) < S(q_{\mathrm{perp}})$) for short durations of relaxation after stretching. It is abnormal at low $q (S(q_\parallel) > S(q_{\mathrm{perp}})$, "butterfly" shape) for long durations. We analyze the influence of the duration of relaxation, the crosslinking ratio, the molecular weight of trapped chains and the elongation ratio. We also show that only interchain scattering is responsible for the butterfly effect. Its origin is not attributed to a phase separation per se. We then discuss the effect of heterogeneities of crosslink density. A model where they would resemble percolation clusters, following random linking of blobs (many blobs per chain in a semidilute solution) is discussed. The structure of such clusters would be revealed by the disinterpenetration associated with a spatial dilution, like by swelling or by stretching, in the parallel direction: in the latter case, this would indeed make parallel scattering grow. But this model does not apply to all situations observed. Another approach simply considers the mismatch between the anisotropy of the displacement of centers of mass of the chains and the weaker anisotropy of their own deformation. This model is still phenomenological.

Key words: Polymer networks; heterogeneities; trapped chains; small-angle neutron scattering; swelling; stretching; decomposition; dynamics

I. Introduction

Small-angle neutron scattering (SANS) studies on crosslinked systems have been performed in the last 10 years on many different crosslinked systems, either containing solvent ("gels") or in dry state. The labeled species were either the solvent or some parts of the gel, crosslinks (difficult to observe), meshes (part of chain between two crosslinks), strings of many meshes following each other (labeled paths). Although the results for undeformed samples (state of preparation) agreed in general with the theoretical view (the scattering is equal or close to the one of the corresponding uncrosslinked systems), discrepancies were found for deformed systems. After a uniaxial extension, for example, one can study the scattering along the different directions with respect with the stretching axis (e.g., parallel and perpendicular) [1, 2]. In this case different problems arose. Let us quote here only the major discrepancy [3, 4]: the deformation along the parallel direction is apparently too weak, which corresponds to a too high scattering. Deformation could also be deswelling of a network synthesized in solution [6], or even swelling of a gel at volume larger than the volume of preparation (overswelling) [6]. In this case, also observed was an abnormal increase of the scattering with swelling. Recently, other strong effects appeared when studying small chains dissolved (but not crosslinked) in a network. Although their orientation appeared very weak at small distances (large scattering vectors q) [7], an abnormal increase of the scattering in the parallel direction was observed at small q vectors [8—10]. This "butterfly" effect was observed in polymer melt (temporary rubber regime) [9, 10] as well as in crosslinked samples, both in molten dry state [8] and swollen by solvent [11]. This paper is a first brief summary of a larger work [12] intending to correlate the response of the same networks to the swelling by a solvent, the inclusion of free chains, and deformation, studied here in presence of these chains. We will see whether, through the butterfly effect, they can be used as probes of the deformation.

II. Experimental section

Preparation of gels

The gels were made in two steps: the first one is to attach some sites to a chain of polystyrene (of given monodisperse molecular weight), at random along the chain. The resulting chains can be considered as "precursor chains" for the second step, in which they are crosslinked by these sites. Two kinds of sites have been used, both kinds being attached on some of the benzene rings. The first kind is a ($-CHCl_2$)-chloromethyl-site obtained by a one step Friedel-Crafts reaction using chloromethylether and $SnCl_4$ (tin chloride) as a catalyst. The number of sites per monomer, S_{Cl}, can be measured by NMR. Chloromethylated precursor chains are crosslinked in semi-dilute solution (concentration C_{prep}) by a second FC reaction without additional crosslinker. In these "CM" gels, the active sites are attached between themselves or to any other benzene ring of a chain.

The second kind of sites is a ($-NH_2$)-aminomethyl group fixed in two steps, a Friedel-Crafts (FC) reaction with chlorophthalimide and $SnCl_4$ as a catalyst, followed by an overnight reaction under reflux with hydrazine. Subsequently, the solution is filtered and the polymer precipitated twice in presence of a base. The number of sites per monomer, S_{NH2}, can be measured by HCl titration. In the second step, sites are linked exclusively together in semidilute solution using benzyl dialdehyde ("AM" gels). If some chains bearing no sites are introduced in the solution before crosslinking, they will not be attached to any other part of chain. The study of these systems is a large part of the paper.

Characterization of the gels

After preparation from a semidilute solution, the system is a gel, i.e., a network containing solvent (toluene). From this state, it can be dipped into the same solvent (or another one) and will usually absorb more solvent and increase in volume, which we call overswelling. The swelling ratio Q will be defined as the ratio of the volume of the gel divided by the volume of the dry gel. Simple networks (free chains absent or removed by washing) are studied in any state between the state of preparation ($Q = Q_{prep}$) and equilibrium swelling ratio ($Q = Q_{eq}$). The latter, Q_{eq}, is the most reproducible characterization for our samples. For example, neither S_{Cl} nor S_{NH2} could not be taken as a good parameter. In particular the AM gels can be dried at increasingly high temperature, from 20° up to 150°C (the last stages under secondary vacuum). This anneals the samples but also modifies the final crosslinking ratio, probably because the reaction is not terminated. For example, the values for three different S_{NH2}, $Q_{eq} = 11,26,35$ if swollen after the preparation will respectively change to 6.5, 15, and 19 if swollen after drying. For convenience, we will define a crosslinking ratio X from Q_{eq}. Using an approximate version of the Flory-Rehner theory, one can assume $M_c \sim Q_{eq}^{5/3}$, where M_c is the average molecular weight between two crosslinks. There, we will use

$$X = Q_{eq}^{-5,3}, \qquad (0)$$

which is, as Q_{eq}, a relevant parameter of our samples (Table 1).

Swelling or drying: gel state or bulk state, and the different labelings

One can label the network at any Q by using deuterated toluene. Another way has been to introduce some uncrosslinkable polystyrene chains of molecular weight M_w in the solution before crosslinking. After crosslinking and drying, we observe

Table 1. Characteristics, cross-section at q tending to zero (S_0) and correlation length (ξ_Q) for representative samples studied at different swelling ratios Q, calculated with respect to the volume in dry state: Q_{prep} in the preparation state (S_{0prep}, ξ_{Qprep}), $Q = 3$ (S_{0Q3}, ξ_{Q3}) for the dried samples, and Q_{eq} at swelling equilibrium (ξ_{Qeq}, S_{0eq}) for all samples

Sample	Q_{prep}	$1/S$	Q_{eq}	X ($Q_{eq}^{-5/3}$)	S_{0prep} cm^{-1}	ξ_{Qprep} Å	ξ_{Qeq} Å	S_{0eq} cm^{-1}	$\xi_Q = Q^a$? a	$S_{0Q} = Q^b$? b
CM		$1/S_{Cl}$								
F78	11.75	[40, 20]	40	0.0021	1	16.8	70	0.95	0.86	0.53
F79	11.75	[40, 20]	37	0.0024	1.08	17.9	65.7	2.283	0.8	0.45
F80	11.75	[40, 20]	17	0.0089	1.73	24	46	2.91	1.54	1.41
AM		$1/S_{NH2}$								
F60	10.5	30	23	0.0054	0.9	15	44	2.35	1.18	0.41
AM, C_{prep} = 10.5, dried, trapped chains.					S_{0Q3}	ξ_{Q3}	ξ_{Qeq}	S_{0eq}		M_{wD}
S1, S2	dried	80	16	0.0093	0.99	19.6	25	2.29		138 000
S4	dried	52	13	0.013						138 000
S5	dried	27	6.3	0.046	0.72	17.7	24.8	2.04		138 000
S3	dried	31	4.5	0.082	1.44	23.5	25.6	2.48		138 000
S6	dried	31	3.9	0.103						138 000
S7A...B	dried	28	7.9	0.032						73 000
S8	dried	47	14.3	0.012						73 000
S9	dried	100	8.2	0.022						138 000
S10	dried	32	9.9	0.030						73 000

that the "dry" networks still contain these chains. If they are deuterated the network is labeled. A third possibility is to crosslink a mixture of deuterated and hydrogeneous precursors. As there are many crosslinks per chains, we obtain strings of several labeled meshes ("labeled paths"): this method will not be used here[1]).

Stretching

The procedure has been described elsewhere [1–4]. After drying, the samples are glassy at room temperature (polystyrene glass transition temperature T_g is 100°C). They are machined in parallelipedic shape, put in an oil bath at a temperature $T > T_g$, and stretched by an elongation ratio λ. They are allowed to relax at constant length for a certain duration $t_R = t_{R1}$. Then they are taken out of the bath, which freezes the deformation in the glassy state, allowing for SANS observation. Later on they can be dipped again into the oil bath, relaxed $t_R = t_{R2}$, etc. The time temperature superposition principle (WLF [5] or Vogel-Fulcher law) is applied in order to reach long t_R by using higher T.

SANS

As described elsewhere [1—3], the sample (1 cm $\times$ 5 cm $\times$ 0.1 cm) is placed on an aperture (7.6 mm diameter). It is flat and the neutron beam is passing through its thickness. The divergence of the beam is low owing to a first aperture of 12 mm at 3 m ahead of the sample aperture. The scattering is recorded on a two-dimensional detector perpendicular to the incident beam (of wavelength W between 5 and 10 Å) at distances D between 1 m and 7 m. Each cell corresponds to an angle θ with the beam. Should it not be recalled in a caption, the setting is $D = 3.2$ m and $W = 10$ Å, giving a range of scattering vector $q = 4\pi/W \sin\theta$ of $[7 \cdot 10^{-3}$ Å^{-1}, $7 \cdot 10^{-2}$ Å$]$. The scattering can be displayed by isointensity lines joining the cells of comparable numbers of counts. One can also radially regroup the cells in parts of rings located within angular sectors of width $\delta\phi = 10$ degrees along directions parallel and perpendicular to the stretching axis of the sample.

---

[1]) The use of labeled paths for these gels showed that their parallel scattering can also be interpreted as a combination of classical anisotropy and the increase of scattering studied here, as proposed earlier ("lozenge effect") [4].

III. Effect of the swelling

From preparation to equilibrium swelling

This section deals with both species of gels, CM and AM, with no labeled chains included in the network. They are labeled by contrast with the solvent, which is deuterated (toluene). The scattering was measured for two values of the swelling ratio $Q = Q_{prep}$ and Q_{eq}. In each case it was compared to the scattering of an equivalent semi-dilute solution of long chains of same polymer volume fraction $\phi_{pol} = 1/Q$. For CM gels at Q_{prep} (Fig. 1), the signal of the gels is rather close to the one of the solution for a low crosslinking ratio ($Q_{eq} = 54$, $X = 1.3 \cdot 10^{-3}$), but increases when increasing X ($2.4 \cdot 10^{-3}$, and $8.9 \cdot 10^{-3}$). At $Q = Q_{eq}$ (Fig. 2), the signal difference between gels and equivalent solutions is in all case larger than in the case of Q_{prep} in Fig. 1. For large X, even if Q_{eq} is smaller, the difference is not smaller. For the two values of Q, the scattering can be fitted to a Lorentzian behavior

$$I(q) = I_{0Q}/(1 + q^2\xi_Q^2), \qquad (1)$$

giving values of I_{0Q} and ξ_Q for $Q = Q_{prep}$ and Q_{eq}. The values are given on Table 1, together with the *apparent* exponents a and b for the Q-dependence of I_{0Q} and ξ_Q. We define here $a = \log(\xi_Q(Q_{eq})/\xi_Q(Q_{prep}))/\log(Q_{eq}/Q_{prep})$ and $b = \log(S_{0Q}(Q_{eq})/S_{0Q}(Q_{prep}))/\log(Q_{eq}/Q_{prep})$, using the *two values* Q_{eq} and Q_{prep} *only* for each sample a lies in between 0.8 and 1.6, b between 0.36 and 1.36. If we now consider the aminomethylated gels in preparation state (Fig. 3), we observe a strong additional signal at low q's. We see also in Fig. 3 that it is possible to fit the data to a sum of two contributions

$$I(q) = I_1(q) + I_2(q)$$
$$= A_Q/(1 + q^2\Xi_Q^2)^2 + I_{0Q}/(1 + q^2\xi_Q^2). \qquad (2)$$

The procedure was to adjust $1/S(q)$ vs q^2 to a straight line at large q giving I_2. The substracted part $I_1(q) = I(q) - I_2(q)$, appears close to $1/q^4$ at medium q. Plotted as $1/\sqrt{(I_1(q))}$ vs q^2, it yields Ξ_Q. In the preparation state, the values of Ξ_Q are difficult to measure precisely in the q range of observation. They lie around 500 Å or more. Passing from Q_{prep} to Q_{eq} (Fig. 4), $I_2(q)$ varies similarly to the case of the chloroaminated gels: I_{0Q} and ξ_Q increase. The subtracted part $I_1(q)$ varies differently from $I_2(q)$: A_Q decreases, and $I_1(q)$ is lowered at all q; The apparent slope (in log-log) is also lowered, probably because Ξ_Q decrease, so that the $1/q^4$ range is shifted at larger q's. Ξ_Q becomes in the same time measurable, around 150 Å. The two scales, Ξ_Q and ξ_Q are however still different enough, even in the swollen case, to allow for separation of $I_1(q)$ and $I_2(q)$ because the ratio $(\Xi_Q/\xi_Q)^2$ is the relevant parameter. Thus, the heterogeneities corresponding to Ξ_Q are *diluted* in the classical way and their size decreases, at variance with the fluctuations responsible for $I_2(q)$.

The values of the exponents a and b as defined above for the CM gels, are here of the same order of magnitude for I_{0Q} and ξ_Q extracted from $I_2(q)$. (Table 1, F60, $a = 1.18$ and $b = 0.41$).

Swelling from the dry state

After a thorough drying of the samples (for reproducibility requirements, of T_g for example) and redipping them into solvent, the equilibrium swelling ratio Q_{eq} is much smaller. Additional crosslinking as well as rearrangement has very likely occured, which makes the state of preparation partially or totally equivalent to bulk state. Reswelling in deuterated toluene has been done on such gels having or not containing free deuterated chains. In that event they were washed, and it was checked on redried pieces that the signal from the small deuterated chains corresponded to $\Phi_D < 1\%$. (Fig. 5) shows the scattering of such dried, washed, and reswollen gel (S1) at $Q = 3$ and $Q \sim Q_{eq} = 8-10$, fitted to Eq. (2). Again, swelling makes I_1 decrease and I_2 increase, more than the equivalent solution.

Interpretation and models

The difference in behavior of I_1 and I_2 is a proof that heterogeneities of different origin may exist inside the samples. The origin of I_1 may be the aggregation of the nonreacted polar amine groups in non polar solvent. The initial solution before crosslinking actually scatters the same at low q [13]. Other heterogeneities of the same type could exist in other gels [14]. It seems that the presence of these aggregates does not influence the effects at large q. We expect the same to hold in the dry state (see next section), but let us now discuss the origin of I_2, which seems more universal.

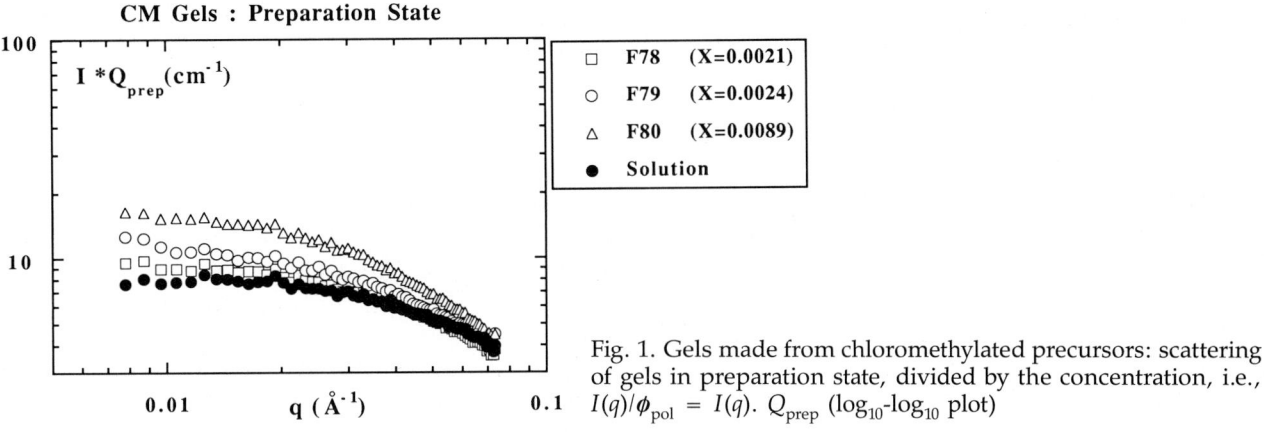

Fig. 1. Gels made from chloromethylated precursors: scattering of gels in preparation state, divided by the concentration, i.e., $I(q)/\phi_{pol} = I(q) \cdot Q_{prep}$ ($\log_{10}$-$\log_{10}$ plot)

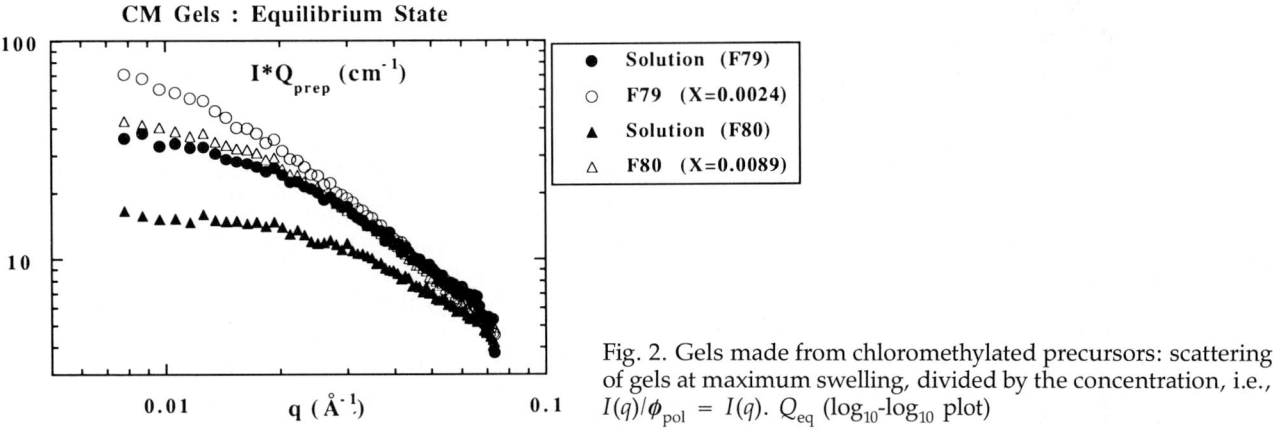

Fig. 2. Gels made from chloromethylated precursors: scattering of gels at maximum swelling, divided by the concentration, i.e., $I(q)/\phi_{pol} = I(q) \cdot Q_{eq}$ ($\log_{10}$-$\log_{10}$ plot)

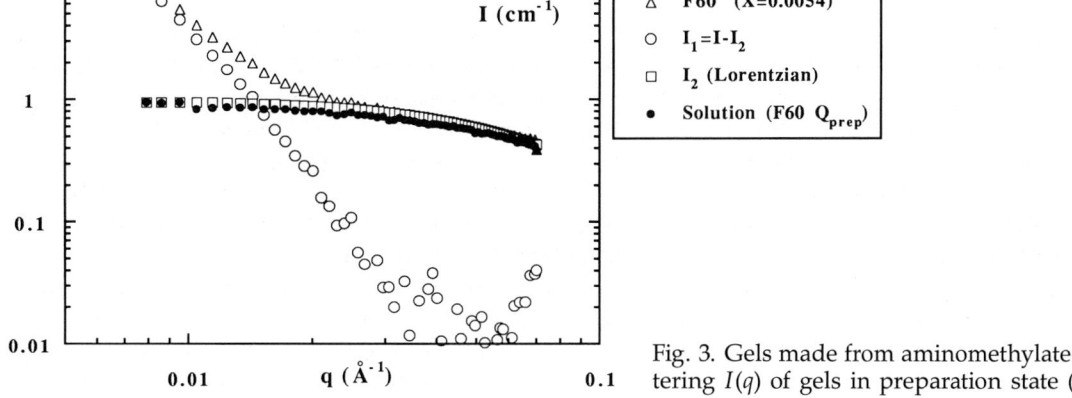

Fig. 3. Gels made from aminomethylated precursors: scattering $I(q)$ of gels in preparation state ($\log_{10}$-$\log_{10}$ plot)

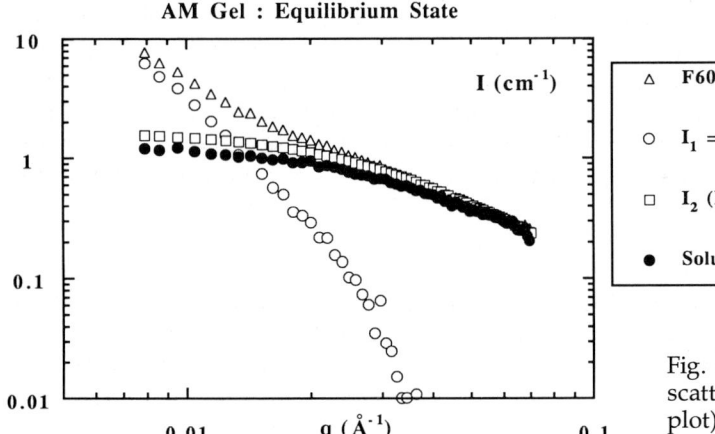

Fig. 4. Gels made from aminomethylated precursors: scattering $I6q$) of gels at maximum swelling ($\log_{10}$-$\log_{10}$ plot)

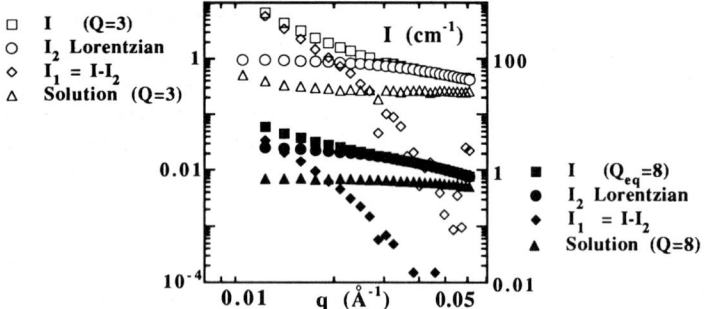

Fig. 5. Gels (S1 and S5, see Table 1) made from aminomethylated precursors, thoroughly dried and washed from their trapped chains: scattering $I(q)$ at two swellings $Q = 3$ and $Q = 8-10$ ($\log_{10}$-$\log_{10}$ plot), and decomposition in two contributions $I_1(q)$ and $I_2(q)$

In [6, 11], some slightly different gels were studied. They were also made of long chains crosslinked in solution, but using a one-shot Friedel-Crafts reaction. They behave like the chloromethylated gels, i.e., they do not crosslink more under drying. In swollen state, on a complete series of Q values between Q_{prep} and Q_{eq}, the scattering, using Eq. (1), yields power laws $\xi_Q = Q^a$ and $S_{0Q} = Q^b$. a and b were found different for the two samples studied, FC1 and FC2, of different crosslink ratio. By analogy, a and b are given here, estimated from Q_{eq} and Q_{prep} only for each sample. For sample FC1 ($Q_{eq} = 24$, $X = 5 \cdot 10^{-3}$), a and b were close to 1.6. For sample F80 studied here, X is close to the value for FC1: we find a and b rather close to 1.6. For sample FC2 ($Q_{eq} = 35$, $X = 2.7\ 10^{-3}$), $a = 1.1$ and $b = 0.9$, meanwhile here for F79 ($Q_{eq} = 37$), $a = 0.8$ and $b = 0.5$. The results are therefore similar in the two experiments using two slightly different ways of crosslinking. The values a and b found for aminomethylated gels are also similar.

As the polymer volume fraction $\phi = 1/Q$, $\xi_Q = \phi^{-a}$ and $S_{0Q} = \phi^{-b}$ increase when ϕ decreases, i.e. when the system is diluted. This effect is actually already observed for *semi-dilute solutions of polymers*: let us call ξ the size of the blob [15], below which the system is a fractal self-avoiding walk (the mass in the blob is $m \sim \xi^{D_f=5/3}$). The average concentration, valid for all scale $>\xi$, is $\phi \sim m/\xi^3 \sim \xi^{D_f-3}$, which leads to $\xi \sim \phi^{-3/4}$: with dilution, the size ξ increases by rearranging small blobs in less numerous larger blobs. This is called *disinterpenetration*. The cross-section limit $S_0(\phi) \sim \phi^2 \xi^3 \sim \phi \xi^3 \sim \phi \xi^{D_f}$ increases with ξ with dilution, if $|3 - D_f| < 3/2$. For $D_f = 5/3$, $S_0(\phi) \sim \phi^{-1/4}$.

Let us first imagine that in crosslinked species, disinterpenetration should not be possible: the swelling would correspond, on the contrary, to a swelling of the blobs with no matter imported from other blobs; $\xi \sim \phi^{-1/3}$ and $S_0(\phi) \sim \phi^2 \xi^3 \sim \phi$ decreases as in a classical dilution. The classical models (which will be better described somewhere else) do not involve any blob, but the philosophy is similar: each chain is swollen separately (by end-to-end pulling) and the scattering decreases due to this expansion. Considering the compressibility leads to similar conclusions, as fluctuations are abated both by the additional shear modulus and the fact that the network is strained by swelling [6, 11].

Another model has been recently proposed [16] by analogy between a percolation problem (linking some objects at random) and the crosslinking at random of a semi-dilute solution, where the objects would now be the blobs. If the chains are very long (number of units $N \gg g$, number of units in a blob) the crosslinking ratio threshold for the blob percolation ($B \sim 1/g$) is much larger than the gel point threshold for chains ($C \sim 1/N$). The effect of dilution of solutions of percolation polymeric clusters has been investigated theoretically. The scattering is $S_0(\phi) \sim \phi \xi^{D_f(3-\tau)} \sim \phi^{-5/3}$ and $\xi \sim \phi^{5/3}$. For $q\xi > 1$, one sees the inside of the clusters, and $I(q)$ should be a power law $\phi/q^{D_f(3-\tau)} \sim q^{-8/5}$. Experiments on polymeric clusters lead to $q^{-8/5}$ law [17], and a maximum in ϕ of $S_0(\phi)$, at a value ϕ^m [18]. However, the power law for the variation of $S_0(\phi)$ with ϕ above ^{m} could not be found with the predicted exponent 5/3. One had to use a master curve to observe a (q, ϕ) superposition [19] in agreement with the theory. Around B, the predictions for dilution could apply for swelling [16]: indeed, the exponent $5/3 \sim 1.6$ is found here surprisingly directly for a few of our samples as well as for the sample FC1 [11] as discussed above.

For other crosslinking ratios, weaker exponents are measured. Theoretically, weaker exponents, which are less universal, could exist far enough from the percolation threshold B (percolation exponents remain valid not too far from B as long as $q > 1/D$, D being the size of the largest cluster). Another consequence of the percolation picture should be a maximum of the effect as a function of the crosslinking ratio: this has not yet been observed and requires further work. The last criticism could be that percolation exponents are observed in cases where the model should not directly apply. This is the case of gels made by end-linking from a solution of much shorter chains [20]. As one is then close to c^*, the blobs percolation threshold and the chains gelation threshold should coincide ($B = C$). For dried gels as well, the percolation model could not apply without redefining the equivalent of the blob size in a melt. However, the behaviors are very similar between the gels made in solution and the gels made from the melt, for PS (here) as well as PDMS, either in the real gelation range [21] or at larger crosslinking ratio [21, 22]. They all belong to the same class, which gives an overscattering after swelling above the volume of preparation. Further experiments are in progress.

IV. Trapping chains

As described above, we have included non crosslinkable chains (mol. weight M_{wD}) in some solutions of AM precursors before crosslinking and checked that it was possible to remove a large fraction of them after crosslinking by washing. We will see below some other signs of their mobility: so, we may call them "free". On the other hand, they are not expelled from the network; thus, we often call them "trapped". We call ϕ_T the total volume fraction of trapped chains with respect to the total volume of polymer. In certain cases, only a fraction $\phi_D = x_D \phi_T$ of trapped chains will be deuterated, the fraction of the non-deuterated chains being $\phi_H = 1 - \phi_D$. If $x_D = 1$, we use indifferently ϕ_D or ϕ_T.

Results for gels with solvent

If the solvent of preparation matches the PS hydrogeneous chains of the network, one can directly see the deuterated chains in the preparation state. Figure 6 displays in $q^2 I(q)/[(\phi_D(1 - \phi_D)]$ = $q^2 S(q)$ a common plateau at high q: the intrachain contribution is visible in this regime and corresponds to a Gaussian string of swollen blobs. The Gaussian behavior is visible because we are at $q < 1/\xi$. At small q interchains contributions appear on Fig. 6; an additional central scattering is visible and increases with M_{wD}. As the system is close to a mixture of two types of chains in a melt, the trapped one and the chain of the network, one may use the well known formula [15]

$$1/I_\chi(q) = 1/(\phi_H S_{1H}(q)) + 1/(\phi_D S_{1D}(q)) - 2\chi . \quad (3)$$

The increase of $q^2 I_\chi \cdot (q)/[(\phi_D(1 - \phi_D)]$ above the plateau value is obtained for χ large enough compared to $\phi_H(S_{1H}(q)$ and $\phi_{1D}(q)$. χ usually represents a chemical enthalpy of mixing, but here χ_{HD} is very small in melt already ($2 \cdot 10^{-4}$), and the presence of solvent decreases its effect even more. Indeed, for the same deuterated chains dispersed in a semidilute solution of long H-chains, no interchain contribution appear. When the H-chains are

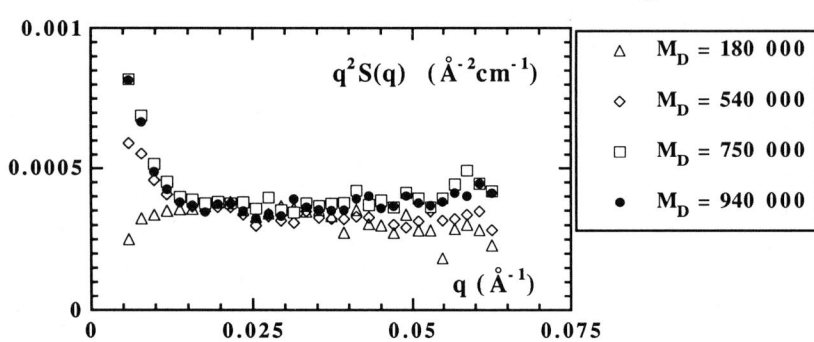

Fig. 6. Gels made from aminomethylated precursors: scattering $I(q)$ from the trapped chains in the gels in preparation state for different molecular weights M_{wD} ($q^2 S(q) = q^2 I(q)/\phi_D (1 - \phi_D)$ representation)

crosslinked, $\phi_H S_{1H}(q)$ may be assumed to become infinite after crosslinking (the network is one infinite chain), or stay close to its value before crosslinking (this would account for the dangling ends of the crosslinked chains). In the first case there would be an increase of $I_\chi(q)$, but we believe that the main effect is an effective χ due to the elasticity of the network, which tends to expell the free chains. The chains are not strongly expelled, but more and more when M_{wD} increases because the entropy of mixing per monomer decreases as $1/M_{wD}$. It is also possible that the free chains are expelled in the soft parts of the gel, as observed above for the solvent itself. However the shape of the curves, within the present accuracy, seem to us easier to describe with Eq. (3) than are the data on swelling and also, as shown below, different from free chains in dried networks, for which a more detailed discussion will be given.

Results for gels dried in the bulk state

Figure 7 shows, for the same molecular weight of trapped chain, a representative variation of $q^2 I(q)/[(\phi_D(1 - \phi_D)] = q^2 S(q)$ with the crosslinking ratio X (related to Q_{eq} by Eq. (0)). For small X (Sample S1), the scattering is the same as for an ideal mixture in a non crosslinked host polymer,

$$1/I_{ideal}(q) = 1/(\phi_H S_{1H}(q)) + 1/(\phi_D S_{1D}(q)), \quad (4)$$

i.e., Eq. (3) with $\chi = 0$ [19]. When X is increased, $I(q)$ increase at small q. Above a certain value of X, one observes *a maximum in the curve* $q^2 S(q)$ at $\tilde{q} \sim$ around $1.5 \cdot 10^{-2}$ Å^{-1} (Note that in the isotropic state no maximum is ever observed for $I(q)$ itself); the position of the maximum $\tilde{q} \sim$ depends on the crosslinking ratio, and probably on the molecular weight M_{wD}. The larger $\tilde{q} \sim$ observed in the series of Fig. 8 (see below) is probably related to the shorter M_{wD}-73 000-involved). At large q, if X remains below a certain value, a plateau in $q^2 S(q)$ is still observed. It corresponds to the Gaussian chain behavior in ideal mixing still visible at large q. For very large X, $q^2 S(q)$ decreases with q at large q (no more plateau), and the maximum of $q^2 S(q)$ can be 10 times higher. For the largest X (Sample 6), a $q^4 I(q)$ representation yields a plateau on a short but significative q range. The scattering becomes similar to a Porod law as scattered by compact objects with sharp interfaces. A second parameter investigated was M_{wD}, between 34 000 and 2 000 000. For the same crosslinking ratio X, the signal in $q^2 S(q)$ is higher for a higher M_{wD}. However, dividing $S(q)$ by the form factor of the corresponding Gaussian single chain $S_{1iso}(q)^{2)}$ quite significantly gathers the data obtained for different M_{wD}. The divided signal depends now only on X. One could actually assume that

$$I(q) = S_{1iso}(q) \cdot g(q), \quad (5)$$

where $g(q)$ would be the Fourier transform of the distribution of the centers of mass of the trapped chains. This would mean that the intrachain scatter-

---
[2]) $S_{iso}(q)$ is calculated from the Debye formula using the molecular weight distribution of the GPC. We actually used $S_{ideal}(q)$ (Eq. (4)), instead of $S_{iso}(q)$ for the determination using Eq. (5), of $g(q)$.

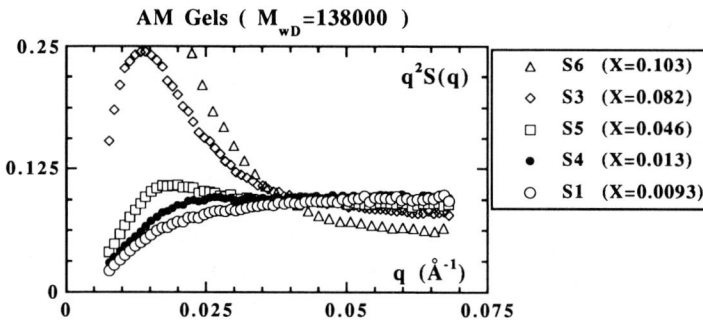

Fig. 7. $q^2 S(q)$ for trapped chains ($M_{wD} = 138\,000$) in a dried networks for increasing crosslinking ratios X (S1: $Q_{eq} = 16$, $X = 0.0093$; S4: $Q_{eq} = 13$, $X = 0.013$; S5: $Q_{eq} = 6.3$, $X = 0.046$; S3: $Q_{eq} = 4.5$, $X = 0.082$; S6: $Q_{eq} = 3.9$, $X = 0.103$)

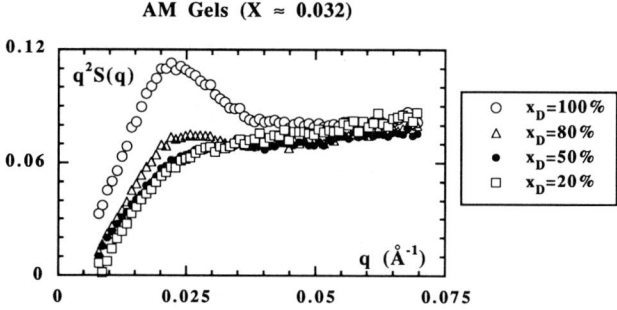

Fig. 8. Variation of $q^2 S(q)$ for different fractions x_D of deuterated trapped chains at constant total fraction of trapped chains: undeformed case (samples S7A,B,C,D, $M_{wD} = 73\,000$, $X = 0.032$)

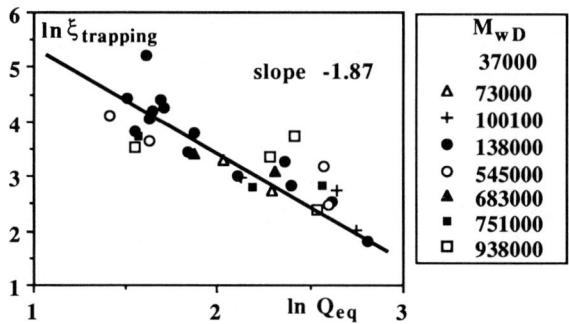

Fig. 9. Correlation length $\xi_{trapping}$ of the interchain scattering: variation with Q_{eq} for different M_{wD}

ing is always equal to $S_{1iso}(q)$. A *proof* of this was obtained by synthesizing several networks from the same precursor, with the same fraction of small chains, but with only a fraction x_D of these deuterated, the rest being hydrogenous, but of same molecular weight distribution. One sees in Fig. 8 the vanishing of the small q interchains contribution as x_D is decreased. Extrapolation at zero of $I(q)/x_D$ at zero x_D yields $S_{1iso}(q)$. This quantity is satisfactorily fitted to the Debye function with the expected radius of gyration ($R_g \sim 0.275\ \sqrt{M_z}$). Linear interpolation of $1/g(q)$ vs q^2 in the q range $[1.6 \cdot 10^{-2}\ \text{Å}^{-1}, 3.2 \cdot 10^{-2}\ \text{Å}^{-1}]$ allows to assume

$$g(q) = g_0/(1 + q^2 \xi_{trapping}^2) . \qquad (6)$$

The variation of $\xi_{trapping}$ with M_{wD} (Fig. 9) is not significant, at variance with its strong variation with the crosslinking ratio ($\xi_{trapping} \sim Q_{eq}^{-1.7+/-.2}$). Our results maybe compared to the one for D-chains trapped in a monomer-grown network (radicalar polymerization of styrene, and divinylbenzene as branching points) [23].

Interpretations

Let us first consider the problem of mixing the guest chains with the network chains. In a classical estimate of the free energy of mixing per monomer, three terms are usually introduced. The first is the translational entropy of mixing, which is kT per chain, or kT/N per monomer for a chain of N monomers. It is always in favor of mixing, but is strongly reduced when increasing N, i.e., M_{wD}, the molecular weight of the trapped chains. It is compensated by two terms. One is the enthalpy of mixing due to the chemical difference between the perdeuterated chains and the network chains, $\Delta E_{HD} = kT \phi_H \phi_D \chi_{HD}$. To this must be added the enthalpy of mixing due to chemical difference between the perdeuterated chains and the crosslinks, which are of different chemical species. We note it $\Delta E_{cross} = kT \phi_D \phi_{cross} \chi_{Dcross}$, where ϕ_{cross} is the volume fraction of crosslinks. The second is the elastic energy ΔE_{el} due to the swelling of the network by the presence of the trapped chains. Following the classical theories, this swelling should be calculated with respect to a "reference state", of swelling ratio Q_0 which we will not try to know here, in view of the two-step crosslinking process involved (in semidilute plus in dry state). We simply assume a variation $\Delta E_{el} \sim \phi_H \nu kT (Q/Q_0)^{2/3}$, where $\phi_H \nu$ is the number of crosslinks per unit volume and, in the dry state, $Q = 1/\phi_H \sim 1/(1 - \phi_{cross} - \phi_D)$. As both ν and ϕ_{cross} increase with X, the two contributions ΔE_{cross} and ΔE_{el} increase with X. Thus, increasing X may lead towards, or achieve phase separation. The picture is oversimplified for our case, because things may occur during the gelation, either between free chains and percolation clusters, or at any time between the gelation threshold and the final state.

The q^{-4} behavior of Sample 6, corresponding to the largest X, can be the signature of a complete separation in two phases with sharp interfaces. The progressive increase of $I(q)$ with X at constant M_{wD} may be better explained by fluctuations prior to separation. Several facts are, however, not explained in this picture. One is the shape of the $q^2 S(q)$ curves for intermediate X. This shape is actually quite different from the one for swollen gels containing free chains (section above). In the classical theory of initial stages of decomposition for chains of two different chemical species, one accounts for the enthalpy of mixing by a χ parameter [15] in Eq. (3). Here, χ would be replaced by a combination of χ_{Dcross}, χ_{HD}, and a contribution corresponding to ΔE_{el}. It is tempting to make this third contribution independent of q, χ_{el}. Thus, to fit our results, we take for $S_{1H}(q))$ and $S_D(q)$ the Gaussian isotropic form factors of the network H-chains and the D-chains prior to crosslinking, and make vary χ in Eq. (3). For sample S5 ($M_{wD} = 138\,000$) we can fit with $\chi = 3 \cdot 10^{-3}$. However, the abscissa of maximum in $q^2 S(q)$ occurs is slightly smaller than in the data. For sample S3, the fit with Eq. (3) will yield a maximum as high as that observed, only *at much smaller q, out of the q range explored here*. Maybe we require a q dependence for the account for ΔE_{el}. Another discrepancy concerns the effect of M_{wD}. Following Eq. (3), it should be strong (the entropy of mixing is much lower). For example, introducing the value of χ fitted for sample S5 in Eq. (3) for chains five times larger in M_{wD} produces a strong phase separation at the same X. Experimentally, this *is not observed* [12]: as said above, the effect of $M_w D$ is not significant. Maybe demixing is not attained because the duration of annealing of the samples (6 h at 150°C) is too short compared to the relaxation and diffusion times. They are already long in a melt [4], and maybe longer if the long chains are more "trapped". We have, however, observed [12] that these long chains are able to relax after deformation in comparable times, at least partly (see Section V).

Another theoretical possibility is that the "demixed" structure is not built by the demixion process, but reflects the heterogeneities in crosslinking. The chains are expulsed from some high crosslink density regions of the network into low density regions. This expulsion effect could be strong enough to be saturated for all values of M_{wD} used here, leading to no variation with M_{wD}. It is therefore interesting a priori to compare the size of heterogeneities in the dry sample with free D-chains and the same sample after removing of the free D-chains and reswelling in D-toluene. The comparison is however biared by the fact that the swelling corresponding to trapping is $Q = 1.1$. Maybe for this reason, one needs a much larger X to get an effect on $\xi_{trapping}$ than on $\xi_{Q=3}$, for example, and for such a large X the system may become close to demixing. This would explain why for low X (S1), $\xi_{trapping} = 0$, but $\xi_{Q=3} = 17$ Å is much larger than the value in semidilute solution (7 Å), meanwhile for large X (S3), $\xi_{Q=3}$ is not much larger (22 Å) but $\xi_{trapping} = 85$ Å is very high. We see that comparisons are delicate.

Finally, it is difficult to exclude an effect of the heterogenities in a swollen sample responsible for $I_1(q)$ (Eq. (2)), if they still exist in bulk. The q^{-4} variation observed for $I_1(q)$ would, however, not lead to a maximum in $q^2 S(q)$ or to an effect at large q, unless the aggregates become much smaller in bulk than in swollen state. $q^2 S(q)$ would better have large values at small q, and a fast, continuous decreasing, leaving at large q the chain contribution alone, i.e., a plateau. Figure 6, for trapped chains in gels in preparation state would better correspond to such a behavior, but not Fig. 7. In addition (see below), the increase under stretching (in parallel direction) of the additional fluctuations in bulk state more resembles the increase under swelling of $I_2(q)$ than the decrease under swelling observed for $I_1(q)$.

V. Elongated state: the butterfly effect

Kinetics for the butterfly effect

The anisotropic intensity contours as well as "parallel" and "perpendicular" intensity regrouped along angular sectors parallel (respectively perpen-

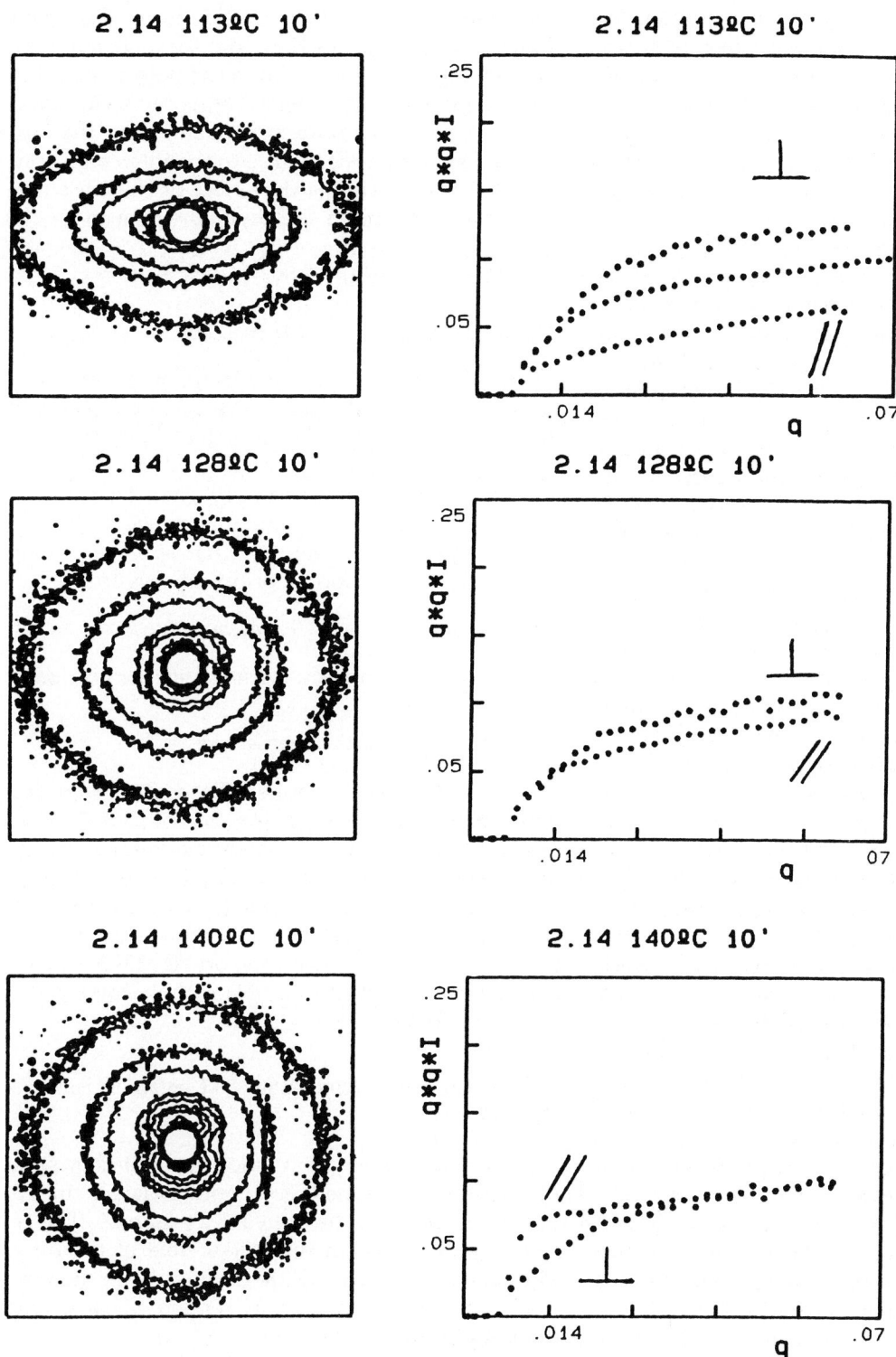

Fig. 10. Kinetics of the butterfly effect: patterns (left-hand side), and $q^2 S(q)$ vs q plots (right-hand side) for scattering parallel ($\parallel$) and perpendicular ($\perp$) to the stretching axis. Sample is S2, $X = 0.0093$, elongation ratio $\lambda = 2.14$, $M_{wD} = 138\,000$; $0 < q < 7 \cdot 10^{-2}$ Å^{-1} in the frame

dicular) to the stretching axis have been studied for several stages of transient relaxation after deformation. In practice, we varied the temperature instead of the time (Fig. 10). At low temperatures (short times), the contours are elliptical, with large axis perpendicular to the stretching axis (vertical here). Correspondingly, the intensity is smaller in parallel than in perpendicular direction. When increasing the time-temperature up to a couple $(t_R, T)_{app}$, a different contribution appears at small angles (close to the edges of the beam catcher), which has the transverse anisotropy. The large axis is now parallel to the stretching axis and the parallel intensity at small q now overtakes the perpendicular intensity. As relaxation is progressing, two lobes on both sides along the stretching axis spread over a larger range of q. The patterns look like the figure "8", or, turning the figure by 90°, like the wings of a butterfly. For the samples of Fig. 10 (M_{wD} = 138 000), we have increased (t_R, T), from (10 min, 140°C) to (10 min, 150°C). This corresponds to a factor 5 in time at constant temperature, but the variation is less than 5%, and zero after 10 min at 160°C. We consider the effect to be stabilized for $(t_R, T)_{stab}$ = (10 min, 140°C). One can estimate the terminal relaxation time $T_{ter}(M_{wD})$ of a melt made from chains of molecular weight M_{wD} [4]. For 34 000 < M_{wD} < 1 000 000, at X values of Fig. 10, $(t, T)_{app} \sim T_{ter}(M_{wD})$. $(t, T)_{stab}$ cannot be measured for long chains, because it is larger than the maximum value before thermal degradation, estimated to (2H, 160°C). In order to extend the range of relaxation, we dissolved into the sample some plasticizer which moved T_g from 100°C to 50°C. No additional evolution after 2H at 160°C was observed for M_{wD} = 700 000 or 1 000 000. A typical effect remains however: the butterfly patterns seem to be contained inside some elliptical envelope (with classical perpendicular great axis). These envelopes are much more anisotropic for long chains than for short chains, even at very long times. This even leads to a lozenge (diamond) shape [4]. It may be due to uncomplete relaxation, or be intrinsic to long trapped chains. Because of their large radius of gyration, they reveal more the anisotropy of $I(q)$. Such an effect is visible even at $qR_g > 1$ for labeled paths [3, 4].

The stabilized butterfly effect
Inter and intra chain correlations

Similarly to undeformed samples, the intrachain contribution $S_1(q)$ of the small chains was extrapolated at zero fraction x_D of deuterated trapped chains, the total fraction of free chains, the crosslinking and the conditions of deformation being as identical as possible. $S_1(q)$ is found identical to $S_{1iso}(q)$, as for the undeformed samples. The butterfly effect then appears due to the interchain contribution (this includes the effect of the host network). We have fitted the divided signal to

$$I(q_\parallel)/S_{1iso}(q) = g(q_\parallel) = g_{0\parallel}/(1 + q^2 \xi^2_{stretch\parallel}) \, . \quad (7)$$

Effect of concentration in trapped chains

Samples synthesized with larger fractions $\phi_D = \phi_T$ = 20%, 24%, 28%, and 32% of free chains (all deuterated) are still "terminated" networks as characterized by low Q_{eq}, undetectable sol fraction and high tensile modulus in dry state. They are far from the gel point. They show *no strong dependence* of the butterfly effect on ϕ_T; that is to say, using the general expression (a priori, always true):

$$I(q) \sim \phi_D S_{1D}(q) + \phi_D^2 S_{2D}(\phi_D, q) \, . \quad (8)$$

$S_{2D}(_D = \phi_T, q)$ shows a *weak dependence* over ϕ_T.

Molecular weight of the trapped chains

A clear effect is a stronger elliptical or even diamond-like of the isointensities for $M_{wD} > 500 000$, but it may be due to an uncomplete relaxation. On the other hand, no effect on the increase of $S(q_\parallel)$, as measured by $\xi_{stretch\parallel}$ and $g_{0\parallel}$ is seen for both stabilized ($M_{wD} < 300 000$) and not stabilized ($M_{wD} > 500 000$) cases. One can see on Fig. 12, for example, that the variation $\xi_{stretch\parallel}(\lambda)$ is the same for M_{wD} = 138 000 and 73 000.

Effect of the crosslinking ratio, and variation with the elongation ratio

The range of elongation ratio was limited by sample breaking. For small X, λ could be larger than 3, but lower than 1.25 or even 1.1 for large X. The effect of λ depends on the initial state of trapping. The different ranges of X determined from the scattering in undeformed state remain valid in the deformed state:

— for $X < 0.01$ ($Q_{eq} > 15$, Sample S1), the undeformed state is ideal mixing. After deformation, the signal in parallel *increases gently* with λ, in perpendicular, it *returns* (after stabilization of the butterfly effect) *very close to the isotropic* one.

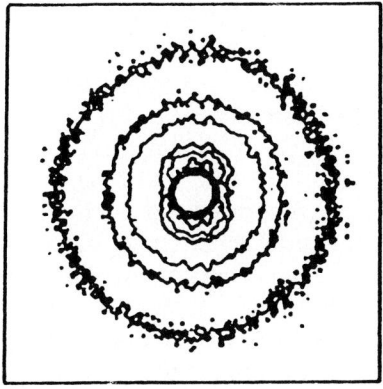

1.6 140°C 10'

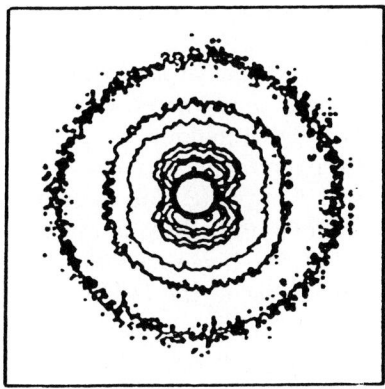

2.2 140°C 10'

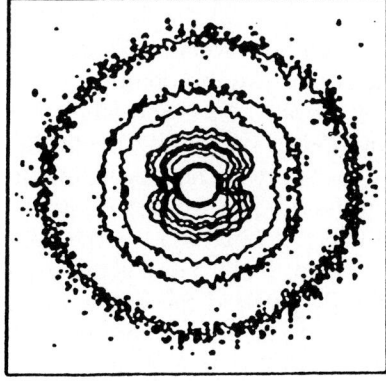

3.2 140°C 10'

Fig. 11. Effect of elongation ratio: patterns for λ = 1.6, 2.2 and 3.2. Sample is S1, X = 0.0093, M_{wD} = 138 000; $0 < q < 7 \cdot 10^{-2}$ Å^{-1} in the frame

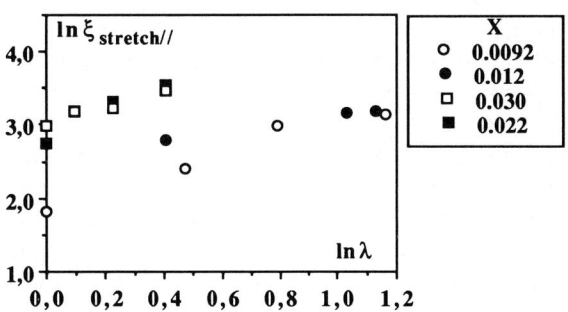

Fig. 12. Correlation length $\xi_{\text{stretch}\|}$ in stretched samples, parallel to the stretching: variation with the elongation ratio λ for two ranges of crosslinking ratios:
— sample S2, $X \sim 0.01$ (M_{wD} = 138 000), and S8, $X \sim$ 0.012 (M_{wD} = 73 000),
— sample S9, $X \sim 0.022$ (M_{wD} = 73 000) and S10, $X \sim$ 0.030 (M_{wD} = 138 000)

— for $0.01 < X < 0.03$ ($8 < Q_{eq} < 15$, Sample S4), the undeformed state scatters more than for ideal mixing. After deformation, the signal in parallel increases *faster* with λ. In perpendicular, the signal *is smaller than the undeformed one*, and *returns towards the ideal mixing* one inasmuch as λ is increased, eventually reaching it.

— for $X > 0.03$ ($Q_{eq} < 8$, as Samples S5 and S3), the scattering in the undeformed state was even larger. After deformation, the signal in parallel may still increase *stronger* with λ. For larger X, it will not increase anymore, giving the impression of a saturation. (But for Sample S3 for example λ and t_R were limited by breaking). In perpendicular, the signal *decreases strongly towards the ideal mixing* one as λ is increased, without being able to reach it.

Thus, the butterfly patterns appear for small X because *the parallel intensity increases* and for large X because *the perpendicular intensity decreases*.

Finally, for $X > 0.1$ ($Q_{eq} < 4$, Sample S6, λ only 1.1) the isotropic scattering q^{-4} associated with a phase separation, the scattering deforms into a signal elliptical in the classical way even for large t_R.

The effect of λ is seen already at very small elongation (1.1), which gives a butterfly effect for any $X < 0.1$. One sees in Fig. 11 (λ = 1.6, 2.2 and 3.2) that the isointensity patterns appear more flat when increasing λ, as if they were contained in an envelope elliptical in the classical direction (great axis perpendicular), following a quasi-affine deformation of the patterns. Meanwhile, the parallel intensity increases with λ, and the perpendicular one decreases, unless it is already overlapping the signal of ideal mixing. Extra patterns corresponding to larger intensities appear in parallel direction. If there was only an affine deformation of the intensity starting from a given situation, say at $\lambda = 1.5$, the patterns would look similar, but the intensity would not increase anymore at small q. Therefore, real data display a combination of two effects, a larger "classical" anisotropy and an increase of the butterfly effect. The latter can also be characterized by the variation of $\xi_{stretch\parallel}$ with λ. In Fig. 12 ($\log_e - \log_e$), the initial slope is around 2, with, at large λ, a slowing down, nearly a plateau. For a soft sample (equivalent to S1), patterns as well as intensities do not vary when passing from $\lambda = 2.8$ to $\lambda = 3.2$ (in the available q range).

Discussion and comparison with models
Chain expulsion and phase separation

In this short paper, we will only *start* the discussion by examining the relation between butterfly effect and expulsion of the free chains. A first question is *what is driving an expulsion*, a second is *where do the expulsed chains go*? The description of the system as close to demixing (see above) in the undeformed state does not give a direct clue for the "butterfly" pattern after deformation. Classical deformation of the undeformed structure will only lead to classical elliptical anisotropy ($S(q_\parallel) < S(q_{perp})$). The completely separated system (Sample S6) follows this trend, giving no butterfly. One thus needs another contribution of the deformation. One could expect it to *induce an expulsion*, a phenomenon observed in solutions and melts. The elastic energy due to swelling by the chains of the temporary or permanent network may increase when the system is elastically deformed [4, 24]. Another elegant explanation [25] would be the difference in monomer orientation of the small chains (isotropic) and of the chains of the network, temporary or permanent, which are still deformed. Differences in dipolar interactions result in a *non zero χ parameter* between chains of *same chemical species*. As for the undeformed systems, these models do not describe the separation process (large fluctuations, spinodal decomposition, or nucleation and growth), neither how the network would limit it, nor the final structure. They do not lead to any butterfly effect. Experimentally, the lack of critical threshold in ϕ_T, M_{wD}, and λ is in any case, in disagreement with any expulsion driven only by deformation.

Clusters separation

A second assumption is that repulsion is also driven by the heterogeneities of crosslink density: free chains invade the regions of low density in the same way as solvent. Indeed, for stretched gels, similar butterfly patterns were observed [11], and explained [26] by analogy between swelling (as described by desinterpenetration [16]) and stretching in the parallel direction. Here, the repulsion of the free chains, for a low X in undeformed state, could be small or, more probably, the clusters of crosslinks are completely interpenetrated, and are thus not visible [16]. This would explain the ideal mixing observed by SANS. After stretching-relaxing, the clusters are revealed in parallel direction either by stress-induced repulsion or, more likely, just through their desinterpenetration, as for swelling, and $I(q_\parallel)$ increases. In perpendicular direction there cannot be more interpenetration and $I(q_{perp}) = I_{ideal}(q)$. For large X, the clusters are already visible in isotropic state, after deformation their signal is enhanced in parallel direction, with a possible saturation. In perpendicular, direction reinterpenetration reduces $I(q_{perp})$ back towards $I_{ideal}(q)$ as in deswelling. This is in large agreement with observations and, thus, is much in favor of the model. $\xi_{stretch\parallel}$ is however difficult to relate to ξ_Q. This is linked first to the fact that it is already difficult to relate $\xi_{trapping}$ to ξ_Q. Also, it is difficult to relate $\xi_{stretch\parallel}$ to $\xi_{trapping}$, because the effect of the deformation varies with the initial state, and because there exists, in addition, a classical effect of the elongation. The variation with λ is smoother than the strongest predicted (as $\lambda \sim c^{-1/3}$, $\xi_{stretch\parallel} \sim c^{-5/3} \sim \lambda_\parallel^5$ [11, 26]). It could be due to a weaker dependence as reported above for $\xi_Q(Q)$ with some of the gels, or because the model does not take the deformation of the clusters into account [26]. Several other weaknesses of the model can be listed. One is that the blobs picture does not apply for heterogeneities in samples crosslinked in the dry state. A second is that no maximum of the effect is observed, in the very large range of X investigated. A third one is the striking similarities of butterfly patterns for simple melts with no crosslinks [9, 10]. We also note that the weak dependence on ϕ_T, M_{wD}, implies that chains of any ϕ_T or M_{wD} are completely expulsed in undeformed state: is this realistic?

Chain separation

Another model [27, 12] does not require expulsion of the chains by the network (see contribution by

Higgs, this volume). In a polymer system like a melt of identical labeled and non-labeled chains, which is uncompressible, there is a balance between the intrachain and the interchain signal

$$S_{1D}(q) = -S_{2D}(q) \tag{9}$$

which makes Eq. (6) become

$$S(q) = (\phi_D - \phi_D^2)S_{1D}(q) . \tag{10}$$

Equation (4) corresponds to Eq. (10) when labeled and non-labeled chains are different; it reflects the same balance. After crosslinking or entanglement, this balance can be maintained. After deformation-relaxation, only the centers of mass of the active meshes, which influences the interchain contribution will be displaced as predicted by classical models: the intrachain contribution of meshes between crosslinks, loose chains, "reptating" chains and, overall, *free chains* will be closer to isotropy. The balance S_1/S_2 is altered: we could call this model chain separation. Equation (4), representing the balance S_1/S_2, can be written for deformed systems

$$1/I_{RPA,\lambda}(q) = 1/(\phi_H S_{1\lambda H}(q)) + 1/(\phi_D S_{1\lambda D}(q)) . \tag{11}$$

In the crossection

$$I_{RPA,\lambda}(q) = g_{RPA,\lambda}(q) \cdot S_{1\lambda}(q) , \tag{12}$$

$g(q)$ obtained through the RPA equation could be *kept*, meanwhile $S_1(q)$ would be *replaced* by a less deformed signal ($\lambda' < \lambda$). Then,

$$I_{unbal}(q) = g_{RPA,\lambda}(q) \cdot S_{1\lambda'}(q) = I_{RPA,\lambda}(q) \cdot (S_{1\lambda'}(q)/S_{1\lambda}(q)) . \tag{13}$$

When passing from Eq. (11) to Eq. (13), $I(q_\parallel)$ increases and $I(q_{perp})$ decreases. Calculated isointensity patterns [27] have elongated butterfly shapes. One disagreement with experiment is the predicted decrease of $I(q_{perp})$ in any case, meanwhile, experimentally, $I(q_{perp})$ decreases only for large X, but remains equal to $I_{ideal}(q)$ for small X. This may be cured by detailed calculations, necessary to establish the model, which are to be overcome. The simplicity and generality of the idea makes it attractive. Note that it could apply to the swelling of gels.

Continuuum mechanics

Very recently, models have arise that have a continuum mechanics formulation. This may still involve frozen-in heterogeneities [28]: this produces modulus heterogeneities and, thus, strain-heterogeneity. It is not yet clear to us whether or not percolation clusters are involved in the process. It is also claimed that *spontaneous* heterogeneities, not frozen in, may just be enhanced by stretching, as in spinodal decomposition ("up-hill diffusion") [29]. This last proposal was motivated by the similarities of effects in melt, with no crosslinking, therefore a more questionable heterogeneous structure. It is also associated with the growing number of observation in flow of similar effects observed by light scattering, which are observed using equivalent rheological models [30—32] as well as disinterpenetration [33]. An active discussion is on going.

VI. Conclusion

We have induced an increase of scattering by three different actions on the same network. The first has been swelling by labeled solvent, the second by incorporating free deuterated chains in a network, and the third by stretching the latter systems. We expected a common origin for these phenomena, and similarity between data using the two labelings, i.e., by the deuterated solvent and by the deuterated free chains. Indeed, we have proved in the latter case that the intrachain contribution is the one of Gaussian isotropic chains in undeformed and deformed systems. Therefore, the trapped chains can effectively be considered as a kind of isotropic polymeric solvent. However, we could not quantitatively relate the correlation lengths extracted from swelling and trapped systems, $\xi_{trapping}$ and ξ_Q. Due to the big difference of size and translational entropy between solvent molecules and free chains, the effect of heterogeneous swelling might be enhanced in the latter case, especially for large crosslinking density. This makes the heterogeneities picture difficult to transfer quantitatively from one system (soft regions are invaded by solvent) to the other (soft regions are invaded by free chains).

These difficulties spread on the study of the deformed system, i.e., the butterfly effect. The response to deforming brought new information. The lack of threshold in total concentration and molecular weight of free chains, and with elongation ratio λ disagrees with a strain-induced decomposition, as described by the classical ingredients of entropy, enthalpy, and Gaussian elasticity. This disagreement is reinforced by the striking observation of decrease of $S(q_{perp})$ with λ.

We hope this work will contribute to the theoretical debate, but nothing is solved yet: the heterogeneities picture is well supported by some features of the results, in particular the decrease, when the undeformed system shows non-ideal mixing of $I(q_{perp})$ towards ideal mixing. There are, nevertheless, weaknesses: no maximum of the effect with the crosslink density (corresponding to a percolation threshold) has been seen yet, and, in practice, there is a larger experimental range of similar effects than in the situation described in the original picture. To make it hold, one should therefore greatly extend its validity domain (attempted recently, but on very qualitative grounds only). Another difficulty is the observation of comparable behaviors in (uncrosslinked) melts of long chains containing small chains: one should, therefore, assume entanglement heterogeneity [34]. Finally, the picture of expulsion of the chains in the soft regions disagrees slightly with the very weak dependence of the effect on total concentration and molecular weight of the free chains. These discrepancies make it worthwhile to consider the more general idea of the chain separation model, not requiring heterogeneities or expulsion, but needing theoretical developments. On a different ground, recently developed continuum mechanics models are even of a more general range of application: they correspond to a kind of stress-induced demixing, different from the usual one, and due to new ingredients in the free energy.

Acknowledgements

The data published here were obtained in the reactor Orphée. We thank C. Thomas, A. Rémy, T. Krebs, and E. Lecosz for their technical help. We thank A. Brulet for improvements of the whole process. We thank A. Lapp, for crucial GPC measurements.

References

1. Boué F, Nierlich M, Jannink G, Ball RC (1982) J Physique Lettres 43:L-589
2. Boué F (1987) Adv Pol Sci 82
3. Boué F, Bastide J, Buzier M, Collette C, Lapp A, Herz J (1987) Prog Colloid Polym Sci 75:152
4. Boué F, Bastide J, Buzier M, Lapp A, Herz J, Vilgis T (1991) Colloid Polym Sci 269:195—218
5. Ferry JD (1987) Viscoelastic Properties of Polymers, J Wiley
6. Bastide J, Boué F, Buzier M (1989) Springer Proc Phys 42
7. Boué F, Farnoux B, Bastide J, Lapp A, Herz J, Picot C (1986) Europhys Lett 1:637
8. Oeser R, Picot C, Herz J (1987) Springer Proc Physics 29:104—111
9. Barea JL, Muller R, Picot C (1987) Springer Proc Physics 29:86-97
10. Bastide J, Boué F, Buzier M (1987) Springer Proc Physics 29:112—120
11. Mendes E (1991) Thesis, Strasbourg and Mendes E, Lindner P, Buzier M, Boué F, Bastide J (1991) Phys Rev Lett 66:1595—1598
12. Hovasse-Zielinski F (1991) Thesis, Paris
13. Ramzi A, current thesis work
14. Mallam S, Hecht AM, Geissler E, Pruvost P (1991) J Chem Phys 10:6447
15. Degennes PG, Scaling Concepts in Polymer Physics, Corn Univ Press, Ithaca
16. Bastide J, Leibler L (1988) Macromolecules 21:2647—2649
17. Bouchaud E, Delsanti M, Adam M, Daoud M, Durand D (1986) J Physique 47:1273
18. Adam M, Delsanti M, Munch JP, Durand D (1987) J Physique 48:1809
19. Munch JP, Delsanti M, Durand D (1992) Europhys Lett 18:577/582
20. Hakiki A (1991) Thesis, Univ Louis Pasteur, Strasbourg, France and Lal J, Bastide J, Bansil R, Boué F, to be submitted
21. Falcoa A, Pedersen I, Mortensen K (to be published)
22. Boué F, Mazan, Leclerc, Moreau, Couarazze, Bastide J, Boué F, unpublished
23. Briber R, Bauer R (1991) Macromolecules 24:1899—1904
24. Nafaile CR, Metzner AB, Wisbrun KF (1984) Macromolecules 17:1187—1195
25. Brochard F, Degennes PG (1988) C.R.A.S. Paris, T. 306, Série II, 699—702
26. Bastide J, Leibler L, Prost J (1990) Macromolecules 6:23
27. Higgs P, this volume
28. Onuki A (1992) J Phys II France 2:45—61
29. Rabin I, Bruinsma R (1992) Europhys Lett 20:79—87
30. Fredrickson GH, Helfand E (1989) Phys Rev Lett 62:2468
31. Milner ST (1991) Phys Rev Lett 66:1477
32. van Egmond JW, Werner DE, Fuller GG (1992) J Chem Phys 96:7742—7757
33. Hasimoto T, Kume T, J Phys Soc Jpn (submitted)
34. Bastide J, Boué F, Mendes E, Buzier M, Zielinski F, Lartigue C (1991) M.R.S. Meeting, Boston (to be published)

Authors' address:

F. Zielinski
Lab. Léon Brillouin
CEA SACLAY
F-91191 Gif sur Yvette, France

Aggregation of free chains within a deformed network: A SANS study

R. Oeser

Institut Laue Langevin, Grenoble, France

Abstract: A reaction, combining labeled oligomers included in an anisotropic matrix, was monitored with small-angle neutron scattering. The initial "butterfly pattern" was strongly enhanced during the reaction, indicating that the growing branched structures are constrained to particular regions within the network. The real-time measurement was continued until the growing clusters had reached their final size. After removal of the constraints the sample completely recovered its original shape. The relaxed isotropic sample exhibits the history of its production by a highly anisotropic scattering pattern. It resembles ellipses whose longer axes are aligned along the stretching direction.

Key words: Small-angle neutron scattering; SANS real-time experiment; PDMS model networks; cluster; butterfly pattern; interpenetrated network; gelation; nonaffine deformation

Introduction

Small-angle neutron scattering experiments on free, i.e. unattached chains in elastomeric polydimethylsiloxane networks showed peculiar contour lines on a two dimensional detector: At small scattering vectors the spectra feature well separated wings, which extend parallel to the stretching direction. Therefore, they were called "butterfly patterns" [1]. Their origin is a nonrandom distribution of the free chains in space, due to the influence of the deformed matrix. Thus, two questions were to be answered:

1) What happens to the spatial distribution of the unattached chains in a uniaxially deformed network when they react among themselves? In other words, does the butterfly pattern remain?
2) Does the relaxed system remember its history, i.e., the way it was produced?

The polydimethylsiloxane system which was used allows to control the crosslinkage and the aggregation chemically. The samples are made in a two-step process, using two independent systems:

- A hydrogenated matrix, which is assumed to be an ideal model network in the classical sense, was made by endlinking from stoichiometric amounts of precursors.
- Bifunctional deuterated chains, which may react among themselves by means of a crosslinker, were allowed to diffuse into the matrix. With an excess of crosslinker the results of the reaction are branched structures, rather than a completely crosslinked network.

The aggregation of small chains within a network corresponds to an intermediate stage towards the build-up of homogeneous, i.e., one-phase, interpenetrated networks (IPN). These networks may have enhanced mechanical properties. If a second network is crosslinked in an anisotropic state, one might create an entangled system whose mechanical properties are direction dependent.

Sample preparation and reaction procedure

To prepare the host network the "Strasbourg recipe" [2] was used, starting from *hydrogenated*

PDMS polymers $\left(M_n = 10\,000, \dfrac{M_w}{M_n} = 1.8\right)$ with a silane group at each end. These were crosslinked with the stochiometric amount of 1,1,2,2-Tetravinyldisiloxane to obtain a tetrafunctional network. To catalyze the hydrosilation reaction, hexachloroplatinic acid in a concentration of $5 \cdot 10^{-4}$ mol per mol silane was added, prior to crosslinking. In order to preserve the catalyst within the network, the sol phase was not extracted.

A reactive mixture was prepared from bifunctional *deuterated* PDMS $\left(M_n = 3000, \dfrac{M_w}{M_n} = 1.7\right)$ and tetravinyldisiloxane. A large excess of crosslinker corresponding to one tetrafunctional molecule per chain end was chosen for the following reasons:

— Branching rather than crosslinking is favored. It is assumed that the small tetrafunctional molecules diffuse freely within the matrix.
— Side reactions of the silane groups [3] with the backbone of the network should be less important, due to the high concentration of double bonds.
— The probability that the silanegroups react with residues of unreacted vinyl groups of the matrix, is reduced.
— Since the sample had to be heated during several hours, the loss of tetravinyldisiloxane due to its volatility (b.p. 66°C/15 mm) had to be compensated.

To make the swelling as homogeneous as possible, the mixture was spread on both surfaces of a stripe of the elastomeric network. After 4 h at room temperature most of the oligomer-crosslinker mixture had been taken up by the sample. The final volume fraction of the deuterated mixture in the host network was 8.6%. The swelling time could not be extended, since the reaction is already going on at room temperature, due to the presence of catalyst in the network.

The swollen sample with a width of 23 mm and a length between clamps of 39 mm was stretched within an oven. A square with 1 cm sides marked with points of chinese in ink in the center of the stretched sample became a rectangle of $7 \cdot 11.5$ mm in the relaxed state corresponding to a uniaxial extension of $\lambda = 1.43$.

The sample was heated to $\approx 65\,°C$ to start the hydrosilation reaction. It was kept at this temperature for 10 days. Afterwards, it was cooled to room temperature and the constraint was removed. The sample relaxed instantaneously and completely recovered the rectangular shape it had before the reaction.

A continuous extraction with cyclohexane in a Soxleth apparatus during 3 days after the experiment resulted in a loss of about 6% of the weight of the sample. This corresponds to the usual sol fractions of this type of model networks, consisting mainly of unreacted monomer.

A complementary observation was made on the remaining reactive mixture which stayed on the support used for swelling: After 2 days at room temperature it had yielded a well crosslinked elastomer, indicating that some catalyst had been taken up from the network.

SANS experiments

Small-angle neutron scattering spectra were recorded on D17 of ILL $\left(\text{sample-detector distance:}\right.$ 3.45 m, wavelength 12 Å, $\left.\dfrac{\Delta\lambda}{\lambda} = 10\%\right)$ with the center of the beam at the right edge of the detector, giving scattering vectors ranging from 0.01 to 0.1 Å^{-1} parallel to the stretching direction of the sample. They were normalized using water as a secondary standard, taking the thickness of the sample and its temperature into account.

Two-dimensional scattering patterns were recorded in real time every 13 min during the first 6 h of the reaction. Another spectrum was taken 20 h after the start of the reaction. After 10 days at 65°C two spectra (Fig. 2b) were recorded at an interval of 5 h. They were identical, indicating the end of the reaction.

After the removal of the constraint the relaxed sample was measured in the same configuration (Fig. 2c) to compare the same contour levels of the stretched with the relaxed state. The elliptical shape of the contours at small angles was checked with the incident neutron beam in the center of the two-dimensional spectrum (Fig. 6).

After the extraction another spectrum of the relaxed sample was recorded at LLB, Saclay. Within experimental uncertainties it was identical with Fig. 6.

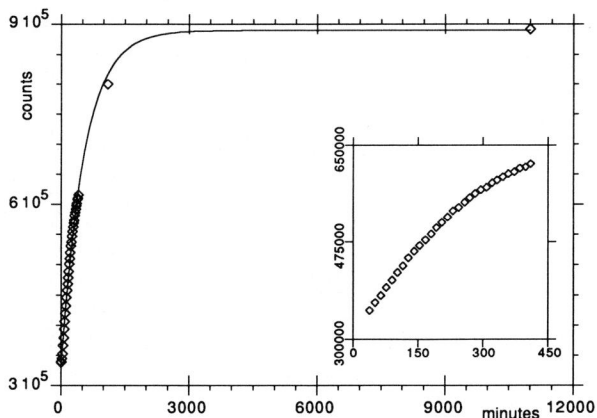

Fig. 1. Total scattered intensity on the whole two-dimensional detector versus reaction time. Insert: real time experiment at the beginning of the reaction

Finally, the single-chain form factor of the deuterated polymer (8% by volume) was measured in a melt of equivalent hydrogenated chains (melt in Fig. 5).

Results of the SANS experiments

The development in time was monitored through the integral scattered intensity on the two-dimensional detector. Immediately after the start of the reaction between the deuterated chains, the total scattered intensity increased linearly with a rate of 1000 counts/min (Fig. 1). After 3 h the rate of increase leveled off and the intensity approached a maximum.

Before the reaction the stretched swollen sample exhibits one-half of a "butterfly pattern" (Fig. 2a), which has been observed for this particular system of labeled unattached chains in a network. During the whole reaction the pattern remains qualitatively the same. The contraction in transverse direction seems to diminish somewhat. However, to obtain

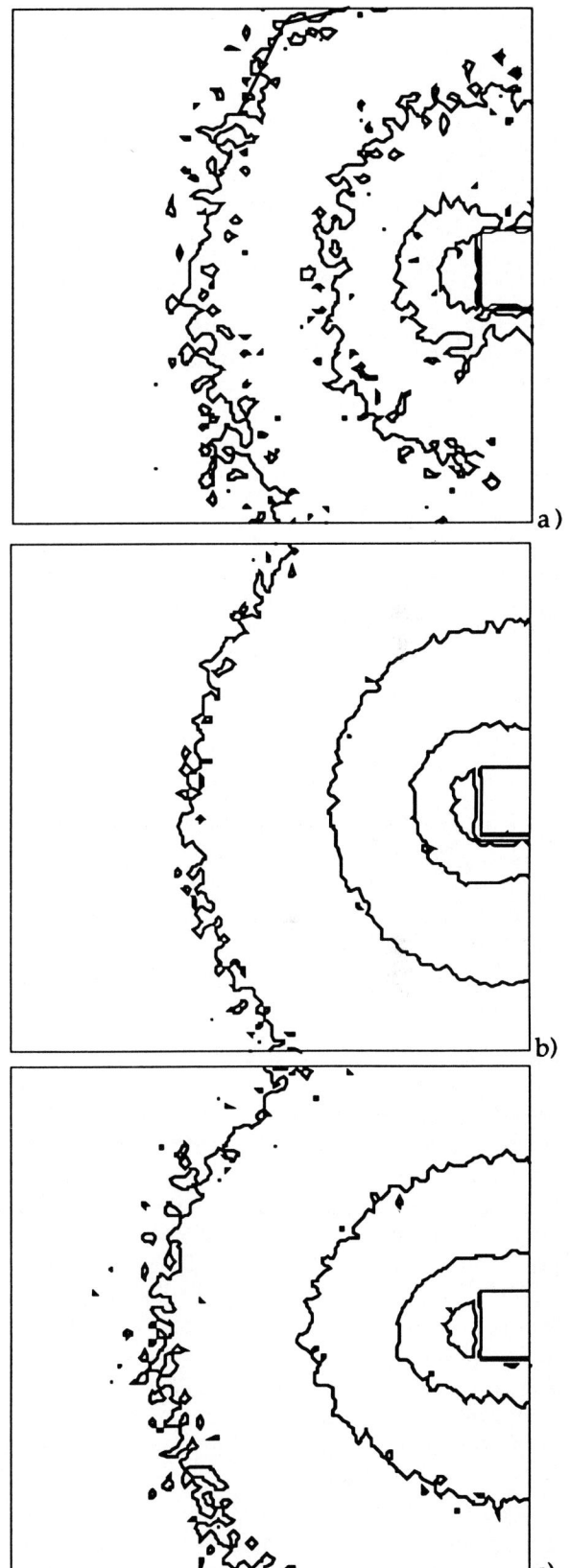

Fig. 2. Normalized two-dimensional spectra from the horizontal stretched sample: a) at the beginning; b) at the very end of the reaction; c) after relaxation. The direct beam has been masked. The isointensity levels are for a): 1.1, 1.8, 2.6, 3.3 cm^{-1} and for b) and c): 1.8, 6.3, 18.3, 31.3 cm^{-1} going from the left towards the beam center. The left edge corresponds in each case to 0.09 Å^{-1}

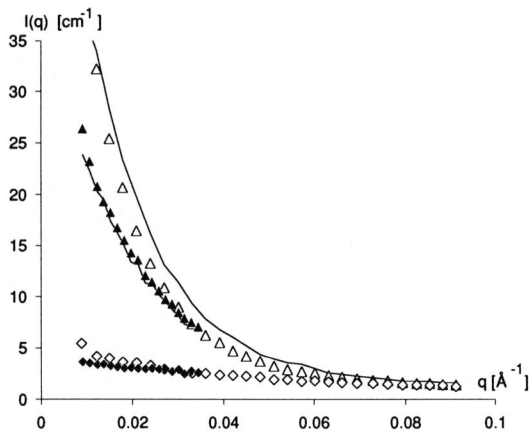

Fig. 3. Characteristic scattering functions parallel and perpendicular to the stretching direction taken from sectors with an opening angle of 10°.
Start of reaction: ◇: parallel ◆: perpendicular.
End of reaction: △: parallel ▲: perpendicular.
After relaxation: continuous lines

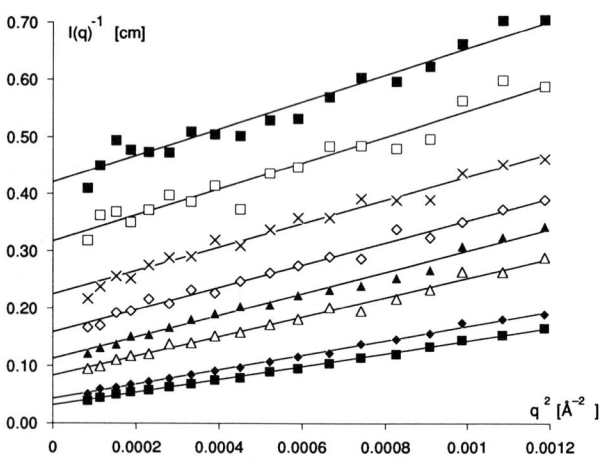

Fig. 4. Zimm representation of scattering functions taken perpendicular to the stretching direction (see ▲, ◆ in Fig. 3) for different times. The lines are fits over the whole range whose results are given in the legend:

Time	R_G	M_w
■ start	41 Å	8 300 [D]
□ 51 min	46 Å	10 900 [D]
× 103 min	52 Å	15 400 [D]
◇ 154 min	61 Å	21 800 [D]
▲ 231 min	71 Å	30 800 [D]
△ 359 min	78 Å	41 500 [D]
◆ 1100 min	93 Å	80 400 [D]
■ 14 000 min	102 Å	109 000 [D]

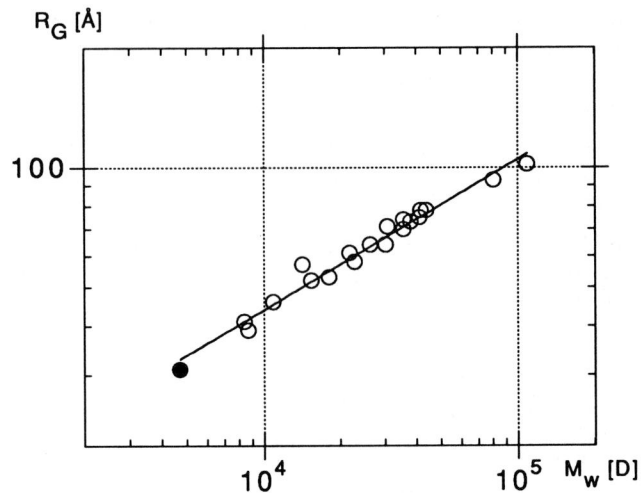

Fig. 5. Double logarithmic plot of apparent values of radius of gyration vs. molecular weight. The fitted power law has a slope of 0.38. (○: during reaction ●: melt)

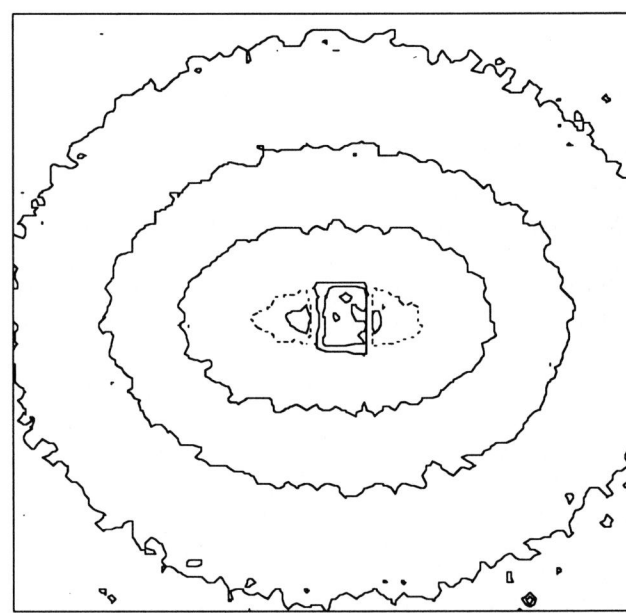

Fig. 6. Spectrum of the relaxed sample taken with beam in the center. The former stretching direction was horizontal. The edge of the detector corresponds to 0.05 Å^{-1}. The contours are 4,8,16,32 (---) and 40 cm^{-1} towards the center

the same shapes on isointensity plots, the levels (Fig. 2a—b) have to be raised drastically.

After relaxation the anisotropy seems to be even more pronounced. The lateral width of the wing has decreased, but its longitudinal extension has increased, resulting in ellipses, whose longer axes are oriented parallel to the stretching direction.

Scattering envelopes (Fig. 3) from sectors with 10° opening angle parallel and perpendicular to the stretching direction were extracted from the two-dimensional spectrum. At small angles both intensities increase by almost an order of magnitude. The scattering functions from the different directions join each other at larger scattering vectors. In general, the scattered intensity at larger q-values does not change during the reaction.

The corresponding scattering envelopes of the relaxed state (Fig. 3) show an enhancement in the parallel direction and a slight decrease perpendicular to the stretching direction.

Experiments on the PDMS system show that the scattering function perpendicular to the stretching direction approaches the form factor of the labeled free chains with increasing deformation. This is used in the determination of an apparent radius of gyration R_G and the apparent molecular mass as a function of time, using Zimm's representation (Fig. 4). The q-range of the fits is the same throughout the whole reaction. The relative range is $0.2 < qR_G < 1$ for the melt and goes up to $0.8 < wR_G < 3.5$ at the end of the reaction. This may be accepted, taking into account the large polydispersity of the clusters. The dependence of the apparent radii of gyration of the clusters on the apparent molecular weights can be described by a power law with an exponent of about 0.38 (Fig. 5).

Discussion

If the hydrogenated chains were of the same nature as the deuterated ones the growing clusters would consist of a random mixture of the two isotopes. One could then measure the form factor of the chains, but the clusters themselves would be invisible [4].

In this experiment the hydrogenated chains are crosslinked and move little during the experiment. Therefore, they serve as a surrounding medium, which provides a contrast for the deuterated chains, without screening the development of the clusters. In other words, the centers of mass of the hydrogenated chains are fixed in space, while the deuterated chains diffuse and combine themselves by means of the crosslinker.

Under the assumption that the form factor can be determined perpendicularly to the stretching direction, the growth of the clusters can be evaluated from the Zimm plots. The difference of the melt (R_G = 31 M_w = 4700) and the first spectrum ("start") in Fig. 4 indicates that the reaction takes place already during the swelling procedure. After an initial rapid rise the maximum values of 102 Å and 109 000 D for the apparent radius of gyration and the molecular weight, respectively, are reached asymptotically. This must be due to the immobility of the large clusters in the stretched matrix, i.e., the diminishing concentration of the small chains determines the kinetics of the process. The finite size is also explained by the fact that the larger clusters are fixed in space and cannot aggregate with themselves. They grow only by the addition of the freely moving small chains.

This may be confirmed by the observed power law for the radius of gyration as a function of the molecular weight. It corresponds to a process of aggregation which is limited by the diffusion of the small chains (DLA) [5]. Computer simulations in three dimensions yield $R_G \propto M^{0.4}$. They also show that the influence of the chemical reactivity tends to compact the aggregates [6] (exponent approaches 1/3) on a length scale above the small chains.

The behavior of the scattering functions at large scattering vectors indicates an undisturbed distribution of the monomeric units. The intensity does not change, since the concentration of deuterated material is constant. It is also isotropic, indicating that neither the clusters themselves nor the deuterated chains are deformed.

The freely moving chains can be thought of as being preferably located in "soft" regions with lower crosslink densities. One explanation for the large fluctuations assumes an inhomogeneous deformation of these regions [7]. This may be the reason for a deviation from the random distribution of the labeled chains, giving rise to the "butterfly patterns" [8].

During the reaction the intensity increases by almost one order of magnitude at small scattering angles in all directions. The change of the form factor of the clusters partially explains the increase. On the other hand, it may be assumed that the proceeding reaction concentrates the labeled chains increasingly more in the "soft" regions. Thus, the anisotropic distribution of their centers of mass becomes more pronounced.

The complete macroscopic recovering of the shape after relaxation indicates that no or only very few elastically effective crosslinks have been formed during the reaction. Because of their compactness and branching structure the clusters are obviously not forming micronetworks which are interpenetrated with the matrix. However, it can be inferred from the measurement done after the extraction of the sol phase that the deuterated particles cannot be removed from the network.

Phenomenologically, the anisotropy of the contours of the relaxed (i.e., macroscopically isotropic) sample is more accentuated than that for the deformed system. This is very different from a system with small chains which returns to an isotropic distribution, accompanied by a strong decrease of the scattered intensity.

The scattering functions parallel and perpendicular to the stretching direction for the relaxed sample are given as lines in Fig. 3. They can be superimposed on the corresponding functions of the deformed sample at the end of the reaction assuming an affine uniaxial deformation with $\lambda = 1.11$. However, this does not include any information from different azimuth angles, i.e., the change of the shape from "butterfly pattern" to ellipses.

Due to the size of the clusters their overall distribution had been fixed, which is equivalent to anchoring them at particular points in the stretched network. Thus, the relaxing matrix determines the distribution of the clusters. The loss of affineness corresponds to the deformation of the meshes in this spatial range.

between the labeled chains. The growth of the clusters corresponds roughly to the conception of diffusion limited aggregation.

There is an anisotropic distribution of the centers of mass of the labeled particles in the beginning which is accentuated throughout the whole reaction. Thus, the fluctuations in the crosslink density of the network govern the repartition of the clusters, which is even conserved when the sample relaxes.

Acknowledgement

The author is indebted to C. Lartigue for improving the manuscript and for recording a spectrum together with colleagues from LLB, Saclay.

References

1. Oeser R, Picot C, Herz J (1988) Polymer motion in dense systems, eds.: D. Richter, T. Springer in Springer proceedings in physics 29:104—111
2. Beltzung M, Picot C, Rempp P, Herz J (1982) Macromolecules 15:1594—1600
3. He X, Lapp A, Herz J (1988) Makromol Chemie 189:1061
4. Beltzung M, Picot C, Herz J (1984) Macromolecules 17:663—669
5. Witten TA, Sander LM (1981) Phys Rev Lett 47:1400
6. Kolb M (1991) in: Gans W, Blumen A, Amann A (eds) "Large Scale Molecular Systems", NATO ASI series B, physics 258:231—251
7. Bastide J, Leibler L, Prost J (1990) Macromolecules 23:1821
8. See also Boué F et al. this volume

Conclusion

The reaction which combines mobile chains in a stretched matrix can be observed directly by means of SANS, since there is a change of correlations only

Author's address:

R. Oeser
Institut Laue Langevin
BP 156x
F-38042 Grenoble Cedex 9

Trapped entanglements in polymer networks and their influence on the stress-strain behavior up to large extensions

M. Klüppel

Deutsches Institut für Kautschuktechnologie e.V., Hannover, FRG

Abstract: The paper demonstrates how a characterization of unfilled, amorphous rubber networks can be evaluated from uniaxial stress-strain measurement data. A network model is proposed that includes finite chain extensibility, topological constraints, and a trapped entanglement constribution to the reduced stress, which does not vanish in the infinite strain limit. Beside their influence on topological constraints, trapped entanglements are assumed to act like additional junctions, which also reduce the limiting extensibility of the network. In this framework a relation between the crosslink contribution and the topological constraint contribution to the reduced stress is derived, which allows the determination of the trapping factor from experimental stress-strain measurement data. The experimental results found for NR-networks cured with thiuram (TMTD) and peroxide (DCP), respectively, confirm the proposed model. It becomes obvious that the low topological constraint contribution found for the DCP-cured networks is related to a lower trapping rate, which results from main-chain scission during the curing procedure.

Key words: Rubber elasticity; trapped entanglements; finite extensibility; topological constraints; network defects

Introduction

The simplest and most convenient way to obtain information about the structure of rubber networks is stress-strain measurements in uniaxial extension. By using Mooney-Rivlin plots of the measured curves, two obvious deviations from ideal rubber elasticity become visible: In the low deformation regime the rubber softens, which gives rise to the phenomenological C_2-term of the Mooney-Rivlin-Equation, and in the upper deformation range it hardens very rapidly. These effects are most naturally explained by topological constraints acting on the network chains, as described by the tube model of Heinrich et al. [1], and the finite extensibility of the chains.

A further deviation from ideal rubber elasticity results from the trapping of interchain entanglements by the crosslinks [2]. These permanently trapped entanglements are assumed to contribute to the equilibrium modulus in a similar way as the crosslinks. In the limit of small deformations this contribution is well established and experimentally confirmed [3—5]. However, it is not quite clear how trapped entanglements influence the modulus on the whole deformation scale. Attempts have been made to relate the constraint modulus of the tube model to the contribution of trapped entanglements found in the small strain limit and, hence, to totally absorb the influence of trapped entanglements on the modulus in the constraining potential of the tube model [1, 6, 7]. Nevertheless, the conclusions drawn from this assumption are contradicting and physically unsatisfactory. For example, in [1], network defects are related to constraint release effects and a correlation between Langleys trapping factor T_e and the tube relaxation parameter β is concluded. This is difficult to understand in view of the original meaning of these quantities. The somewhat different conclusions drawn in [6] and [7] contradict the scaling relations found between the plateau modulus of the uncrosslinked melt and the topological constraint moduli of the related networks.

Theory

Due to the difficulties in relating the influence of trapped entanglements on the modulus to the constraint modulus of the tube model and in view of the close similarity between trapped entanglement and crosslinks, we conclude that the constraint contribution of trapped entanglement cannot totally be included into the constraining potential, used in different tube models. Instead, trapped entanglements give rise to the introduction of additional network junctions. A distinction between trapped entanglements and crosslinks results from their possible difference in elastic efficiency, only.

This concept assumes three different influences on a single network strand between two crosslinks. The first influence results from the two crosslinks, the second is due to trapped entanglements of the chain under consideration, which act like additional junctions, and the third results from entanglements and crosslinks of the surrounding chains, which is taken into account in the constraining potential of a tube model [1]. By using a result of Kästner [8], who found that all influences on network strands can be assumed to be decoupled and additive, we conclude that trapped entanglements, crosslinks, and topological constraints contribute to the stress additively. It follows that the reduced stress, measured in the low deformation regime, consists of three terms:

$$\sigma_{\text{red}} = G_c + G_e T_e + G_N \varphi(\lambda)$$
$$\cong G_c + G_e T_e + G_N \frac{1}{\lambda} \quad \text{for } \lambda \geqslant 1, \quad (1)$$

where:

$$\sigma_{\text{red}} = \frac{\sigma_0}{\lambda - \lambda^{-2}} \quad (2)$$

$$G_c = A_c \nu_c RT = \frac{A_c \hat{\varrho}_p l^2 RT}{M_s R_c^2} \quad (3)$$

$$G_N = \frac{\beta^2 \hat{\varrho}_p l^2 RT}{4\sqrt{6} M_s r_0^2} \quad (4)$$

$$\varphi(\lambda) = \frac{2(\lambda^{\beta/2} - \lambda^{-\beta})}{\beta(\lambda^2 - \lambda^{-1})} \quad (5)$$

$$G_e = A_e \nu_e RT \quad (6)$$

$$\nu_e = \frac{5 G_N^0}{4 RT}, \quad (7)$$

λ is the deformation ratio, σ_0 is the stress related to the cross-section of the undeformed sample, R is the gas constant, T the absolute temperature, ν_c and ν_e are the densities of elastically effective strands related to crosslinks and entanglements in the melt before crosslinking, G_N^0 is the plateau modulus of the melt, A_c and A_e are structure factors related to crosslinks and trapped entanglements, T_e is the trapping factor of Langley, $\hat{\varrho}_p$ is the mass density of the elastically effective network, l and M_s are the length and the molar mass of the statistical chain segments, r_0 is the mean fluctuation radius of the chain segments (tube radius), and R_c is the mean end-to-end distance of the network chains between two crosslinks.

β is a tube relaxation parameter, which takes into account constraint release effects. These can result from the simultaneous relaxation of many chains, which build the constraining tube arround a single chain. We assume that, in dry rubber networks, this relaxation can be neglected and β equals 1. Only in swollen rubbers does the tube relaxation becomes significant and β is used to describe the strong swelling dependence of the constraint modulus G_N. In highly swollen rubbers, where G_N vanishes, a total relaxation of the tube deformation takes place and β equals zero. In contrast to the original tube model of Heinrich et al., Eq. (1) predicts an influence of trapped entanglements on the reduced stress also in highly swollen rubbers. Due to the nature of trapped entanglements, this influence seems reasonable. An experimental test of (1) should be possible by performing stress-strain measurements at highly swollen rubbers where G_N vanishes.

The structure factors A_c and A_e are related to the fluctuation range of the crosslinks and trapped entanglements. They determine the elastic efficiency of the crosslinks and trapped entanglements, respectively. By using heuristic arguments, Kästner found an expression for A_c from the joint introduction of restricted junction fluctuations and the mean field picture of the tube model [8]:

$$A_c = 1 - \frac{2}{f} \left\{ 1 - \frac{2}{\sqrt{\pi}} \frac{K \cdot e^{-K^2}}{\text{erf} K} \right\}, \quad (8)$$

where

$$K = \sqrt{\frac{3f}{2}} \frac{r_c}{R_c}, \quad (9)$$

and f is the functionality and r_c is the fluctuation radius of the crosslinks. As $r_c \cong r_0$, (9) can be rearranged by using Eqs. (3) and (4):

$$K \cong \sqrt{\frac{3f\beta^2}{8\sqrt{6}A_c} \frac{G_c}{G_N}}. \quad (10)$$

Because of the higher mobility of trapped entanglements compared to the crosslinks, we expect the elastic efficiency of trapped entanglement to be smaller than the efficiency of crosslinks, i.e., $A_e < A_c$. However, a determinative equation for A_e remains unknown at this stage.

So far, the theory is only applicable for small deformations where Gaussian statistics are valid. At larger values of deformation the finite extensibility of network chains has to be taken into account. This is done by considering the conformational entropy contribution resulting from the orientation of chain segments. Instead of the Gaussian distribution function, it leads to an expression involving the Inverse Langevin Function for the end-to-end distance of network chains [9]. This distribution function has to be applied to all three terms of Eq. (1).

For the first two terms of (1) an equation derived by Treloar, who used a series expansion for the Inverse Langevin Function, is applied [9]. By relating the distribution function for the end-to-end distance of network strands to the mean position of crosslinks and trapped entanglements, the concept of junction fluctuation, expressed by the structure factors A_c and A_e, is included.

For the third term of Eq. (1), we expect a small change in its contribution to the reduced stress only, because the influence of topological constraints rapidly goes to zero at large deformations, where significant differences between Gaussian- and inverse Langevin statistics occur. Hence, a decrease of tube constraints due to finite extensibility, which was assumed by other authors [10], is neglected in the following representation:

$$\sigma_{\text{red}} \cong (G_c + G_e T_e) \left\{ 1 + \frac{3}{25n} \left(3\lambda^2 + \frac{4}{\lambda} \right) \right.$$
$$\left. + \frac{297}{6125 n^2} \left(5\lambda^4 + 8\lambda + \frac{8}{\lambda^2} \right) + \ldots \right\}$$
$$+ G_N \frac{1}{\lambda}. \quad (11)$$

Here, n is the mean number of statistical segments between successive junction. According to the assumption that trapped entanglements contribute to the finite extensibility as well as do crosslinks [10], we conclude:

$$n = \frac{\hat{\varrho}_p}{(v_c + v_e T_e) M_s}, \quad (12)$$

or, by using (3) and (7):

$$n = \frac{\hat{\varrho}_p RT}{\left(\dfrac{G_c}{A_c} + \dfrac{5}{4} G_N^0 T_e \right) M_s}, \quad (13)$$

n determines the limiting extensibility of the network by the relation $\lambda_{\max} = \sqrt{n}$, where the full series expansion (11) approaches infinity. In the limit $n \to \infty$ Gaussian statistics, i.e., Eq. (1), results.

An experimental test of Eq. (11) by using (13) requires the knowledge of the trapping factor T_e. Unfortunately, it cannot be extracted from the previous theory, because the value of the structure factor A_e remains unknown. However, a determination of the trapping factor is possible if the scaling arguments concearning the plateau modulus of the uncrosslinked melt are generalized to entangled networks. In the melt case, the right scaling predictions for the plateau modulus are found by relating the tube radius r_0^{melt} to the step length of the primitive path, i.e., the mean spacing of successive entanglements [1]. It yields:

$$G_N^0 = \frac{4\varrho_p l^2 RT}{5 M_s (\xi r_0^{\text{melt}})^2} \quad (14)$$

$$\xi = \frac{\sqrt{n^{\text{melt}}}\, l}{r_0^{\text{melt}}} \quad (15)$$

$$n^{\text{melt}} = \frac{\varrho_p}{v_e M_s}. \quad (16)$$

ϱ_p is the mass density of the polymer melt and ξ is a scaling factor, which fixes the proportionality between the tube radius and the mean spacing of entanglements in the melt.

In the network case is seems reasonable to assume a similar relation between the tube radius r_0 and the mean spacing of successive junctions, i.e., crosslinks or trapped entanglements:

$$\xi = \frac{\sqrt{n}\, l}{r_0}. \quad (17)$$

This means that the constraining potential of the tube model is governed by crosslinks and entanglements that are trapped, only. Entanglements of dangling chain ends are assumed to give no contribution to the tube constraints.

From the principles of the tube model it is clear that deviations from Eq. (17) should appear for slightly crosslinked networks, with multiple entangled long chain ends. However, (17) is expected to be fullfilled for moderately but almost completely crosslinked networks, where the trapping factor has reached its limiting value and no more trapping takes place. Networks defects, resulting from main-chain scission during the crosslinking procedure, are assumed to have a strong influence on the constraint modulus G_N if (17) is used, because entanglements which are freed by main-chain scission and are not trapped do not contribute to the tube constraints. A scaling relation between G_N, G_N^0 and G_c is found from Eq. (17) by inserting Eqs. (13) and (4):

$$\frac{G_c}{A_c} = \frac{4\sqrt{6}\, G_N}{(\xi \beta)^2} - \frac{5}{4} G_N^0 T_e. \quad (18)$$

Hence, a plot of G_c/A_c against G_N should result in a linear dependence for sufficiently high crosslinked networks, where the trapping factor has reached its limiting value. From the axis intersection of the regression line the limiting value $T_{e,\max}$ is extractable, while the slope determines the value of the scaling factor ξ.

The elastically ineffective dangling chain ends are closely related to the trapping of entanglements, because $\sqrt{T_e}$ equals the probability that a randomly choosen chain segment is connected to the gel by both paths. By using Scanlan's concept of counting elastically effective strands, as explaned in the appendix of reference [4], an equation between the density of crosslinks μ_c and the density of elastically effective strands ν_c results. A similar relation is found between the mass density $\hat{\varrho}_p$ of the elastically effective network and the mass density ϱ_p of the rubber samples:

$$\nu_c = (3\sqrt{T_e}\, w_g - T_e)\mu_c \quad (19)$$

$$\hat{\varrho}_p = \left(\frac{3}{2}\sqrt{T_e}\, w_g - \frac{T_e}{2}\right)\varrho_p. \quad (20)$$

These equations hold for tetrafunctional crosslinks. w_g is the gel fraction of the networks, which has a value very close to 1 in moderately crosslinked highly extendable rubber samples.

Experimental results

Natural rubber samples (SMR CV 50) were cured with two different crosslinking systems at 155°C up to 90% of the maximum torque found in vulcameter curves. In one case, peroxid (DCP), and in the other case thiuram (TMTD) together with the same amount of ZnO were used for the crosslinking procedure. The amount of crosslinking agent was varied systematically. The uniaxial stress-strain behavior of the rubbers up to large extensions was measured by using strip-samples. The extension ratio was determined by an optical system (Universalprüfmaschine Zwick 1445). To avoid stress-induced crystalization, the measurements were performed at 100°C. The deformation velocity was choosen to be small (10 mm/min) in order to evade dynamical contributions to the measured stress.

Figures 1 and 2 show the measured reduced stress of samples cured with different amounts of TMTD and DCP, respectively, plotted against the inverse deformation ratio. In addition, the calculated Gaussian contribution to the reduced stress, as defined in Eq. (1), is shown for every stress-strain measurement point. The calculation was performed by using a graphical itteration sheme for the values of G_c, $G_e T_e$, G_N, A_c and T_e, involving Eqs. (8), (10), (11), (13), (18), and (20) together with the plots shown in Figs. 3—5.

The functionality of the crosslinks was taken to be $f = 4$. As stated above, $\beta = 1$ was assumed for unswollen rubber samples and the gel fraction was taken to be $w_g = 1$. The mass density of the samples was measured as $\varrho_p = 900$ kg/m³. The molar mass of the statistical segments was taken

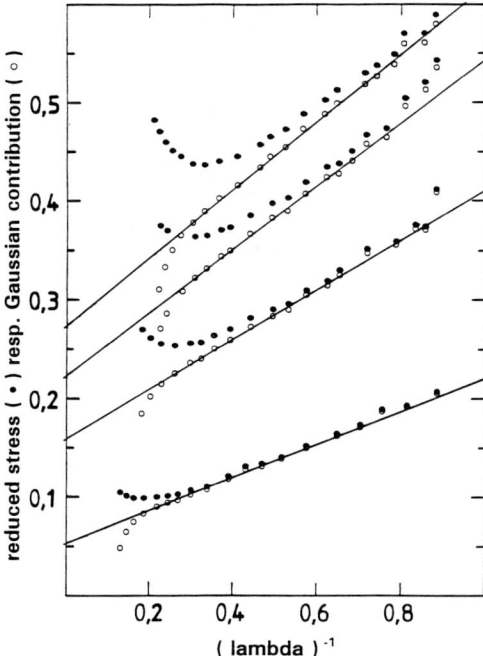

Fig. 1. Mooney-Rivlin plot of the reduced stress (●) at 100°C and the calculated Gaussian contribution (○) for NR-networks cured with 1, 2, 3, 4 wt% thiuram (TMTD)

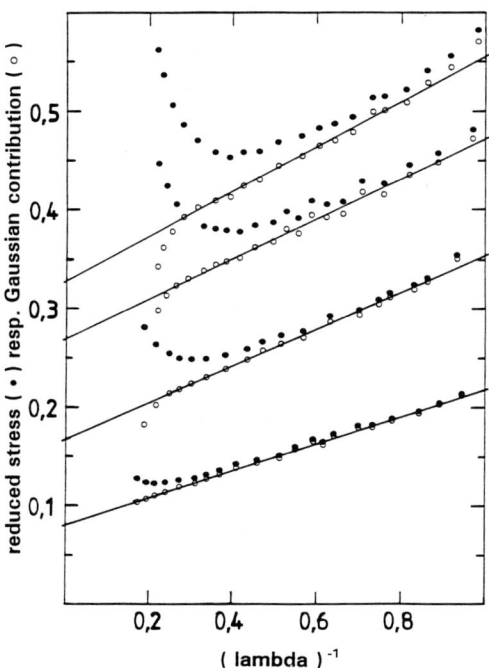

Fig. 2. Mooney-Rivlin plot of the reduced stress (●) at 100°C and the calculated Gaussian contribution (○) for NR-networks cured with 0.5/1/1.5/2 wt% peroxid (DCP)

from reference [11] to be $M_s = 130$ g/mol. The plateau modulus was found from dynamical measurements to be $G_N^0 = 0.53$ N/mm² at 100°C.

The series expansion of (11) was taken into account up to 4th order in n, as given in [9]. This corresponds to a good approximation of the inverse Langevin Function up to $\lambda/\sqrt{n} \cong 0.7$.

The regression lines inserted in Figs. 1 and 2 are related to the observed linear behavior of the Gaussian contribution on a large deformation scale. Deviations occur at small and very large extensions only. According to Eq. (1) the axis intersections of the regression lines determine the values of $G_c + G_e T_e$ and G_N. Figures 3 and 4 show the infinite strain contribution $G_c + G_e T_e$, found from several measurements, as plotted against the concentration of crosslinking agent TMTD and DCP, respectively. Starting from a gel-point (GP), in the lower concentration regime a non-linear dependence is observed for both network types. However, for higher concentrations of crosslink agent a linear dependence results. An extrapolation of the corresponding regression lines to the concentration of crosslinking agent at the gel-point (dotted lines) gives a non-zero contribution for the value $G_c + G_e T_e$ in both cases. As G_c is assumed to vanish at the gel point, this contribution is identified with the limiting amount of the trapped entanglement contribution $G_e T_{e,\max}$. The crosslink contribution G_c is taken to be that part of $G_c + G_e T_e$ which changes in direct proportionality to the concentration of crosslinking agent and equals zero at the gel point (dashed lines).

The gel point was constructed by using the cycle rank concept of Flory, which assumes that the number of crosslinks equals the number of primary chains n_p at the gel point. n_p was found from the measured mean number molecular weight $M_n = 1.62 \cdot 10^5$ g/mol for the uncrosslinked melt.

Figure 5 shows a plot of the crosslink contribution G_c, divided by the corresponding structure factor A_c, against the constraint contribution G_N. For both network types a linear dependence is observed in the regime of higher concentrations of crosslinking agent. The corresponding regression lines show nearly the same slope, but different axis intersections. According to Eq. (18), the scaling factor ξ is extracted from the slope of the regression lines and the limiting value of the trapping factor $T_{e,\max}$ is determined by the axis intersection for both crosslinking systems. The characteristic network parameters, which are found from the plots in Figs. 3—5, are summarized in Table 1.

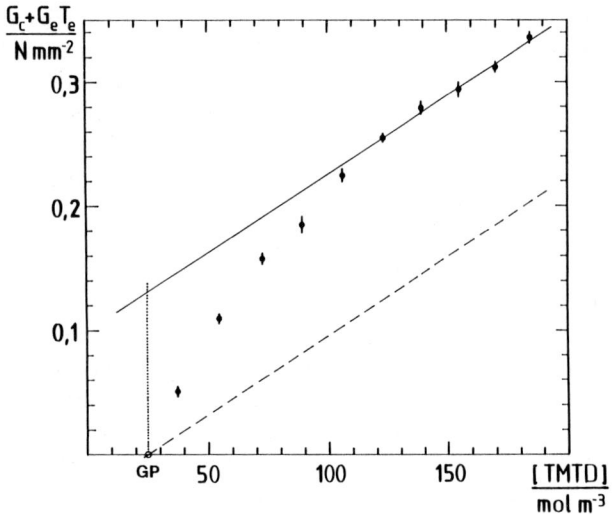

Fig. 3. Plot of the Gaussian infinite strain contribution $G_c + G_e T_e$ to the reduced stress vs. concentration of crosslinking agent TMTD. (GP: Gel point; (——): G_c — contribution)

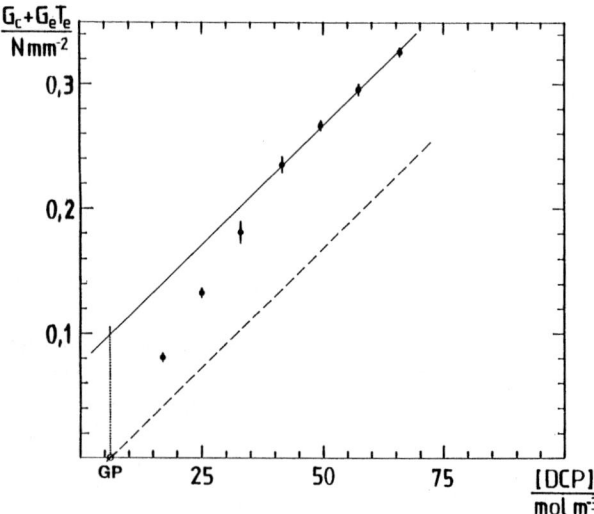

Fig. 4. Plot of the Gaussian infinite strain contribution $G_c + G_e T_e$ to the reduced stress vs concentration of crosslinking agent DCP. (GP: Gel point; (——): G_c — contribution)

They are all related to the higher crosslinking regime, where the trapping factor has reached its limiting value $T_{e,max}$. The limiting amount of elastically effective network strands is calculated from Eq. (20). The crosslinking efficiencies

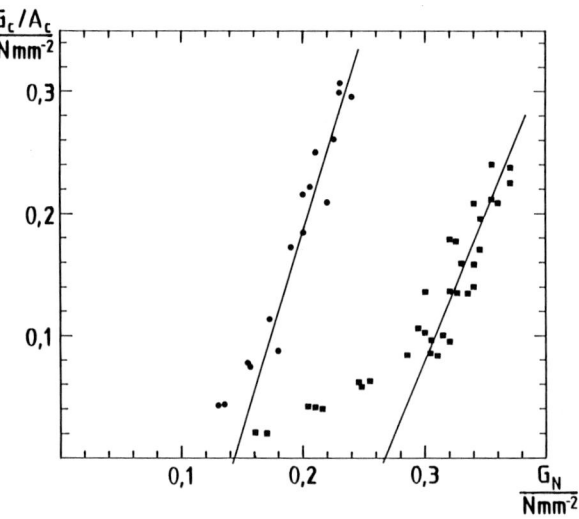

Fig. 5. Plot of the Gaussian crosslink contribution G_c to the reduced stress devided by the corresponding structure factor A_c vs the constraint contribution G_N. ((■): TMTD; (●): DCP)

Table 1. Characteristic network parameters which are independent of crosslinking densities

	$T_{e,max}$	ξ	$\dfrac{\widehat{\varrho}_{p,max}}{\varrho_p}$	A_e	$\dfrac{\mu_c}{[\text{TMTD/DCP}]}$
NR/TMTD	0.98	2.01	0.99	0.20	0.23
NR/DCP	0.67	1.73	0.89	0.23	0.85

$\mu_c/[\text{TMTD}]$ and $\mu_c/[\text{DCP}]$ are found from the slope of the dashed lines in Figs. 3 and 4 together with Eqs. (3) and (19). The values of the structure factor A_e, which corresponds to the elastic efficiency of trapped entanglements, are calculated from the ratio between $G_e T_{e,max}$ and $T_{e,max}$ by using Eqs. (6) and (7). The values for the structure factor A_c, which are related to the elastic efficiency of crosslinks, are plotted in Fig. 6 as a function of concentrtion of crosslinking agent TMTD and DCP, respectively. They are calculated from Eq. (8) and (10).

Conclusions

The experimental results confirm the proposed model. The calculated Gaussian contribution to the

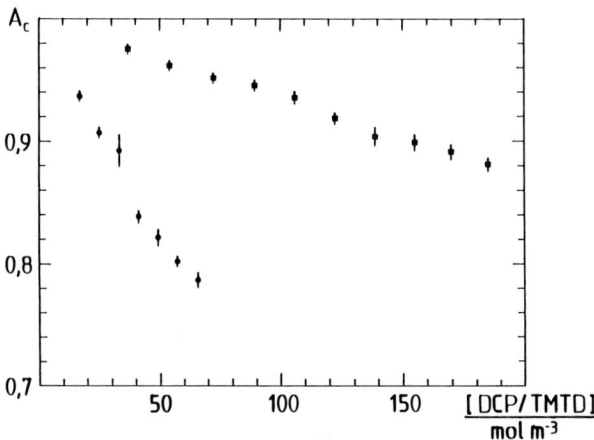

Fig. 6. Plot of the calculated structure factor A_c vs the concentration of crosslinking agent TMTD (■) and DCP (●)

reduced stress shows the required λ^{-1}-dependence on a large deformation scale (Figs. 1 and 2). The deviations at very large extensions are explained by stress-induced scission of chains, while the deviations at small extensions are probably due to the insufficient measurement conditions in this range. A determination of the reduced stress in the small strain regime requires an exact association between the zero of strain and the zero of stress, which is difficult to realize if the measurements are performed on a large stress scale. The proposed non-vanishing influence of trapped entanglements on the Gaussian infinite strain contribution to the reduced stress (Eq. (1)) is found if the finite extensibility of network chains is taken into account (Figs. 3 and 4). The low elastic efficiency of trapped entanglements, which is found to be about four times smaller than the elastic efficiency of crosslinks, is related to the higher mobility of trapped entanglements compared to the crosslinks.

The proposed linear scaling relation (17) between the tube radius and the mean spacing of successive junctions leads to reasonable results. The different limiting values $T_{e,max}$ of the trapping factor (Table 1) indicate that, in the TMTD-curing procedure, nearly all entanglements are trapped at higher crosslinking densities and nearly perfect networks result, while about one-third of the entanglements in the melt are not trapped at all if DCP is used in the crosslinking procedure. By using a result of Langley [2], who found that the limiting value of the trapping factor depends on the relative scission level only, this can be related to a main-chain scission probability of about 10%. The somewhat lower value of the scaling factor ξ for the DCP-cured networks, compared to the nearly perfect networks resulting from the TMTD-curing procedure (Table 1), can be related to the dilution of the network by the dangling chain-ends. Hence, we conclude that the different scaling behavior of the constraint contribution G_N for both networks types is well understood in the proposed model. The consideration of tube relaxation effects, as assumed in [1], is not necessary for understanding the experimental results.

References

1. Heinrich G, Straube E, Helmis G (1988) Adv Polym Sci 85:33
2. Langley NR (1968) Macromolecules 1:348
3. Dossin LM, Graessley WW (1979) Macromolecules 12:123
4. Langley NR, Polmanteer KE (1974) J Polym Sci Polym Phys 12:1023
5. Oppermann W, Rennar N (1987) Prog Colloid Polym Sci 75:49
6. Gaylord RJ (1983) Polym Bull 9:186
7. Marrucci G (1981) Macromolecules 14:434
8. Kästner S (1981) Colloid Polym Sci 259:499
9. Treloar LRG (1975) The Physics of Rubber Elasticity, 3rd Ed, Clarendon Press, Oxford
10. Edwards SF, Vilgis TA (1986) Polymer 27:483
11. Bandrup J, Immergut EH (Eds) (1975) Polymer Handbook, 2nd Ed, Wiley Interscience, New York

Author's address:

Dr. Manfred Klüppel
Deutsches Institut
für Kautschuktechnologie e.V.
Eupener Straße 33
3000 Hannover 81, FRG

Physical and chemical network effects in polyurethane elastomers

L. Apekis[1]), P. Pissis[1]), C. Christodoulides[1]), G. Spathis[2]), M. Niaounakis[2]), E. Kontou[2]),
E. Schlosser[3]), A. Schönhals[3]), and H. Goering[3])

[1]) Department of Physics, National Technical University of Athens, Greece
[2]) Department of Theoretical and Applied Mechanics, National Technical University of Athens, Greece
[3]) Zentralinstitut für Organische Chemie, Berlin, FRG

Abstract: The effects of the physical and chemical networks on the relaxation processes and on the morphology of polyester-based polyurethane elastomers were studied by thermal, thermomechanical, and dielectric methods. Two series of polyester-based polyurethane elastomers were prepared by varying the NCO/OH ratio during the second step of polymerization while the hard segment content was maintained almost constant, at about 30% in the first series and at about 39% in the second. Differential Scanning Calorimetry (DSC), dielectric ac and Thermally Stimulated Depolarization Current (TSDC) measurements and Dynamic Mechanical Analysis show a main relaxation related to the glass transition of the soft parts of the polyurethane copolymers. The results of all the techniques used have shown that the increase of the NCO/OH ratio produces an increase in the glass transition temperature and a more homogeneous morphology. Dielectric ac and TSDC relaxation spectroscopy provided more information about the dynamics of the glass transition.

Key words: Polyurethane; physical and chemical network; glass transition; molecular mobility; dielectric; mechanical relaxation

Introduction

The microdomain structure, the phase mixing, and the properties of polyurethane elastomers are affected by many factors such as the types of hard and soft segments, molecular weight of the components, the chain extender, the stoichiometric composition, and the polymerization method [1—4]. Aspects of the effects of physical and chemical cross-linking on the structure and properties of polyurethane elastomers have been investigated by means of dynamic mechanical [5, 6], calorimetric, mechanical and spectroscopic [7] and dielectric [8, 9] methods. A theory for the network formation in polyurethane systems due to side reactions has been developed [10].

In this work the effect of the physical and chemical networks on the morphology and on the relaxation processes of polyester-based polyurethane elastomers were studied by thermal, thermomechanical, and dielectric methods. Two series of polyester-based polyurethane elastomers were studied, which were prepared by varying the NCO/OH ratio during the second step of polymerization while the hard segment content was maintained almost constant, at about 30% in the first group and at about 39% in the second.

Materials

The materials were prepared in a two-step polymerization process. The details of the procedure have been described elsewhere [7]. In the first group of elastomers the polyester used was poly(ethylene adipate) (Desmophen 2000, Bayern AG) with an average molecular weight of 2000 (PEA). In the second group the polyester was poly(ethanediol-1,4-butanediol adipate) (Desmophen 2001, Bayer AG) with an average molecular weight of 2000 (PEBA).

This mixed polyester was used in order to reduce crystallinity in the soft phase of the copolymers. In both groups the diphenylmethane diisocyanate (MDI) was Desmodur 44 (Bayer AG) and the chain extender 1,4-butanediol (BDO). The prepolymer composition was maintained constant with a molar ratio of MDI/PEA of about 2.8/1.0 for the first group of elastomers and MDI/PEBA of about 3.9/1.0 for the second group. The NCO/OH ratio, during the second step of polymerization, was 0.9, 1.0, 1.1, 1.2, and 1.3 for the first group of samples, and 0.9, 1.0, 1.1, and 1.2 for the second group. The samples were designated by the NCO/OH ratio values (EPU09, EPU10, ...).

Experimental

Differential Scanning Calorimetry (DSC)

The DSC measurements on the first series of samples were made using a Du Pont 9000 Thermal Analyzer with a heating rate of 20 K/min [7]. The measurements on the second series of samples were made using a Perkin Elmer DSC-7 Calorimeter with a heating rate of 10 K/min [15].

Dynamic Mechanical Analysis (DMA)

The temperature dependence of the dynamic mechanical properties of samples from the second series of elastomers was determined with a recording Torsion Pendulum at a frequency of 1 Hz with a heating rate of 2 K/min [15].

Dielectric ac measurements

The complex permittivity of samples from the second series of elastomers was measured in the frequency range of 10 Hz—100 kHz at various temperatures using the ac-bridge method [13].

Thermally Stimulated Depolarization Current (TSDC)

The TSDC measurements were made in the temperature range of 77—300 K. The TSDC method, its principles, and the procedures followed for evaluating the activation energy W, the pre-exponential factor τ_0, and the contribution of a relaxation mechanism to the static permittivity $\Delta\varepsilon$, as well as the experimental apparatus used, have been described elsewhere [11, 12].

Results and discussion

Figure 1 shows typical DSC thermograms from the four types of elastomers studied (the series with

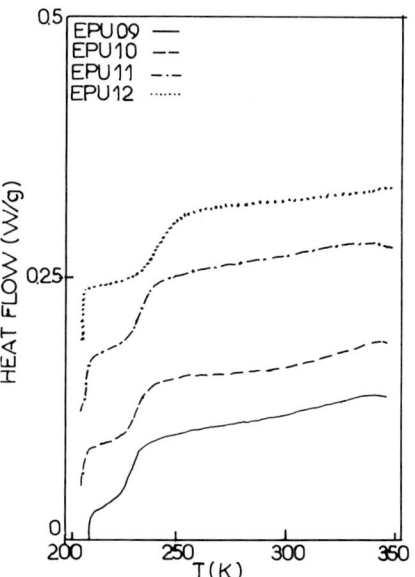

Fig. 1. DSC thermograms of the four different types of polyurethane studied (to make the plots clearer the different thermograms are shifted vertically)

MDI/PEBA about 3.9/1.0 or hard segment content about 40%). The thermograms show a main transition in this temperature region, which according to our previous work on polyurethane elastomers is related to the glass transition of the soft part of the copolymers [4]. The DSC results show that the glass transition temperature T_g is shifted to higher temperatures as the NCO/OH ratio is increased. The data concerning the main transition (α-relaxation) from DSC measurements as well as those from the other methods were summarized in Table 1 and plotted in Fig. 2.

In Fig. 3 typical global TSDC spectra from the same four types of polyurethane are shown. The plots show a broad dispersion with two maxima in the low-temperature region and two peaks, whose positions depend on the NCO/OH ratio, in the high-temperature region. In a previous work we studied a series of polyurethane copolymers with the same three components (polyester, MDI, BDO), but with stoichiometric compositions and constant NCO/OH ratio [4]. TSDC spectra from those materials were similar to the spectra shown in Fig. 1. According to those results, the broad low-temperature dispersion with the two maxima is due to secondary local (β and γ) relaxations observed in many other polymers [14]. The dielectric relaxations

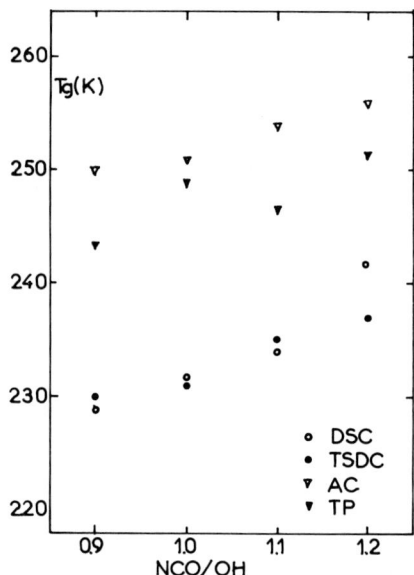

Fig. 2. T_g-values versus NCO/OH ratio as determined by the four different methods used (see Table 1)

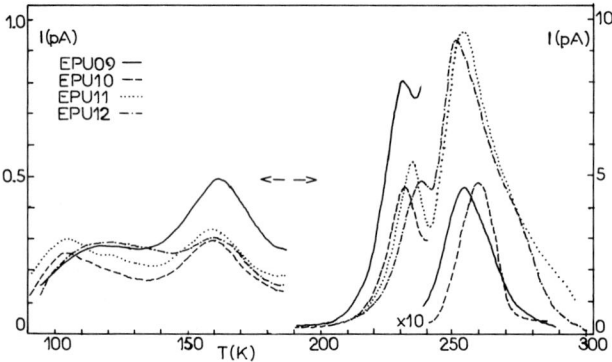

Fig. 3. TSDC plots for the four types of polyurethane studied

Table 1. α-relaxation, EPU (MDI/PEBA/BDO, MDI/PEBA. 3.8/1.0)

Sample	DSC T_g(°C)	TSDC T_m(°C)		ac 1 kHz T_m(°C)	TP T_m(°C)
		dipolar	MWS		
EPU 09	−44.2	−43	−19	−23	−29.5
EPU 10	−41.2	−42	−13	−22	−24.0
EPU 11	−39.2	−38	−19	−19	−26.5
EPU 12	−31.2	−36	−22	−17	−21.5

causing the two high-temperature peaks were found to be related to the glass transition (α-relaxation) of the polyurethane samples. The peak at lower temperature is due to molecular mobility of soft polyester parts at the glass transition of the copolymers and is located at the glass transition temperature T_g as measured by DSC, while the peak at higher temperature was attributed mainly to interfacial Maxwell-Wagner-Sillars (MWS) polarization, which takes place at the interface between soft and hard regions and reflects a drastic change in the mobility of soft polyester parts [4]. In the present work, our interest is focused on the high-temperature relaxations related to the glass transition of the elastomers. Following our previous results, it can be concluded from the TSDC data presented in Fig. 3 that the increase of the NCO/OH ratio from 0.9 to 1.2 causes a shift of the glass-transition temperature T_g from 231 to 237 K (see also Fig. 2).

Figure 4 shows the dielectric loss as a function of temperature, measured at 1 kHz for the same four types of polyurethane elastomers. These results show a dielectric behavior similar to that observed in the TSDC plots, i.e., a broad dispersion at the low-temperature region whose position and magnitude allow us to attribute it to secondary (β and γ) relaxations and a main dispersion at the high-temperature region whose position and magnitude allow us to relate it to the main glass transition (α-relaxation) of the copolymers. Figure 4 shows the same shift of the main relaxation as that of the glass transition from TSDC data.

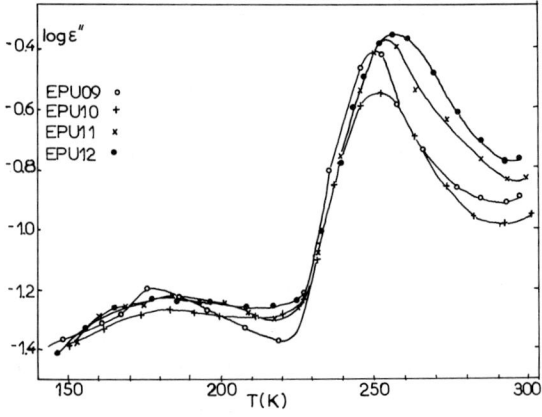

Fig. 4. Temperature dependence of the dielectric loss for the four types of polyurethane measured at 1 kHz

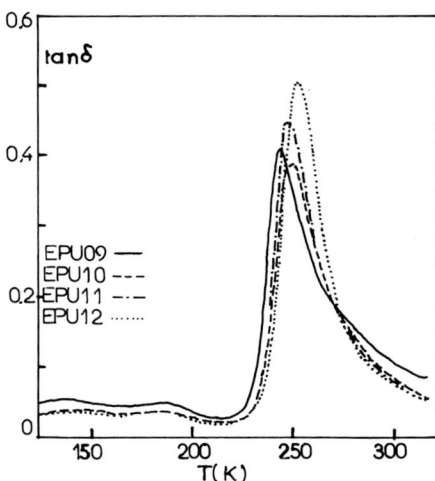

Fig. 5. Temperature dependence of the dynamic mechanical loss tangent of the four polyurethanes measured at 1 Hz

the other series with hard segment content of about 30%. The results of these measurements concerning the glass transition (a-relaxation) of the elastomers are summarized in Table 2. These data show the same influence of the NCO/OH ratio on T_g.

Table 2. α-relaxation EPU (MDI/PEA/BDO, MDI/PEA: 2.8/1.0)

Sample	DSC T_g(°C)	TSDC T_m(°C)	
		dipolar	MWS
EPU 09	−30.3	−36	−9
EPU 10	−29.0	−34	−13
EPU 11	−26.0	−32	−17
EPU 12	−24.0	−28	−16
EPU 13	−23.5	−31	−16

In Fig. 5 the dynamic mechanical loss (tan δ) of the same four types of elastomers is shown as a function of temperature, measured at the frequency of 1 Hz. These results show a dynamic behavior similar to that observed with dielectric measurements in the same temperature region, i.e., a broad dispersion with two maxima in the low-temperature region which may be attributed to secondary (β and γ) relaxations and a main relaxation at the high-temperature region which may be related to the glass transition of the samples. Although the tan δ plots are presented in Fig. 5, in Fig. 2 the values of the maximum loss temperature T_m are those obtained from the measurements of the dynamic loss modulus G'' as directly expressing the dynamic mechanical losses. To make the comparison easier between dielectric (TSDC and ac) and mechanical measurements, it is useful to mention that the equivalent frequencies of the TSDC measurements are of the order of mHz, while the ac plots of Fig. 4 were obtained at a frequency of 1 kHz and those of Fig. 5 at a frequency of 1 Hz.

The comparison of these results allows us to assume that they express different aspects of the molecular mobility during the transition from the glass to the rubber state of the elastomers as well as the influence of the NCO/OH ratio on the dynamics of mechanical and dielectric relaxations of the glass transition. DSC and TSDC measurements have also been performed on five types of elastomers from

In both series of polyurethanes, the same increase in the NCO/OH ratio results in a shift of T_g to higher values of approximately 7 degrees, with the exception of the abrupt increase observed for EPU12 in DSC measurements, seen in Fig. 2.

The values of T_g are lower in the second series of polyurethanes studied (Table 1) than those of the first series (Table 2), although their hard segment content is higher than that of the first series. This difference of about 10 K in T_g may be mainly due to the polyester used in the two series. PEBA (used for the second series) was expected to give lower T_g's because its composition prevents crystallinity into the soft part of the copolymer while PEA (used for the first series) does not prevent it [7, 15].

This fact provides more evidence that, during glass transition, mainly the soft polyester segments are involved and that the results concerning molecular mobility at glass transition refer mainly to the soft polyester segments mobility.

The mangitude of the MWS relaxation accompanying the glass transition relaxation in TSDC measurements was found to be reduced by the increase of the NCO/OH ratio, i.e., by the increase of the degree of cross-linking, as can be seen in Fig. 3. This MWS relaxation is considered as reflecting a drastic change in the mobility of soft polyester parts by the activation of space charges trapped in the interface between soft and hard regions [4]. The reduction of its magnitude may therefore be seen as a result of the interface reduction caused by the increase of the degree of cross-linking.

The increase of the NCO/OH ratio introduces in the structure of the copolymers an increase in density of chemical cross-links which restricts the formation of hard domains, changes the morphology, and causes a mixing of hard and soft segments [7, 15, 17]. The increase of T_g with increasing NCO/OH ratio observed in our measurements may reflect the influence of the hard segments, dissolved in the soft phase matrix from the hard domain disruption, on the soft segments mobility. This interpretation is consistent with the results of Mahboubian Jones et al. [9] obtained from segmented polyurethane elastomers with different 1,3 and 1,4 Butanediol concentrations. Similar results concerning the influence of the degree of cross-linking on the dynamics of mechanical and dielectric relaxations of the glass transition and on the microphase separation of polyurethanes have been reported by Havranek et al. [5] and Zielinski et al. [8]. The relatively small shift of the T_g of our results is in agreement with the results of Delides and Pethrick [3] regarding the importance of the molecular weight of the ester block and those of Schlosser and Schönhals [13] regarding the importance of the urethane concentration in defining the relaxation behavior of the polyurethane systems.

More information concerning the glass transition dynamics of the elastomers has been provided by dielectric ac and TSDC spectroscopy.

In Figs. 6 and 7 are plotted the results of the measurements of the real, ε', and the imaginary, ε'', parts of the complex permittivity as a function of frequency at different temperatures for two types of

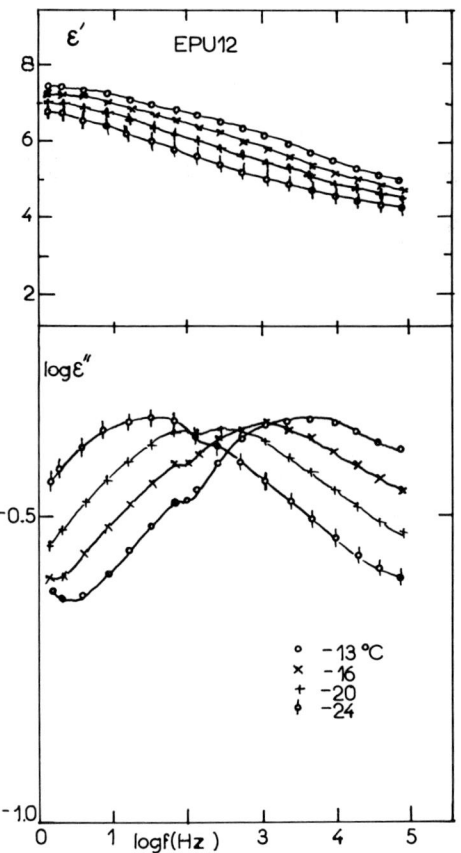

Fig. 7. Frequency dependence of the real and the imaginary parts of the complex dielectric permittivity of the sample EPU 12 measured at different temperatures

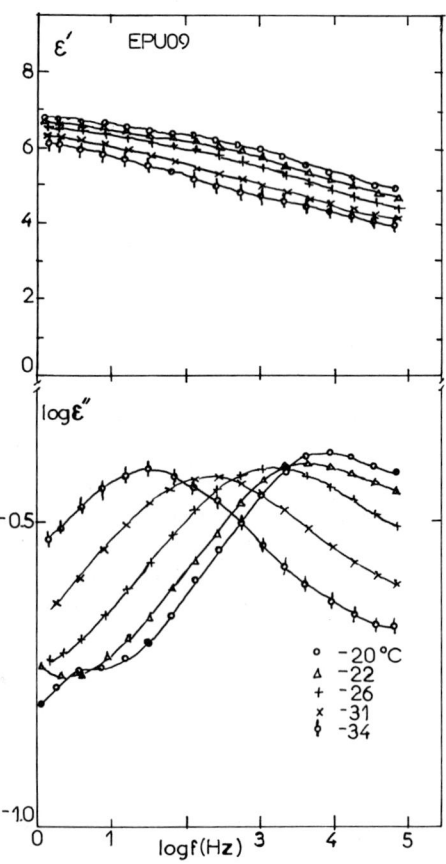

Fig. 6. Frequency dependence of the real and the imaginary parts of the complex dielectric permittivity of the sample EPU 09 measured at different temperatures

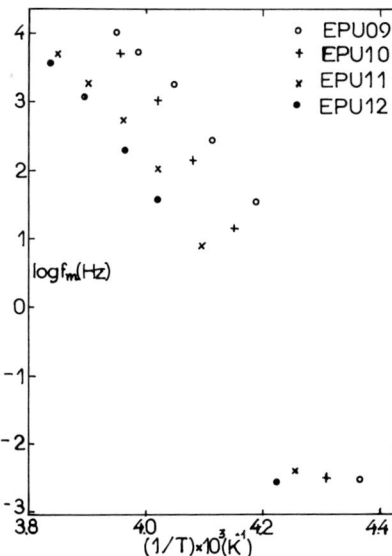

Fig. 8. Plots of the log of the maximum dielectric loss frequency, f_m, versus the reciprocal temperature for ac and TSDC data, concerning the glass transition

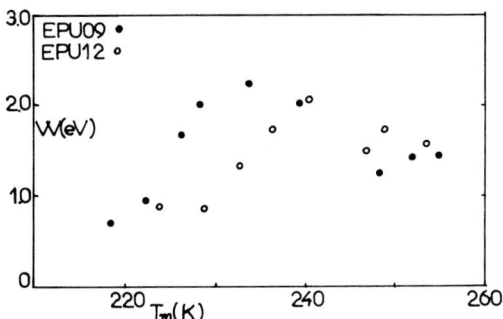

Fig. 9. Activation energy, W, of the thermal sampling responses isolated in the high-temperature region of the TSDC spectra for the samples EPU 09 and EPU 12

elastomers. The results of this kind of measurements for all types of samples are presented in Fig. 8 where the log of the maximum dielectric loss frequency, f_m, is plotted as a function of temperature. The TSDC peak temperatures are included and correspond to equivalent frequencies [11] calculated from the relaxation parameters obtained with the thermal sampling technique mentioned below. The observed deviation from the simple Arrhenius behavior of the ac plots is common for relaxations related to the glass transition of polymers [14].

The Thermal Sampling (TS) technique of the TSDC method has also been used. The method consists of sampling a relaxation process within a narrow temperature window [11, 16]. In our measurements the temperature window was 5 K and the polarizing time about 80 s. In Fig. 9 the plots of W of the TS repsonses versus their peak temperatures T_m are presented for the sample EPU09 and EPU12. These results show that, although the increase of the ratio NCO/OH results in an increase of the relaxation times (shifting T_g to higher values) it does not significantly affect the spectrum of the activation energies of the relaxations. The high values of the activation energy (and the calculated corresponding very low values of the pre-exponential factor τ_0) suggest that both relaxations are related to cooperative type of motion. Both spectra present a wide distribution of W in the temperature region of the dipolar peak and a narrow distribution of about 1.5 eV in the region of the MWS peak. These results and the fact that the two peaks overlap in the TSDC spectrum make possible a dipolar contribution to the MWS peak [9].

Acknowledgements

Support from the General Secretariat for Research and Technology of the Ministry of Industry, Energy and Technology of Greece (GSRT) and from the Bundesministerium für Forschung und Technologie (BMFT) is gratefully acknowledged.

References

1. Seymour RW, Cooper SL (1973) Macromolecules 6:48
2. Dev SB, North AM, Reid JC (1972) In: Karatz FE (ed) Dielectric Properties of Polymers, Plenum, New York p 217
3. Delides C, Pethrick RA (1981) Eur Polym J 17:675—681
4. Spathis G, Kontou E, Kefalas V, Apekis L, Christodoulides C, Pissis P, Olivon M, Quinquenet S (1990) J Macromol Sci Phys B29(1):31—48
5. Havranek A, Ilavsky M, Nedbal J, Bohm M, Soden MV, Stoll B (1987) Colloid Polym Sci 265:8
6. Cooper SL, Tobolski AV (1967) J Appl Polym Sci 11:1361
7. Kontou E, Spathis G, Niaounakis M, Kefalas V (1990) Colloid Polym Sci 268:636—644
8. Zielinski R, Rutkowska M (1986) J Appl Polym Sci 31:1111
9. Mahboubian Jones MGB, Hayward D, Pethrich RA (1987) Eur Polym J 23:855—860
10. Dusek K, Spirkova M, Havlicek I (1990) Macromolecules 23:1774—1781

11. Van Turnhout J (1980) In: Sessler GM (ed) Topics in Applied Physics, Vol 33: Electrets, Springer, Berlin 1980, p 81
12. Pissis P, Anagnostopoulou-Konsta A, Apekis L (1987) J Exp Bot 38:1528
13. Sclosser E, Schonhals A (1989) Colloid Polym Sci 267:133
14. Hedvig P (1977) In: Dielectric Spectroscopy of Polymers, Hilger, Bristol
15. Spathis G, Niaounakis M, Kontou E, Apekis L, Pissis P, Christodoulides C, Zscuppe M (submitted)
16. Apekis L, Pissis P (1987) J Physique C1, 48:127—133
17. Georing H, Pohl G, Joel D, Carius HE (1985) Plaste und Kautschuk, 32 Jahrgang, 1:14

Authors' address:

L. Apekis
Dept. of Physics
National Technical University of Athens
Zogrfou Campus
GR-15773 Athens, Greece

Influence of the thermodynamical quality of the solvent on the properties of polydimethylsiloxane networks in swollen and dry states

L. Rogovina, V. Vasiliev, and G. Slonimsky

Institute of Organoelement Compounds of Russian Academy of Sciences, Moscow, Russia

Abstract: The dependence of the elasticity modulus and the equilibrium swelling of the endlinked PDMS networks on the thermodynamical quality of the solvent in which the networks are formed and in which they swell is studied. The molar masses of crosslinked oligomers are 800 and 11000. The solvent quality is changed by the increase in the amount of nonsolvent in binary liquids mixture. The correlation between ternary phase diagrams and properties of swollen and dry networks (elasticity modulus, swelling, density) is established. The study of network deswelling in solvents of worsening quality shows that the abrupt collapse proceeds for the networks formed in good solvents at high dilution. The collapse results in an increase of the exponent value of the power law of elasticity modulus concentration dependence.

Key words: Chemical networks; poly(dimethylsiloxane); thermodynamical quality of the solvent; elasticity modulus; swelling

In our earlier studies, we established the influence of polymer concentration during chemical network formation in solution on the elasticity modulus and equilibrium swelling of the initial gels and the dry networks obtained from these gels, these networks being exemplified by poly(dimethylsiloxanes) and poly(isocyanurates) [1—3].

The aim of the present study is to investigate the dependence of poly(dimethylsiloxane) (PDMS) networks properties on the thermodynamical quality of the solvents in which these networks were formed and in which they swell.

The thermodynamical quality of the binary solvent was systematically changed by enriching the latter with the nonsolvent and the ternary phase diagrams of polymer-solvent-nonsolvent systems were compared with the change of the elasticity modulus and the swelling of the gels formed in these solvents [4].

Figure 1a shows these phase diagrams for linear PDMS of different molar masses ($\bar{M}_n$) and Fig. 1b shows the same diagrams in quasibinary form for both linear and crosslinked PDMS. The networks were obtained by endlinking PDMS oligomers in solution at the concentration 5—20% at room temperature, $\bar{M}_n$ of the oligomer being M_c of the network.

The thermodynamical quality of the binary solvent, characterized by χ-value, deteriorates proportionally to the volume fraction γ of the nonsolvent (acetonitrile) in the mixture with chloroform (Fig. 2). The value of χ was calculated from the measurements of the elasticity modulus and the equilibrium swelling [5] of the networks with $M_c = 11000$ at which χ is independent of the degree of crosslinking [1].

As the molar mass of linear PDMS decreases, the one-phase state of the solution remains in a wider range of solvent compositions and the phase separation occurs in poorer solvent. However, the endlinking of the oligomer of smaller molar mass leads to the formation of more dense network and this results in syneresis of the gels in better solvents than the solvent in which the phase separation of the same linear PDMS had occurred.

Thus, the difference in phase diagrams for linear and crosslinked polymers is much more pronounced for the oligomer with smaller $\bar{M}_n$.

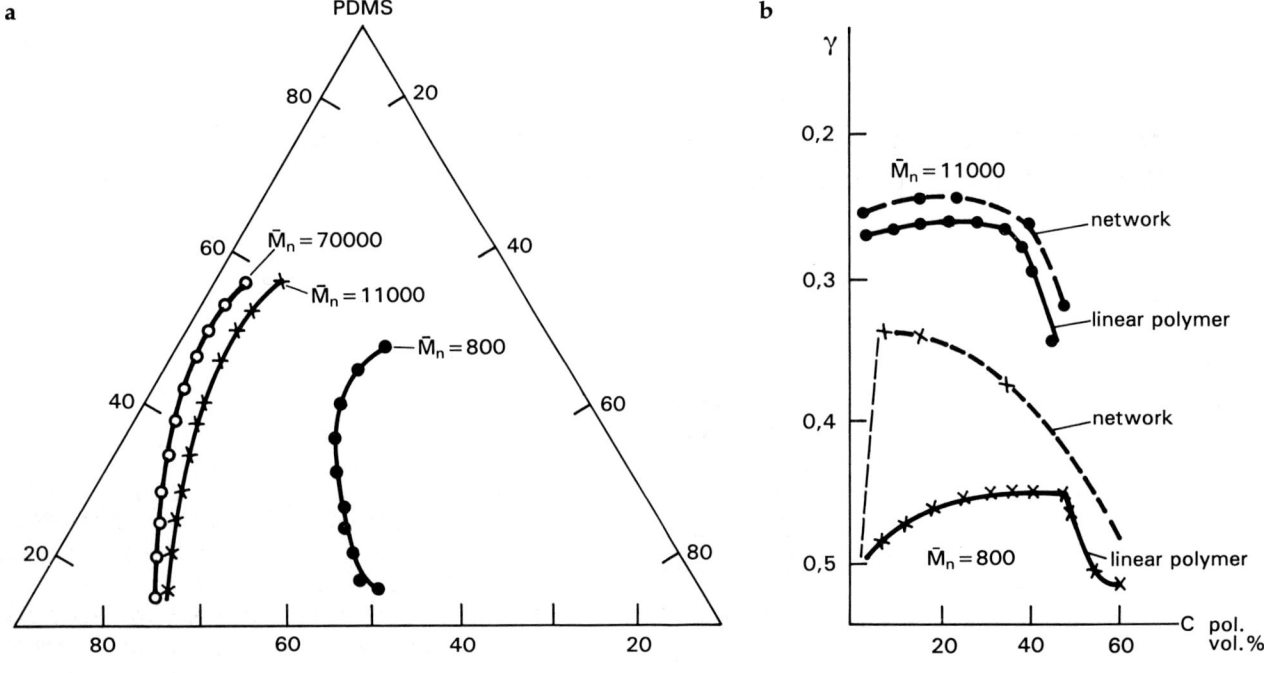

Fig. 1. Ternary phase diagrams PDMS-chloroform-acetonitrile (a) and these diagrams in quasi-binary form (b). γ is the volume fracture of acetonitrile in its mixture with chloroform

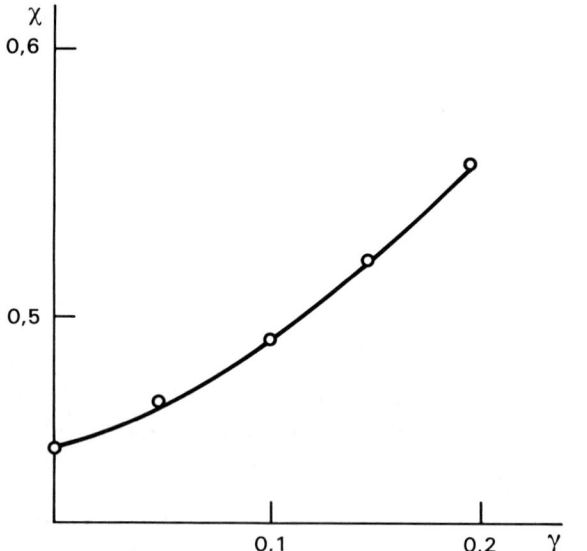

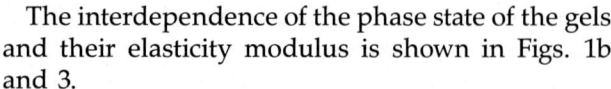

Fig. 2. The dependence of the polymer-solvent interaction parameter χ on the solvent composition

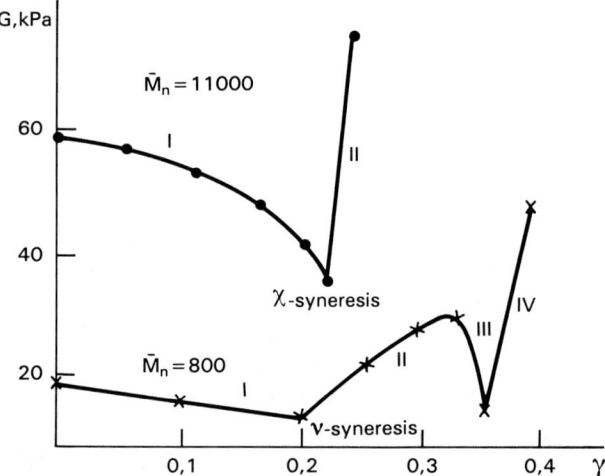

Fig. 3. The dependence of elasticity modulus G of the gels formed in binary solvents on their composition γ

The interdependence of the phase state of the gels and their elasticity modulus is shown in Figs. 1b and 3.

1) *PDMS with* $\bar{M}_n = 11000$: In one-phase state (I) the solvent deterioration gives rise to the tightening of PDMS molecular coil, which is confirmed by decreasing $[\eta]$ of the relevant dilute solutions (Fig. 4). This results in increasing intramolecular cyclization during crosslinking and, consequently, in a decreasing number of elastically active chains of the network. As a result, the elasticity modulus decreases in this region, as is shown in Fig. 3, region 1.

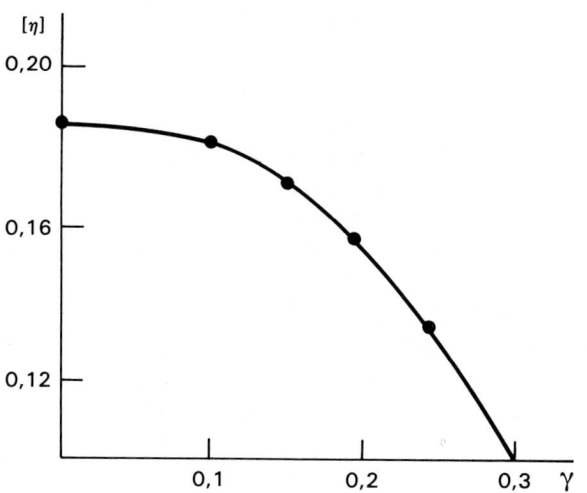

Fig. 4. The dependence of the intrinsic viscosity $[\eta]$ of linear PDMS on the solvent composition γ

As the gel transits to the two-phase state, the syneresis occurs. In [6], Dušek distinguishs two types of syneresis:

χ-syneresis caused by incompatibility between network and solvent, and v-syneresis caused by increase in the number of crosslinks. In the region II which corresponds to phase separation on phase diagram the χ-syneresis occurs. As a result, the turbid porous gels are formed. The polymer concentration inside the gels increases and, subsequently, the elasticity modulus grows considerably.

2) *PDMS with* $\bar{M}_n = 800$. In the case of dense networks based on short oligomers two types of syneresis occurring during crosslinking differently determine the properties of gel.

After the decrease of the elasticity modulus in the one-phase region (I) the v-syneresis, caused by a great number of crosslinks, develops long before the phase separation (region II on curve 2 in Fig. 3). As a result, transparent nonporous gels are formed, where their modulus increases due to removal of the solvent.

However, the further solvent deterioration leads to the prevailing decrease of the number of elastically active chains and, as a result, v-syneresis ceases

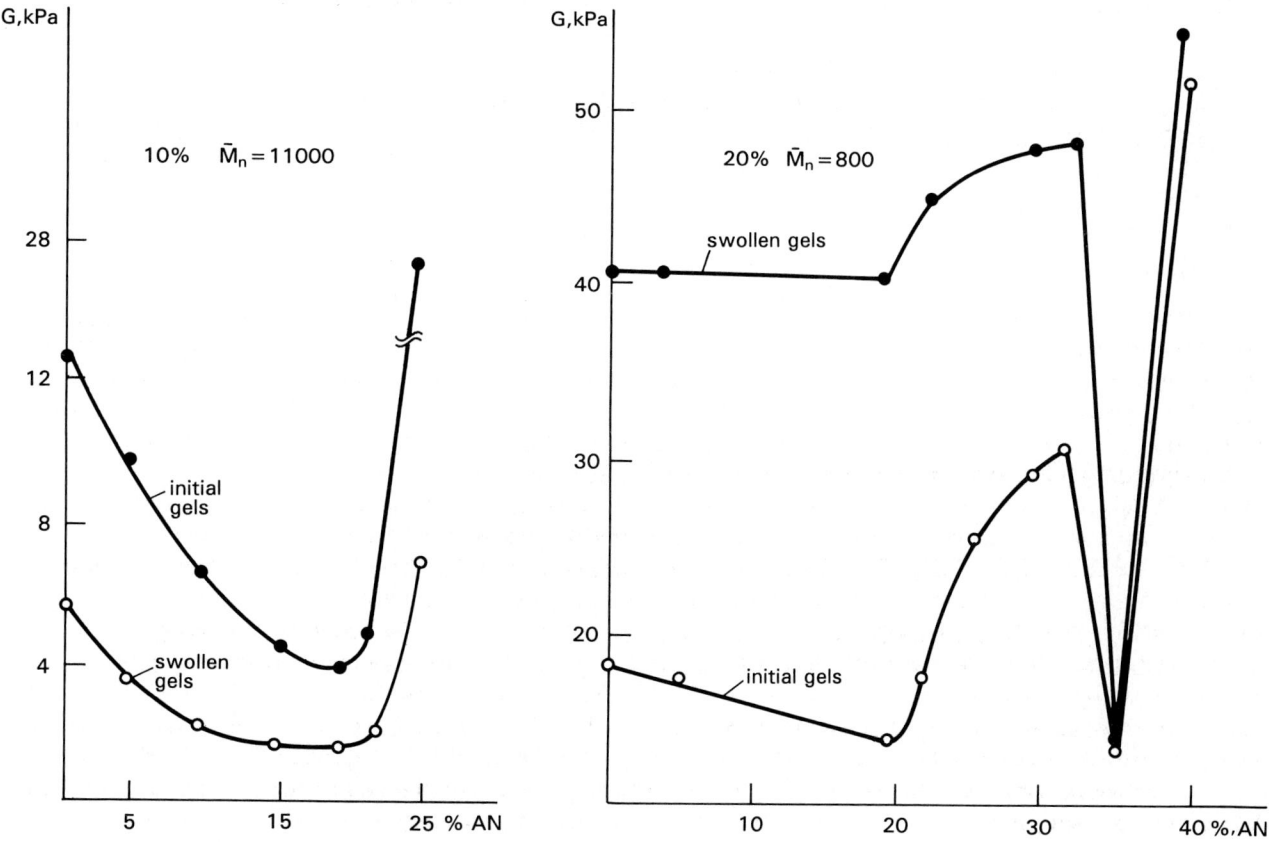

Fig. 5. Influence of the swelling on the elasticity modulus G of the networks with different M_c

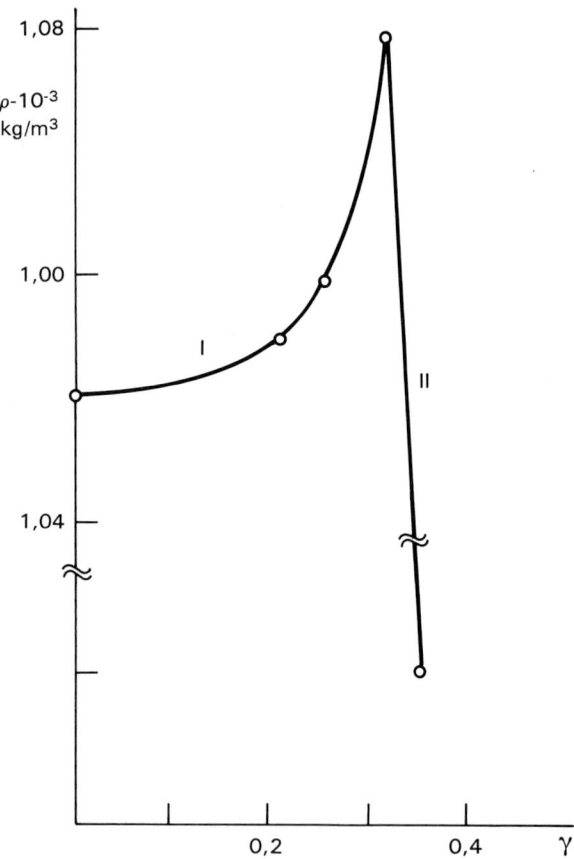

Fig. 6. The dependence of the density ρ of dry networks on the composition γ of the solvents in which they are formed

and the modulus decreases (III). Upon further solvent worsening the χ-syneresis develops as a result of phase separation according to the phase diagram (IV). In this case the elasticity modulus increases sharply.

The swelling of these gels does not effect the character of the dependence of the elasticity modulus on solvent quality. However, Fig. 5 shows that the swelling decreases all the moduli of the networks with $\bar{M}_n = 11000$ and increases the moduli of the networks with $\bar{M}_n = 800$.

This last, unusual phenomenon is the result of a non-Gaussian behavior of these networks due to the short network chains, which determines their extension during swelling.

The density of dry networks obtained upon solvent removal from the gels grows in region I due to the considerable compactization of the highly imperfect networks. Figure 6 shows that the density decreases in region II due to retaining the porous structure formed in this region in the initial gel state. The same kind of compactization of the gels structure with a great number of topological defects was observed for the gels obtained in good solvent, but at very high dilution [1]. We believe this is typical of defect networks in general, and this phenomenon is closely related to the transition of network chains from the coil to crumpled globula conformation [7].

The network properties are determined by both the nature of the solvent in which they are formed and that of the solvent they swell in.

In the first case, the network topology changes and the equilibrium swelling of these gels in a good solvent (toluene) is determined by the decrease of the number of crosslinks and it varies oppositely to the elasticity modulus change (Fig. 7a).

The different states of the swelling of the networks with the same topology synthesized at the same conditions may be obtained by their deswelling in the different solvents upon the gradual solvent deterioration until the pure nonsolvent is used. Nonporous gels are always obtained at these conditions.

Figure 7b shows that, for networks synthesized without solvent or in concentrated solution, the deswelling proceeds gradually. However, for the network obtained at high dilution (5%) the abrupt deswelling occurs at the composition of the solvent corresponding to that of the phase separation during network formation.

This network collapse is of the same nature as the network compactization during drying.

The variation of network formation conditions results in different values of the power law exponents for the dependence of the elasticity modulus on the polymer concentration in gels (v_2).

For the networks obtained at different concentrations in a good solvent (toluene), we obtained $G \sim v_2^{1.7}$ and for the same networks after their swelling in toluene to equilibrium $G \sim v_2^{2.25}$.

For the networks obtained at different concentrations in poor solvent (chloroform/acetonitril = 80/20), we obtained $G \sim v_2^{2.9}$, and for the network obtained without solvent and swollen in the solvents of different quality $G \sim v_2^{0.4}$. Only in this latter case did all swollen states of the network have the same topology.

For the networks obtained in the good solvent at high dilution (5%) and deswollen in the solvents of

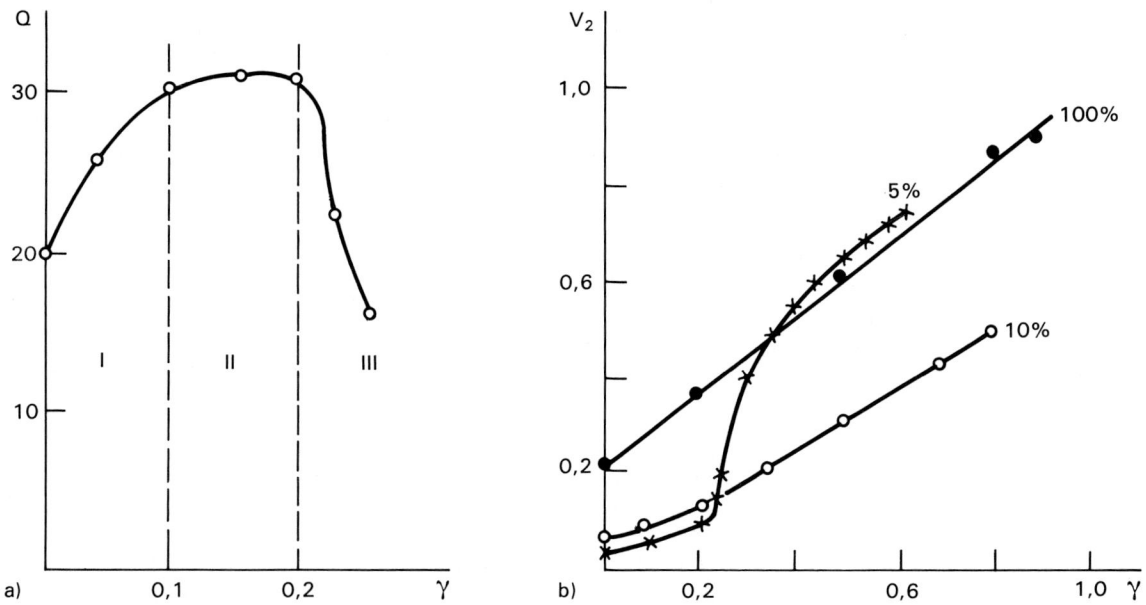

Fig. 7. The dependence of equilibrium swelling of the gels on the composition of the solvents in which they are formed (a) and in which they are swollen (b)

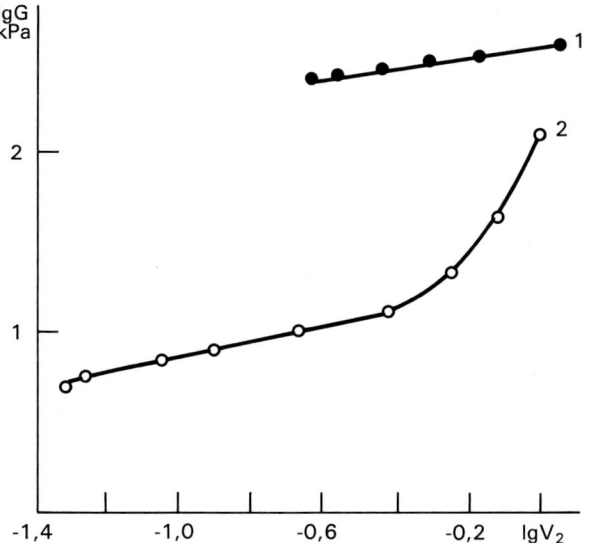

Fig. 8. The elasticity modulus-concentration dependence for the networks ($\bar{M}_n = 11000$) swollen in the solvents of worsening quality:
1) network obtained without solvent; 2) network obtained at $v_0 = 0.05$, that being the volume fraction of polymer in solution during network formation

worsening quality the exponent is not constant, as is shown in Fig. 8. Before beginning of the collapse it is 0.5, in the collapse region it is 1.3, and after collapse up to the dry state it is 2.9—3.0. This means that the collapse of the polydimethylsiloxane networks, although it is not so pronounced as in charged networks, results in a completely new structure.

Thus, the controlled variation of the solvent nature at network formation and swelling allows the purposeful regulation of the network structure and properties.

References

1. Vasiliev VG, Rogovina LZ, Slonimsky GL (1985) Polymer 26:1667—1676
2. Rogovina LZ, Vasieliev VG, Frenkel CM, Pankratov VA, Slonimsky GL (1984) Visokomol Sojed (Polym Sci USSR) 26:182—188
3. Rogovina LZ, Vasieliev VG, Slonimsky GL (1985) Polym Sci USSR 27:803—804
4. Rogovina LZ, Vasieliev VG, Slonimsky GL (1990) Polym Sci USSR 32:2086—2092
5. Rietsch F, Froelich D (1979) Europ Pol J 15:349—359
6. Dušek K (1965) J Pol Sci B3:209—212
7. Grossberg AY, Nechaev SK (1988) J de Phys 49:2095—2100

Authors' address:

Dr. L. Rogovina
Institut of Organoelement Compounds
of Russian Academy of Sci
28 Vavilova Str
Moscow 117813, Russia

Triblock copolymers, mesogels and deformation behavior in poor solvents

A. Halperin*) and E. B. Zhulina+)

*) Max Planck Institut für Polymerforschung, Mainz, FRG
+) Institut für Physik, Johannes Gutenberg University Mainz, FRG

Abstract: A fundamental distinction between the mesophases formed by ABA triblock copolymers and by AB diblock copolymers is the bridging of A domains by B blocks. The ABA mesophases form physically crosslinked networks characterized by non-uniform spatial distribution of high-functionality crosslinks. The swelling of these networks by selective solvents gives rise to novel "mesogels". Three theoretical aspects of these systems, focusing on the lamellar case, are considered: i) The SCF theory of the equilibrium fraction of bridging chains in mesophases formed by a melt of ABA triblock copolymers. ii) The swelling equilibrium and the deformation behavior of mesogels swollen by a selective solvent for the B blocks. iii) The deformation behavior of lamellar gels immersed in a poor solvent for the B blocks.

Key words: Collapse; grafted chains; scaling; SCF bridging; phase transition

Introduction

Among the important factors determining the behavior of polymeric networks are the elasticity of the constituting chains and the manner of their crosslinking. The elastic response of the chains involves two familiar scenarios: The deformation of flexible chains in a melt or in solution and the deformation of rigid, rod-like polymers. In the first case the restoring force is entropic in origin, while in the second it is due to the "mechanical" rigidity of the polymers. The crosslinks are typically of low functionality and uniform distribution in space. In the following, we explore, theoretically, situations incorporating different ingredients. One is the *deformation behavior of flexible chains immersed in a poor solvent* [1, 2]. In particular, we focus on the stretching of dense polymer brushes, i.e., chains which are terminally bound to a flat surface so that the distance between grafting sites, d, is small in comparison to the radius of a collapsed free chain, $R \sim N^{1/3}$, where N is the degree of polymerization. As we shall see, the stretching of polymer brushes in a poor solvent proceed via a first-order phase transition involving a coexistence of a dense, weakly deformed layer and a dilute, strongly stretched layer. The coexistence region is associated with a novel force law, $f \sim L^0$ where L is the overall thickness of the brush, i.e., the stress in independent of the strain. The second ingredient is *crosslinks of high functionality and non-uniform distribution in space* [3]. We will mostly focus on lamellar configurations where the crosslinks are constrained to aligned, equally spaced planes. It is possible, however, to imagine other interesting situations such as crosslinks which are constrained to parallel lines. These situations may be realized in networks based on *mesophases of ABA triblock copolymers*. In these systems some of the B chains bridge different A domains. This crosslinking is expected to be long lived if the A domains are glassy or crystaline in the regime of interest. Their stability may be enhanced further by selective crosslinking of the A domains. This may be accomplished, for example, by radiation-induced crosslinking. *Mesogels* is the term we use for the gels obtained by swelling such networks by a selective solvent for the B chains [3]. As we shall see, the behavior of mesogels exhibits novel features: The swelling is anistropic; the swelling equilibrium is described by

c^* theorems of distinctive forms; asymmetry of compression and stretching and non-affine deformation behavior in good solvents. An especially interesting feature is the deformation behavior of mesogels immersed in a poor solvent. For lamellar gels one expects the $f \sim L^0$ force law to characterize the stretching behavior in poor solvents.

We begin with a discussion of the factors determining the fraction of bridging B chains in mesophases formed by a melt of flexible ABA triblock copolymer. The quantitative conclusions of this section concern the equilibrium distribution of bridging and non-bridging B blocks in a lamellar mesophase. The next section concerns the scaling behavior of mesogels swollen by a good solvent. It is mostly concerned with aligned lamellar gels. Finally, we apply scaling analysis to the deformation of lamellar mesogels in poor selective solvents, i.e., a precipitant for the A blocks and a poor solvent for the B blocks. In all cases, we ignore polydispersity effects. The ABA triblocks copolymers are assumed to be monodispersed, comprising two identical A blocks consisting each of N_A monomers and a flexible B block incorporating $2N_B$ monomers. The discussion is further limited to the strong segregation regime, i.e., the A and B blocks are strongly incompatible and the solvent, when present, is highly selective. In this regime the AB interface is narrow, of width comparable to monomer size. This choice is convenient since, in this limit, the AB surface tension, γ, is independent of N_A and N_B. Furthermore, in this case the B blocks are, in effect, grafted [4] to the AB interface: The AB junctions are constrained to a narrow interfacial region and the B blocks are thus anchored to the interface by their end-groups. Densely grafted chains exhibit interesting properties traceable to their stretched configurations. Mesogels are interesting from this perspective because their elementary structural units are grafted layers of B chains. Micellar gels are obtained from micellar mesophases and are comprised of close-packed, spherical, grafted layers; lamellar gels may be envisioned as stacks of flat "brushes" of densely grafted layers, etc. The study of mesogels may thus afford a novel approach to the investigation of grafted layers. Furthermore, mesogels may allow simple experimental probes of otherwise inaccessible facets of grafted layers. Such is the case, for example, with grafted layers subjected to stretching. Little is known of these systems [5] experimentally and our theroetical discussion is accordingly rough.

Copolymeric mesophases: Triblocks vs. diblocks

Mesophases formed by ABA triblock copolymers exhibit bridging of A domains by B blocks [6, 7]. This is an important feature since it yields physically crosslinked networks. In turn, these are of interest as a source of industrially useful thermotropic elastomers [7]. The phenomenon is also of fundamental interest as a qualitative difference between the mesophases formed by diblock and triblock copolymers. These mesophases are otherwise quite similar, both with regard to symmetry and to the characteristic dimensions of the domains [8]. A theoretical model for this phenomenon must account for the various options available to the B blocks. If each of the A blocks belongs to a different domain the B blocks form a bridge. When the two A blocks are incorporated into a single domain the B block forms a loop. In this case, one must allow for the possibility of bridging due to entanglements between loops anchored to different domains. As we shall see, the theory does not distinguish between entangled loops and direct bridges for the case of a perfectly monodispersed sample. Finally, one of the A blocks may be embedded in a B domain, giving rise to a "dangling end". This last option may be neglected in the strong segregation limit considered here. In the following, we aim to obtain the equilibrium ratio of bridges and loops in a strongly segregated lamellar phase.

Lamellae formed by flexible $A_N B_{2N} A_N$ triblock copolymers are often considered equivalent to two lamellae comprising $A_N B_N$ diblock copolymers. The characteristic dimensions of the lamellae are obtainable via an Alexander-type argument [4, 9, 10]. It is focused on the interplay between the intrafacial energy associated with the AB boundary and the deformation penalties due to the stretching of the grafted A and B blocks. The intrafacial term is $\sigma \gamma$ where σ is the surface are per block. All chains are assumed to be uniformly stretched and the deformation term is given by the elastic free energy of a Gaussian coil, $kTL^2/Na^2 \approx kTNa^2/\sigma^2$. Minimization with respect to σ yields the equilibrium characteristics of the layer

$$L/a \approx (\gamma a^2/kT)^{1/3} N^{2/3} \qquad (1)$$

and

$$\sigma/a^2 \approx (kT/\gamma a^2)^{1/3} N^{1/3} \ . \qquad (2)$$

The Alexander model can account for certain properties of triblock copolymer mesophases such as the characteristic dimensions of the domains [10]. However, it is clearly too restrictive. The uniform stretching assumption places an undue constraint on the chains and the resulting free energy is thus overestimated. A more realistic approach, allowing the chain ends to be anywhere within the layer, was recently devised by Semenov [11]. This self-consistent field (SCF) analysis recovers the results obtained by the Alexander approach apart from slightly different multiplicative constants. While the SCF approach did not yield qualitatively new results for mesophases of diblock copolymers it is, as we shall see, crucial for a proper analysis of mesophases formed by triblock copolymers. It suggests that the B layer consists of three regions: i) a central region where only bridging chains are present. In this "dead zone" the chains are uniformly stretched; ii) two boundary layers which contain both bridging and non-bridging chains. In these layers the stretching is non-uniform. In contrast to the Alexander model the chains are not uniformly stretched and the free energy of the bridging chains is higher than that of non-bridging chains. This picture was first reported by Johner and Joanny [12] for the case of a swollen brush adsorbed to a wall by its free ends. It was independently developed by Zhulina and Pakula [13] for the melt case and by Ligoure et al. for the adsorption of solvated telechelics between two walls [14]. Our discussion follows the Semenov approach as modified in [13]. Consider a B layer comprising blocks consisting of $2N$ monomoers such that the area per anchored end group is σ. Since the layer is symmetric with respect to the midplane it is sufficient to consider one half of the layer, of thickness $L = Na^3/\sigma$. The thickness of the boundary layer is denoted by h and the fraction of bridging chains by q. Each bridging chain deposits N' monomers within the boundary layer. The remaining chains are non-bridging loops. In the following a loop is treated as two linear chains comprising each of N monomers. We aim to characterize the equilibrium structure of the B layer for given values of q and N'. These determine h since a volume of $(L-h)\sigma$ in the central region contains $q(N-N')$ monomers of volume a^3 thus leading to

$$h/L = 1 - q(N-N')/N . \tag{3}$$

The bridging chains in the central region are uniformly stretched. The elastic free energy per copolymer due to the chains segments embedded in the central layer is thus

$$F_c/kT = 3q(L-h)^2/2(N-N')a^2 . \tag{4}$$

The corresponding tension is

$$f_c/kT = 3(L-h)/(N-N')a^2 . \tag{5}$$

The chains in the boundary layer are not uniformly stretched, and its description is accordingly more complex. The average elastic free energy per chain in the boundary layer is

$$F_b/kT = (1-q)(3/2a^2) \int_0^h dy\, g(y) \int_0^y E_n(x,y)dx$$

$$+ q(3/2a^2) \int_0^h E_b(x,h)dx . \tag{6}$$

$E(x,y) = dx/dn$ is, essentially, the local tension in a chain whose endpoint is located at height y. The tension is actually given by $f_b = kT(3/a^2)E$. The representation of the local elastic free energy as $E(r,r')dr$ is identical to the familiar $E^2(r,r')dn$. However, in our case $E(r,r')dr$ is replaced by a one-dimensional form, $E(x,y)dx$, because of the stretched configurations adopted by the grafted B blocks. These are also reflected in the requirement that $x < y$. These simplifications are possible because the stretching reduces the number of configurations accessible to the coil. The possible configurations may be considered as weak fluctuations around a single dominant trajectory. $E_b(x,h) = E_b(x)$ is the local tension in bridging chains whose end points are always positioned at the periphery of the boundary layer. $E_n(x,y)$ is the local tension in a non-bridging chain whose end is located at height $y \leq h$. The end-points of such chains are distributed throughout the boundary layer and $g(q)$ is the corresponding height distribution function.

In order to find the equilibrium state of the boundary layer, the free energy given by (6) must by minimized subject to the following constraints

$$\int_0^y E_n^{-1}(x,y)dx = N \tag{7}$$

$$\int_0^h E_b^{-1}(x)dx = N' \tag{8}$$

and

$$\Phi(x) = (1-q)(a^3/\sigma) \int_0^h g(y)E_n^{-1}(x,y)dy$$

$$+ qa^2/\sigma E_b(x) = 1 . \tag{9}$$

Constraints (7) and (8) assure the conservation of monomers for bridging and non-bridging chains. The constant melt density within the layer is reflected in Eq. (9). Note the choice of integration domain in the first term of (9), stating that monomers at height x may only originate in non-bridging chains with end points at a greater height x, $x < y < h$. This is another manifestation of the stretched configurations of the grafted B coils. These constraints are supplemented by the requirements of mechanical equilibrium which impose equality of the tension at the edge of the boundary layer to the tension in the central region

$$E_b(h) = (L - h)/(N - N') . \quad (10)$$

The minimizations utilizes the Langrange method of undetermined multipliers as explained in [11]. The equilibrium structure of the boundary layer, for a given set of σ, N' and q, is specified by the following

$$E_n(x,y) = (\pi/2N)(y^2 - x^2)^{1/2} \quad (11)$$

$$E_b(x,h) = (\pi/2N)(\Lambda^2 - x^2)^{1/2} , \quad (12)$$

where $\Lambda = h/\cos(\pi\tau/2)$ and $\tau = (N - N')/N$ is the fraction of the monomers, belonging to a bridging chain, which are deposited in the central layer. The distribution function for the free chain ends is

$$g(y) = \frac{1}{(1-q)L} \frac{y(h^2 - y^2)^{1/2}}{\Lambda^2 - y^2} . \quad (13)$$

For $q = \tau = 0$ (13) reduces to the $g(y)$ of a dense brush of free chains. The combination of (3), (12) and the equality of tensions at h specifies the interedependence of q and τ:

$$2q/\pi \tan(\pi\tau/2) = 1 - q\tau . \quad (14)$$

Knowing $E_n(x,y)$, $E_b(x)$ and $g(y)$, it is possible to obtain the elastic energy per chain, $F_{el} = F_b + F_c$

$$F_{el}/kT = (F_{el}^0/kT)[(1 - \tau q)^3 + (12/\pi^2)q^2] , \quad (15)$$

where $F_{el}^0/kT = (\pi^2/8)(Na^4/\sigma^2)$ is the elastic free energy of a free ($q = 0$), dense brush. By taking (14) into account, one obtains F_{el} for small q values, $q \ll 1$,

$$F_{el}/kT = (F_{el}^0/kT)[1 + (2q/\pi)^4] . \quad (16)$$

Apart from the elastic free energy it is necessary to account for the mixing entropy of bridging and non-bridging chains. The free energy per chain, F_{chain}, assuming random mixing is

$$F_{chain}/kT = F_{el}/kT + q\ln q + (1 - q)\ln(1 - q) . \quad (17)$$

Minimization with respect to q yields the equilibrium value of q, q_{eq}, for a given σ

$$q_{eq} \approx (kT/F_{el}^0)^{1/3} \approx (\sigma/a^2)^{2/3}N^{-1/3} . \quad (18)$$

This suggest that q_{eq} is weakly dependent of N. However, (18) is a rough estimate. The characteristic dimensions of the lamellae, L and σ, were obtained from an Alexander-type argument; the two-dimensional entropy of the endgroup was not taken into account; polydispersity effects were ignored. Also, the Semenov theory provides a satisfactory description in the limit of long, strongly stretched chains. Furthermore, (18) applies to systems in thermodynamical equilibrium while real systems may fail to equilibrate. Nevertheless, this rough picture may provide some insight on the factors determining the equilibrium fraction of bridging chains. In the following, we mostly ingore this detailed description. Rather, we focus on the consequences, i.e., the qualitative features of the behavior of mesogels. For simplicity, we will assume that all B blocks form bridges.

Mesogels

Mesogels are obtained by swelling any of the mesophases formed by ABA triblock copolymers. It is thus possible to envision lamellar, cylindrical and micellar gels. Their detailed structure depends on the nature of the A blocks. These may be flexible, rigid or crystallizable. The characteristic dimensions of the A domains are somewhat different for each case [4]. We mostly consider lamellar gels formed from crystalline-coil-crystalline ABA triblock copolymers. The special features of cylindrical and lamellar gels are most evident in "single crystal" samples [5], i.e., aligned mesogels where the A domains are parallel. The discussion is limited to such samples. It is concerned, primarily, with their macroscopic characteristics: swelling equilibrium and deformation behavior. For simplicity, the A domains are assumed to be *perfectly rigid*. In particular, they are assumed to retain their equilibrium dimensions as attained in the melt mesophase. We mostly focus on the idealized limit of $q = 1$ obtained when all B blocks bridge different A domains. Finally, we ignore, at this early stage, complications due to entanglements, defects, nonuniformities, etc.

The swelling behavior of mesogels is qualitatively distinct [3]. While micellar gels swell uniformly, the swelling of aligned lamellar and cylindrical gels is anisotropic. Lamellar gels swell unaxially, along the lamellar normal. Biaxial swelling is characteristic of cylindrical gels. Furthermore, the swelling equilibrium of mesogels is quantitatively distinct as seen by comparing the c^* theorems applying to swollen mesogels with the theorem which concerns simple gels. The c^* theorems [15, 16] specify the average monomer concentration, c_e, within a gel in equilibrium with a reservoir of solvent, i.e., when the chemical potentials of the solvent inside and outside the gel are equal. For simple gels the c^* theorem states that c_e is maintained at c^*, the overlap threshold for semidilute solution of free chains comprising N monomers where N is also the average number of monomers between adjacent crosslinks [15, 16]

$$c_e \approx c^* \sim N^{-4/5} . \tag{19}$$

In a good solvent the gel is thus envisioned as an array of close packed blobs of size $\xi \approx R_F \sim N^{3/5}$ where R_F is the Flory radius of the free chains. This formulation, as the Flory theory of gels [6], specifies an equilibrium state reflecting a balance between osmotic and pressures. It does not allow for the role of entanglements, etc. The c^* theorems for mesogels retain this physical foundation. However, the overlap threshold now applies to the highly branched structural subunits. Thus, a swollen micellar gel may be viewed as an array of star polymers at the overlap threshold. For a lamellar gel the subunit is a flat brush of densely grafted chains. Similarly, the equilibrium state of a swollen cylindrical gel is described in terms of close packed, grafted cylinders. To illustrate the point consider the swollen state of a lamellar gel formed by ABA crystalline-coil-crystalline triblock copolymers. In this system the area per grafted end-group, σ, scales as $\sigma \sim N^{1/3}$ where $2N$ is the polymerization degree of the flexible B block [17]. The grafting density is high, $\sigma \ll R_F^2$, and the chains crowd each other. Since $q = 1$ is assumed, the monomer volume fraction, Φ, is constant, $\Phi \approx Na^3/L\sigma$ where L is the thickness of the layer. Furthermore, all chains are uniformly stretched. Accordingly, the Alexander model [4, 9] for a flat brush of densely grafted chains accurately describes the system: A slab of semidilute solution envisioned as an array of close packed concentration blobs of size $\xi \approx \Phi^{-3/4}a$ comprising $g \sim \Phi^{-5/4}$ monomers such that $\xi \approx g^{3/5}a$. The free energy per chain, F_{chain}, consists of two terms: An elastic penalty due to the stretching of a Gaussian string of blobs from $R_0 \approx (N/g)^{1/2}\xi$ to L. A second term accounting for monomer-monomer interactions by the kT per blob ansatz, i.e., $(N/g)kT$.

$$\begin{aligned}F_{chain}/kT &\approx (L^2/Na^2)\Phi^{1/4} + N\Phi^{5/4} \\ &\approx \Phi_e^{9/4}(\sigma/a^2)(L_e/a)[(L/L_e)^{7/4} + (L_e/L)^{5/4}] .\end{aligned} \tag{20}$$

The chemical potential is given by $\mu = F - \Phi\partial F/\partial\Phi$, where $F = F_{chain}/L\sigma$ is the free energy density. The equilibrium condition thus leads to [3]

$$\Phi_e \approx (a^2/\sigma)^{2/3} \sim N^{-2/9} \tag{21}$$

as compared to $\Phi_e \sim N^{-4/5}$ for simple gels. For this Φ_e the brush is characterized by $\xi \approx \sigma^{1/2}$ and $L_e \approx (N/g)\xi \approx N(a^2/\sigma)^{1/3}\sigma$. This corresponds to the equilibrium state of an isolated brush. Thus, the c^* theorem for lamellar gels yields Φ_e corresponding to the overlap threshold, c^*, of aligned lamellar brushes. The precise N dependence of Φ_e varies somewhat with the nature of the A block but the $\Phi_e \approx (\sigma/a^2)^{2/3}$ behavior is general. Similar considerations yield c^* theorems for micellar and cylindrical gels. Thus, for gels comprising grafted cylinders of length H such that each incorporates f block copolymers, we have

$$\Phi_e \approx (a/H)^{1/2}f^{1/2}N^{-1/2} . \tag{22}$$

The swelling equilibrium of micellar gels incorporating star-like micelles comprising each of f block copolymers is characterized by

$$\Phi_e \approx f^{2/5}N^{-4/5} . \tag{23}$$

In both cases the f dependence was retained because the scaling behavior of f varies with the nature of the A blocks. Of the various mesogels, micellar gels are the most similar to simple gels. This is apparent in the predicted scaling behavior of Φ_e for the two systems. The underlying factor is amount of chain stretching in the swollen gel. The constituting chains of a simple gel, within this picture, are undeformed. In micellar gels, the deformation of the arms emanating from each star-like junction is rather weak. The degree of stretching and the deviation from simple gel behavior grow as we pass to cylindrical and to lamellar gels. This is a consequence of the geometry of the grafted structures forming the elementary subunits of these mesogels [4]. The mutual crowding, and the resulting stretching diminish along the sequence flat, cylindrical, spherical as the volume available to each chain grows.

The similarity between micellar gels and simple gels extends to the qualitative aspects of their deformation behavior. Micellar gels, as simple gels, are expected to have isotropic compressibility. Also, the compression is expected to occur with a concomitant change in the cross-section of the sample. When the deformation rate is high and the solvent content cannot readjust, the volume of the sample remains constant [15, 16]. Different behavior is expected from aligned cylindrical and lamellar gels [3]. The compressibility is no longer isotropic since the gel may be compressed only by forces acting along the lamellar normal. Furthermore, the lateral dimensions of the sample are fixed. Accordingly, compression or extension of an aligned lamellar gel must be accompanied by a change in the volume of the sample irrespective of the time scale. As was noted above, the constituing chains in a mesogel are stretched in equilibrium while the chains forming a simple gel are not. This, together with the geometrical constraints on the cross-section of the samples, give rise to another distinctive feature of mesogels: The deformation behavior of cylindrical and lamellar gels is qualitatively different for compression and extension. While the response to compression is dominated by the osmotic pressure, the restoring force associated with extension is due to the elasticity of the chains. The asymmetry between compression and extension is also manifested in the role of non-bridging chains, i.e., non-entangled B loops. These affect the elastic response of a stretched mesogel only weakly. In marked contrast, all chains, irrespective of their bridging role, contribute to the osmotic restoring force arising in compressed mesogels. This point is of particular interest since non-entangled loops are analogous to dangling chains in simple gels. Finally, the configurations of individual chains in a deformed simple gel are often specified by the assumption of affine deformation: The relative deformations of an individual chain and of the sample are assumed to be identical. As we shall see, lamellar gels in good solvents are expected to exhibit non-affine deformation behavior.

For concreteness, we now consider the deformation behavior of lamellar gels in greater detail. As before, we base the discussion on the Alexander model for flat grafted layers. In the equilibrium state of the gel, as specified by the c^* theorem, the chains are strongly stretched. The blob size is $\xi \approx \sigma^{1/2}$ and the layer thickness, $L/a \approx N(a^2/\sigma)^{1/3}$, corresponds to a fully stretched chain of blobs of length $(N/g)\xi$.

Compression increases the concentration within the layer. Consequently, the blobs are smaller and the chains are less stretched. F_{chain} as given by (20) describes this regime. The force law per chain, $f = -\partial F_{\text{chain}}/\partial L$, yields the pressure $p = f/\sigma$

$$p/kT \approx \Phi_e^{9/4} a^{-3}[(L_e/L)^{9/4} - (L/L_e)^{3/4}] \ . \tag{24}$$

The compression is accompanied by a decrease in the lateral dimensions of the chains, $\langle \Delta r_\perp^2 \rangle = (N/g)\xi^2$, from $N(\sigma/a^2)^{1/6}a^2$ to $N(\sigma/a^2)^{1/6}(L/L_e)^{1/4}a^2$. Since the lateral dimensions of the sample remain constant, the chains undergo non-affine deformations. Within this picture the compression of a lamellar gel is indistinguishable from the compression of two independent brushes. In particular, the roles of bridging and non-bridging chains are identical. A different approach is necessary for the description of the stretching behavior. Since the chains in equilibrium state are envisioned as fully stretched strings of blobs, further extension is only possible upon rearrangment of the blob structure. A description of this regime was suggested recently by Rabin and Alexander [18] who generalized an earlier approach due to Pincus [15, 19]. Each chain is pictured as a fully stretched string of Pincus blobs of size ξ_p determined by $\xi_p f \approx kT$, where f is the tension in the chain. For sufficiently weak deformations the blobs exhibit self avoidance and each blob incorporates $g_p \approx (\xi_p/a)^{5/3}$ monomers. Acccordingly, the chain length is $L \approx (N/g_p)\xi_p \approx N(fa/kT)^{5/3}a$ and its elastic free energy, as given by the kT per blob prescription, is $(N/g_p)kT \approx N(fa/kt)^{5/3}kT$. However, $\xi_p < \sigma^{1/2}$ while the volume per chain, σL, is larger then σL_e. Accordingly, the blobs are not close packed and the familiar kT/ξ^3 ansatz for the interaction free energy density fails. An alternative approach [18] utilizes a Flory type argument allowing for blob-blob interactions: Each blob experiences an interaction energy of $kT\Phi_p$ where $\Phi_p \approx \xi_p^2/\sigma$ is the blob volume fraction. The repulsive interaction energy in a good solvent is thus $(N/g_p)(\xi_p^2/\sigma)kT$ per chain and the total free energy per stretched chain is

$$F_{\text{chain}}/kT \approx \Phi_e^{9/4}(\sigma/a^2)(L_e/a)[(L/L_e)^{5/2} + (L_e/L)^{1/2}] \ . \tag{25}$$

The corresponding restoring pressure is

$$p/kT \approx \Phi_e^{9/4} a^{-3}[(L/L_e)^{3/2} - (L_e/L)^{3/2}] \ . \tag{26}$$

The deformation of the individual chains is, again, non-affine. Their lateral dimensions decrease from $\langle \Delta r_\perp^2 \rangle \approx N(\sigma/a^2)^{1/6} a^2$ to $N(\sigma/a)^{1/6}(L_e/L)^{1/2} a^2$. This non-Gaussian regime occurs in a limited domain. For strong extensions the blobs are too small to experience excluded volume interactions. As a result, a new regime, characterized by Gaussian-Pincus blobs, appears. The boundary is specified by $\xi_p \approx r_b$ where $r_b \approx (a^3/v)a$ is the size of a Gaussian blob comprising g_b monomers such that the monomer-monomer interactions are of order kT [15] i.e., $vg_b^2/r_b^3 \approx vg_b^{1/2}/a^3 \approx 1$. The free energy per chain in the very strong stretching regime is

$$F_{chain}/kT \approx R_0^2/\sigma + (L/R_0)^2 , \qquad (28)$$

where the first term accounts for the interaction free energy $(N/g_p)(\xi_p^2/\sigma)kT$ and the second for the deformation penalty of a Gaussian coil. In this regime the deformation behavior of the coils is affine, as expected.

Deformation behavior in poor solvents

It is difficult to study the deformation behavior of free collapsed globules. The observation of free chains in a poor solvent is hampered by the precipitation of the polymers. This problem may be circumvented by grafting the chains to a surface. It is then possible to probe their deformation behavior by using a force measurement apparatus capable of lateral motion. Depending on the grafting density, one may study the behavior of a sense, collapsed brush or of isolated globules. However, the analysis of such shear-induced deformation is complicated by dynamical effects. Another possible route involves the study of binary gels incorporating a minority of insoluble chains. This may allow the study of isolated globules under stress. A related approach involves the stretching of aligned lamellar gels immersed in a poor solvent for the B blocks. In this case the experiment probes the stretching behavior of a collapsed brush while avoiding dynamical complications. As we shall see, this system is expected to exhibit novel strain-stress relationship [2]. These features occur because the stretching involves a first-order phase transition associated with a coexistence of a dense, weakly deformed layer with a dilute layer of strongly stretched chains.

The deformation behavior of a single globule [1] may provide a useful introduction. A stretched globule exhibits an instability reminiscent of a first-order phase transition. This can be seen if we imagine the deformation taking place as a continuous process. In other words, assume that the spherical globule is deformed continuously into ellipsoids of growing eccentricity. Linear response is expected for weak deformations. This $f \sim L$ regime occurs when the longer semi-axis, of length L is, comparable to the radius of the collapsed globule, $r_c \sim N^{1/3}$. When L is much longer the deformed globule is reminiscent of a cylinder of diameter d. Since the globule is a dense object conservation of volume, $d^2 L \approx r_c$, determines d. In this regime the lateral surface area, $\sim r_c^{3/2} L^{1/2}$, is dominant and the associated surface energy gives rise to a nonlinear restoring force, $f \sim L^{-1/2}$. For stronger deformations, when $L \sim N$, a Gaussian-restoring force, $f \sim L$, is expected. The sequence $f \sim L$, $f \sim L^{-1/2}$, $f \sim L$ describes a van der Waals loop in the fL diagram. The interpretation in terms of a Maxwell construction suggests a first-order phase transition involving "tadpole"-like chain configurations, i.e., a strongly stretched chain segment emerging from a dense globule. However, the globule is a finite system incapable of undergoing proper phase transitions. One must allow for finite size effects. In this case their origin is the change in the surface area of the globule upon repartitioning of the monomers between the two "phases". The details are beyond the scope of this paper. Nevertheless, it is important to note that this effect is abent if the "head" of the "tadpole" is embedded in a collapsed brush. This suggests that a stretched brush may indeed undergo a proper phase transition.

A similar approach can be applied to collapsed brushes. That is, one assumes that the stretching takes place as a continuous process and search for the signature of an instability. Proceeding, again, within the framework of the Alexander model, we write the free energy per chain as

$$F_{chain}/kT \approx L^2/Na^2 + N(-|v|\phi + w\phi^2) , \qquad (29)$$

where $\phi = Na^3/\sigma L$ is the monomer fraction while v and w are, respectfully, the second and third viral coefficients. Here, and in the following, w is taken to be of order unity. The first term accounts for the Gaussian elasticity of the chains and the second reflects binary as well as ternary interactions between the monomers. Since we aim to characterize the fL diagram it is helpful to express F_{chains} in terms of the tension $f/kT \approx L/Na^2 \equiv a^{-2}E$.

This leads to

$$F_{chain}/kT \approx E^2 - |v| a^3/\sigma E + wa^6/\sigma^2 E^2 . \quad (30)$$

The condition $\partial F_{chain}^2/\partial^2 E = \partial F_{chain}^3/\partial E^3 = 0$ reveals a critical point at

$$|v|_c \approx aw^{3/4}/\sigma^{1/2} ; \quad E_c \approx aw^{1/4}(a^2/\sigma)^{1/2} , \quad (31)$$

suggesting that a first-order phase transition occurs for $v < -|v|_c$.

A more detailed description of the "stretching transition" is desirable. This requires a detailed picture of the collapsed brush. A convenient description is possible in terms of blobs. In a poor solvent the discussion is based on collapsed blobs [15, 20, 21] of size $\xi_c \approx g_c^{1/2} a$. Each blob incorporates g_c monomers exhibiting Gaussian statistics. g_c is determined by the requirement that the attractive interactions within the blobs are comparable to kT, i.e., $(v/kT)(g_c^2/\xi_c^3) \approx g_c^{1/2} |\Delta T|/\Theta \approx 1$, where $v/kT \approx v_0(\Delta T/\Theta)$ is the second viral coefficient, Θ is the theta temperature, and $\Delta T = T - \Theta$. Limiting ourselves to the case of $\sigma > \xi_c^2$, we may picture the layer as a close packed array of ξ_c blobs. The layer thickness, L_c, is given by

$$L_c \approx N(a^2/\sigma)\xi_c . \quad (31)$$

As in the good solvent case, the chains are, in effect, confined to capillaris of cross-section σ. However, in a brush swollen by a good solvent the blob size is $\xi \approx \sigma^{1/2}$ while in the poor solvent case ξ_c is independent of σ. The expression for the thickness of the layer thus corresponds to a string of N/g_c close-packed ξ_c blobs within a cylinder of diameter $\sigma^{1/2}$. The density within the collapsed brush, $g_c a^3/\xi_c^3$, is that of the dense phase resulting from polymer precipitation. The interface of the layer is associated with a surface tension of $\gamma \approx kT/\xi_c^2$.

The stretching of a collapsed brush involves three stages: i) For small extensions, $L = L_c + h \approx L_c$, a linear response is expected. The initially flat brush boundary develops "goose pimples," hemispherical deformations of height h and basal area σ. The linear restoring force, $f \sim h$, is due to the associated increase in surface energy. ii) The following stage is a first-order phase transition. It involves a coexistence of a dense, weakly deformed layer and a dilute layer of strongly stretched chains. The coexistence curve is obtained by equating the chemical potentials of monomers in the two phases for a given L. The free energy per chain due to the N_d monomers embedded in the dense layer is

$$F_d/kT \approx -N_d/g_c + (d^2 + h^2)/\xi_c^2 . \quad (32)$$

The first term assigns an attractive energy of $-kT$ to each ξ_c blob. For $N_d = N/F_d$ is the free energy of the weakly deformed layer described in i). The corresponding chemical potential, $\mu_d = \partial F_d/\partial N_d$, is

$$\mu_d/kT \approx -g_c^{-1} . \quad (33)$$

The remaining $N_s = N - N_d$ monomers form a stretched string of Pincus blobs of size ξ_s. In a poor solvent the blobs are Gaussian, i.e., $\xi_s \approx g_s^{1/2} a$ where g_s is the number of monomers in a blob. The lateral fluctuations of these strings, $\langle\Delta r_\perp^2\rangle \approx Na^2$ are larger than σ and chain-chain interactions occur. The free energy per chain due to the stretched segment is

$$F_s/kT \approx L_s^2/N_s - (N_s/g_s)\xi_s^2/\sigma . \quad (34)$$

The first term, $L_s^2/N_s \approx N_s/g_s$, where $L_s \approx (N_s/g_s)\xi_s$ is the thickness of the dilute layer, accounts for the elastic free energy by assigning kT to each Pincus blob. Chain-chain interactions give rise to the second term. Following Rabin and Alexander each blob is assumed to experience interaction energy of order $kT\Phi_b$ where $\Phi_b \approx \xi_s^2/\sigma$ is the volume fraction of blobs. However, in the poor solvent case the blob-blob interactions are attractive rather than repulsive. The chemical potential, $\mu_s = (\partial F_s/\partial N_s)_L$, is

$$\mu_s/kT \approx -g_s^{-1} - a^2/\sigma . \quad (35)$$

The equality of the chemical potentials, $\mu_s = \mu_d$ 3 μ leads to $g_s^{-1} = g_c^{-1} - a^2/\sigma$ and to $\xi_s = \xi_c(1 - \xi_c^2/\sigma)^{-1/2}$. This specifies F_d and the tension, f, in the stretched chains

$$f/kT \approx (1 - \xi_c^2/\sigma)^{1/2}/\xi_c \quad L_c + h_m < L < L_m . \quad (36)$$

Matching f in regions i) and ii) determines $h_m \approx \xi_c(1 - \xi_c^2/\sigma)^{-1/2}$. The second boundary of the coexistence region, L_m, corresponds to the incorporation of all monomers into the stretched layer, i.e., $L_m \approx (N/g_s)\xi_s \approx N(a/\xi_c)(1 - \xi_c^2/\sigma)^{1/2}a$. The osmotic interactions reinforce the restoring force. As a result, the Pincus blobs are slightly larger than ξ_c while the tension in the chains is somewhat lower than kT/ξ_c, the tension in the absence of chain-chain interactions. The main feature is the $f \sim L^0$ dependence in this regime. iii) A third, final, regime occurs when all monomers are incorporated into the stretched brush. The chains are pictured as stretched strings of Gaussian Pincus blobs. However, as opposed to the good solvent case considered in section III, the blob-blob interactions in a poor solvent are attractive. The tension in the chain is due, primarily, to the Gaussian elasticity with small correction accounting for the contribution of blob-blob interactions. Althogether, we have

$$f/kT \approx L^2/R_0^2 + (R_0^2/\sigma)L^{-1} . \qquad (37)$$

Some support for this picture was found in numerical SCF calculations [22]. However, neither our analysis nor the SCF calculations allow for the possibility of lateral aggregation of the stretched chains. Such an effect may modify the behavior of the system.

Discussion

Macroscopic samples of aligned, "single crystal," mesophases are required for the experimental study of mesogels. The production of such samples was reported by Keller and Odell [5] and by Hadzioannou et al. [23]. It is based on the application of shear fields and thermal annealing. Also, the samples must be robust enough to withstand swelling and deformation with minimal disruption of the A domains. Thus, Keller et al. reported [5] reversible biaxial swelling of cylindrical mesogels up to ~20% gain in total volume while further swelling was associated with break up of the cylindrical domains. It may be possible to enhance the robustness of the samples by using lamellar rather than cylindrical mesophases, by increasing the size of the A domains and by inducing selective crosslinking in the A domains.

The behavior of mesogels is of interest in its own right. However, experiments involving mesogels afford other opportunities: As noted earlier, only the extension behavior is sensitive to the fraction of bridging chains. Thus, a comparison between the stretching and compression behavior is expected to provide information on the actual q value in a mesophase. This probe does not distinguish between direct bridges and entangled loops. Also, mesogels may allow the study of strongly stretched flexible chains in good as well as in poor solvents. If the deformation is attained by shearing the sample it is possible to avoid complications due to deswelling.

Acknowledgemets

The auhtors would like to acknowledge with thanks the hospitality and support which made this work possible. A. H. the hospitality of Professor E. W. Fischer and the Max-Planck Institut für Polymerforschung. E. B. Z. the hospitality of Professor K. Binder and the financial support provided by the A. v. Humbold Foundation. It is also a pleasure to acknowledge instructive correspondence with L. I. Klushin.

References

1. Halperin A, Zhulina EB (1991) Europhys Lett 15:417
2. Halperin A, Zhulina EB (1991) Macromolecules 24:5393
3. Halperin A, Zhulina EB (1991) Europhys Lett 16:337
4. Halperin A, Tirrell M, Lodge TP (1992) Adv Polym Sci 100:31
5. Keller A, Odell JA (1985) In: Folkes MJ (Ed) Processing, Structure and Properties of Block Copolymers. Elsevier, New York
6. Mark JE, Erman B (1988) Rubberlike Elasticity A Molecular Primer, John Wiley, New York
7. Legge NR, Holden G, Schroeder HE (Eds) (1987) Thermoplastic Elastomers, Hanser Publishers, Munich
8. Benoit H, Hadziioannou G (1988) Macromolecules 21:1449
9. Alexander S (1977) J Phys (France) 38:977
10. Birshtein TM, Zhulina EB (1985) Polym Sci USSR 27:1613
11. Semenov AN (1985) Sov Phys JETP 61:731
12. Johner A, Joanny JF (1991) Europhys Lett 15:265
13. Zhulina EB, Pakula T, Macromolecules (in press)
14. Ligoure C, Leibler L, Rubinstein M, Macromolecules (in press)
15. de Gennes PG (1979) Scaling concepts in polymer physics, Cornell University Press, Ithaca
16. Candau S, Bastide J, Delsanti M (1982) Adv Polym Sci 44:30
17. Birshtein TM, Zhulina EB (1990) Polymer 31:1312
18. Rabin Y, Alexander S (1990) Europhys Lett 13:49
19. Pincus P (1986) Macromolecules 9:386
20. Lifshits IM, Grosberg AY, Khokhlov AR (1978) Rev Mod Phys 50:683
21. Williams C, Brochard F, Frisch HL (1981) Ann Rev Phys Chem 32:433
22. Klushin LI (private communication)
23. Hadziioannou G, Mathis A, Skoulios A (1979) Colloid Polym Sci 257:136

Authors' address:

A. Halperin
Materials Department, UCSB
Santa Barbara, CA 93106, USA

Microemulsion mediated polymer networks*)

H.-F. Eicke, U. Hofmeier, C. Quellet, and U. Zölzer

Institut für Physikalische Chemie, Universität Basel, Switzerland

Abstract: An interesting case of soft condensed matter are microemulsion-mediated polymer networks. Depending on the strength of the polymer-water or polymer-oil domain interactions, a variety of networks is formed extending from stable gels to fluids, displaying purely viscous flow. Details of the formation and properties of the networks formed will be discussed. Of particular interest is the regime of viscoelastic networks. It is tempting to compare experimental data from rheometric oscillatory measurements of such systems with predictions of the mode coupling theory. Preliminary results are presented.

Key words: Networks; microemulsions; gels; soft condensed matter; dynamic light scattering; rheology

Introduction

The subject of this contribution may be approached by considering polymer or copolymer "solubilities" in complex fluids, i.e., here the so-called microemulsions. Microemulsions are isotropic, thermodynamically stable mixtures of oil, water, and surfactant (= amphiphilic molecules). In contrast to a mixture of three miscible components which are homogeneous down to a molecular scale, the particular feature of microemulsions lies in their molecular (colloidal) heterogeneity, i.e., they are composed of water and oil domains separated by a monomolecular surfactant layer. Figure 1 shows such a three-component (Gibbs) phase triangle and where the water-oil microemulsion is to be located (open circles in dotted square). It is essential to point out that the polymers do not interact directly, but rather via the colloidal domains of water or oil (frequently called nanodroplets because of the nanometer size range of the correlation length [1—4]) which offers, simultaneously, good solvents to the functional groups and the hydrocarbon moieties of the polymers. Of course, the polymeric substrates may be one-sidedly restricted only to the aqueous or hydrocarbon domains (as, e.g., with gelatin) or to both as with block copolymers of the ABA type. In the latter case this "selective solubility" leads by itself to a network formation where the functionality of the entanglement points, i.e., the nanodroplets with their dissolved polymer blocks determine the dimensionality of the network. However, in the former an intrinsic property of this biopolymer, the coil-helix transition, must be operative between the aqueous domains, in order to produce network (gel) formation.

The new materials created by the dissolution of copolymers in these complex fluids have recently met considerable interest; they are described as dynamically disordered systems with no tendency to crystallize. They are expected to show a thermorheologic simple behavior which greatly facilitates — at least in principle — the analysis of the rheometric results. Finally, they might be promising examples of the recently advocated mode coupling theory [5].

Experimental

The preparation of the $H_2O/AOT/i-C_8H_{18}$ (AOT = sodium bis[2-ethylhexyl]sulfosuccinate) microemulsions and the synthesis of the copolymers has been

---

*) Dedicated to Professor Bernhard Wunderlich on the occasion of his 60th birthday.

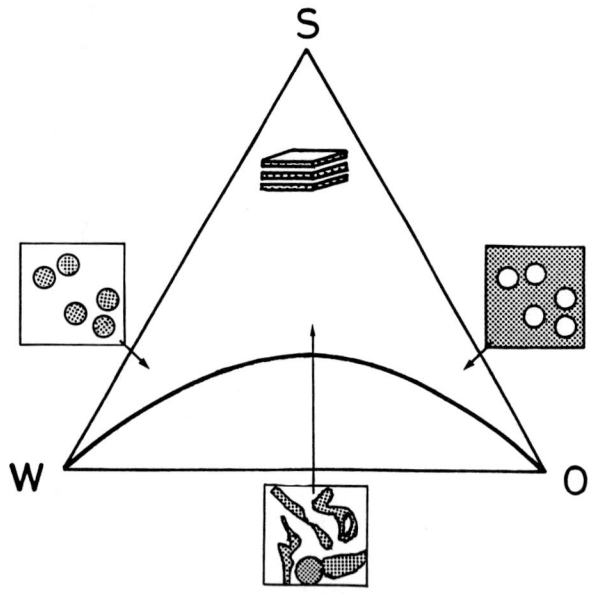

Fig. 1. Gibbs phase triangle: w/o microemulsion (open circles in dotted square); Schulman line: constant correlation length, $\xi = \Phi_w \Phi_0 / \Phi_s$

described extensively elsewhere [6, 7]. Also, suitable experimental techniques for the investigations of these systems were described in detail; see, e.g., [7—10]. Oscillatory shear investigations were carried out with a CARRIMED CSR H 100 with a homemade thermostatic control unit, sample holder, and adjustment facilities. The measuring system was a parallel plate geometry (acrylic material) of 6 cm diameter and gap width of 400—900 μm. In this range the results were independent of the gap width. Also, no time dependence of the moduli was observed. The frequency window of oscillatory measurements was sample-dependent and mostly covered the range between $4 \, 10^{-2}$ and 40 Hz. The software of the instrument allows one to perform the measurements under constant applied strain, in our case 0.4%, and ensures to remain within the system's linear viscoelastic region. A more detailed description of the apparatus will be published elsewhere.

A commercial quasi-elstic light-scattering goniometer (SP-86 ALV Langen, FRG) was used, equipped with a 3000 ALV digital correlator employing 23 simultaneous sampling times and thus allowing monitoring of widely spaced decays in the same experiment; additionally, 32 real-time, fast channels with a delay time of 0.8 μs were used. The measurements were made in the homodyne mode using an argon ion laser operated at 496.5 nm. The temperature of the sample cell was set to 298.2 ± 0.05 K.

Results and discussion

There exists an obvious relationship between the structure of the polymer and the properties of the network formed (Fig. 2): the latter will depend upon the size of the hydrophilic and hydrophobic moieties of the polymer. Thus, a mostly hydrophilic biopolymer like a polypeptide (gelatin) will be predominantly confined within the aqueous nanodroplets with some apolar residues adsorbed onto the water/surfactant interface. Observation of a correlated percolation (Fig. 3), i.e. the shift of the percolation of the pure microemulsion to lower temperatures, indicates that the "nanogels" (nanodroplets containing gelatin) exhibit incremental attractive interactions resulting in nanogel clusters. The final gel state in the presence of solubilized polypeptide is particularly stable due to the coil-helix transition of the gelatin and of the helical linkages between the nanogels. This coil-helix transition becomes apparent in the present system essentially in two phenomena with increasing gelatin concentration (see Fig. 3): the first process describes a percolation transition of the nanogels which — at about the critical gelatin concentration — is accompanied by a strong increase of the electric conductivity by several orders of magnitude (Fig. 4), indicating the formation of an infinite fractal cluster (see, e.g., [1—4]); the second corresponds to the formation of helical linkages between the nanogels. They are responsible for the considerable rigidity of the gelatin gels. Its stability is destroyed if the temperature is increased above the upper helix stability temperature ($T_H \sim 300$ K) where the helices melt. This may be considered a strong argument in favor of the above claimed role

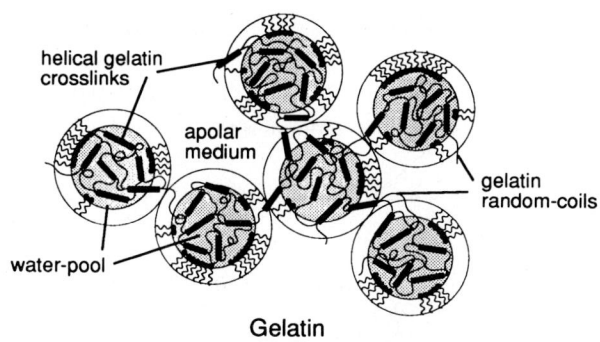
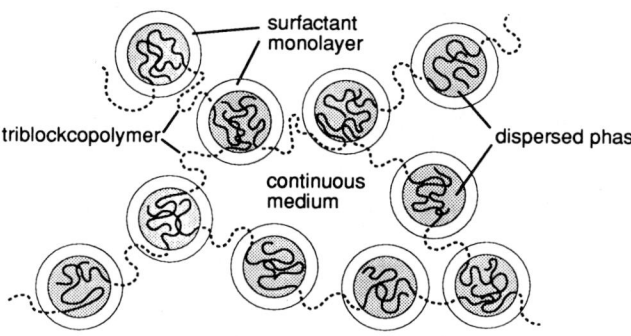
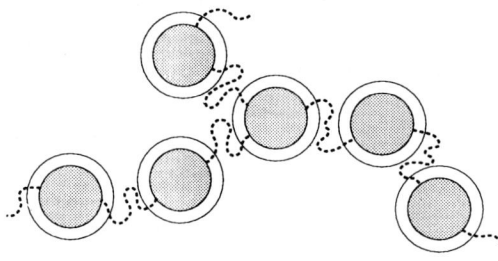

Fig. 2. Complex fluid mediated polymer networks (schematic all in $H_2O/AOT/i-C_8H_{18}$; w/o microemulsion). Top: Gelatin Middle: POE-b-PI-b-POE; Bottom: $PS(SO_3Li)_2$

of the tropocollagen-like helices in interconnecting the nanodroplets. The specific electric conductivity plots in the percolation regime for varying gelatin concentrations are mutually parallel. This indicates, probably, that the transitions are topologically unaffected by the amount of added gelatin. The pronounced temperature dependence of the electric conductivity in the gel state mirrors the potential role of the polypeptide helices as spacers between the droplets in the percolated state (see Fig. 4).

The steady-state shear viscosity does not diverge, neither at the percolation threshold of the pure microemulsion, nor at low solubilized gelatin concentrations, but does so only above a certain gelatin concentration where the "equivalent" temperature of the percolation threshold (i.e., corresponding to this particular gelatin concentration) drops below T_H (Fig. 5). Only then sufficient inter-droplet helices can form and percolate through the sol. The steady-state viscosity traces the degree of inter-droplet helices within typical nanogel clusters. One has to be aware, however, that only the pre-percolative state is recorded in this way, since at higher temperatures (at and above the percolation transition of interdroplet linkages) the network is affected by the experimental procedure (decomposition of linkages by shearing stress). In the reliable experimental region of this viscosity study the slopes of the viscosity plots are independent of gelatin concentration; they may thus be collapsed into a single master plot from which a tentative critical exponent of $k = 0.6 \pm 0.1$ can be determined. This result is consistent with data obtained from so-called strong gels [11]. Such a critical exponent of the scalar elasticity percolation of permanent bonds through a set of monomers was proposed by de Gennes [12]. The latter suggested an analogy between viscosity and the electric conductivity of a random (static) network of resistors and superconductors. It is of particular interest that the critical exponent of the electric conductivity, which was also measured in this study ($s = 0.9 \pm 0.3$), is in agreement with this proposition.

Information on the fractal geometry of (finite) nanogel clusters below the percolation threshold can be inferred from electric birefringence measurements in the sol-gel transition state. If the typical cluster size s_{typ} (which makes the largest contribution to the electric birefringence and which is proportional to the Kerr constant) is plotted against concentration of gelatin, a strong increase of s_{typ} is detected. This indicates an increasing clustering of the nanogels. Utilizing a relationship [13] between the steady-state shear viscosity and the apparent volume fraction Φ_{app} of the nanogels in the nanogel clusters, Φ_{app} is shown to increase faster with increasing gelatin concentration than with the volume fraction of the gelatin containing nanodroplets. Such a behavior is to be expected for fractal clusters [14]. Φ_{app} and s_{typ} are correlated by a simple scaling law, i.e.,

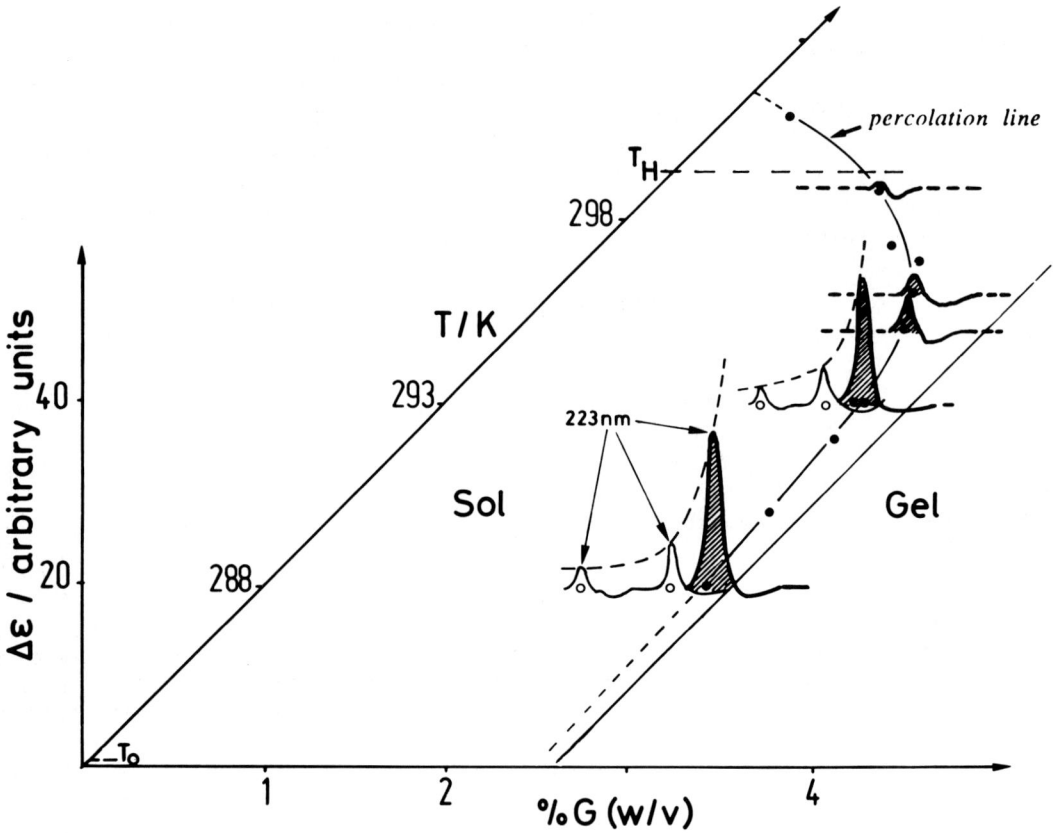

Fig. 3. Percolation thresholds (determined by electric conductivity measurements) and dichroitic absorption against gelatin concentration, $w_0 = [H_2O]/[AOT] = 60$, $c_{AOT} = 0.1$ mol/dm^3, from [7]; for further information see [4]

$$\Phi_{app} \sim s_{typ}^{(d-D)/D} \Phi_0 , \qquad (1)$$

where Φ_0 is the volume fraction of nanogels. Figure 7 demonstrates the linear relationship of $\log(\Phi_{app}/\Phi_0)$ against $\log s_{typ}$. From the slope a fractal dimension of $D = 2.0 \pm 0.1$ is derived in good agreement with the value predicted from (finite) clusters below percolation.

Light-scattering data [8] in the gel phase (i.e., above percolation) reveal that the fractal dimension coincides with the value found for finite nanogel clusters below percolation. This coincidence prompts us to assume that the topology of the clusters below and above percolation is the same. The experimental information so far available on microemulsion-mediated gelatin gels supports our view of a swollen, tenuous network expanded in an oil continuous medium.

In relation to their structure, block-copolymers of the ABA type with long hydrophilic A- and hydrophobic B-blocks are another example of well-suited polymers for interlinking nanodroplets. (In order to avoid misunderstandings, the formation of such networks is traditionally to be described by a reaction between nanodroplets and copolymers via, e.g., the mass action law, i.e., only a part of the copolymer is involved in linking different droplets.) The A-block is a polyoxyethylene and the B-block is a polyisoprene which meet excellent solvent properties in their respective domains. This is seen in the case of polyoxyethylene in the water domains from the considerable heat of mixing [15]. The "functionality" of the entanglement points (i.e., the water domains) is determined by the number of A-blocks from different block copolymers per nanodroplet. In contrast to the microemulsion — gelatin network, this system does not exhibit form-stable networks at comparable polymer concentration, but starts to flow if exposed to gravitational force. Only at very high polyoxyethylene concentra-

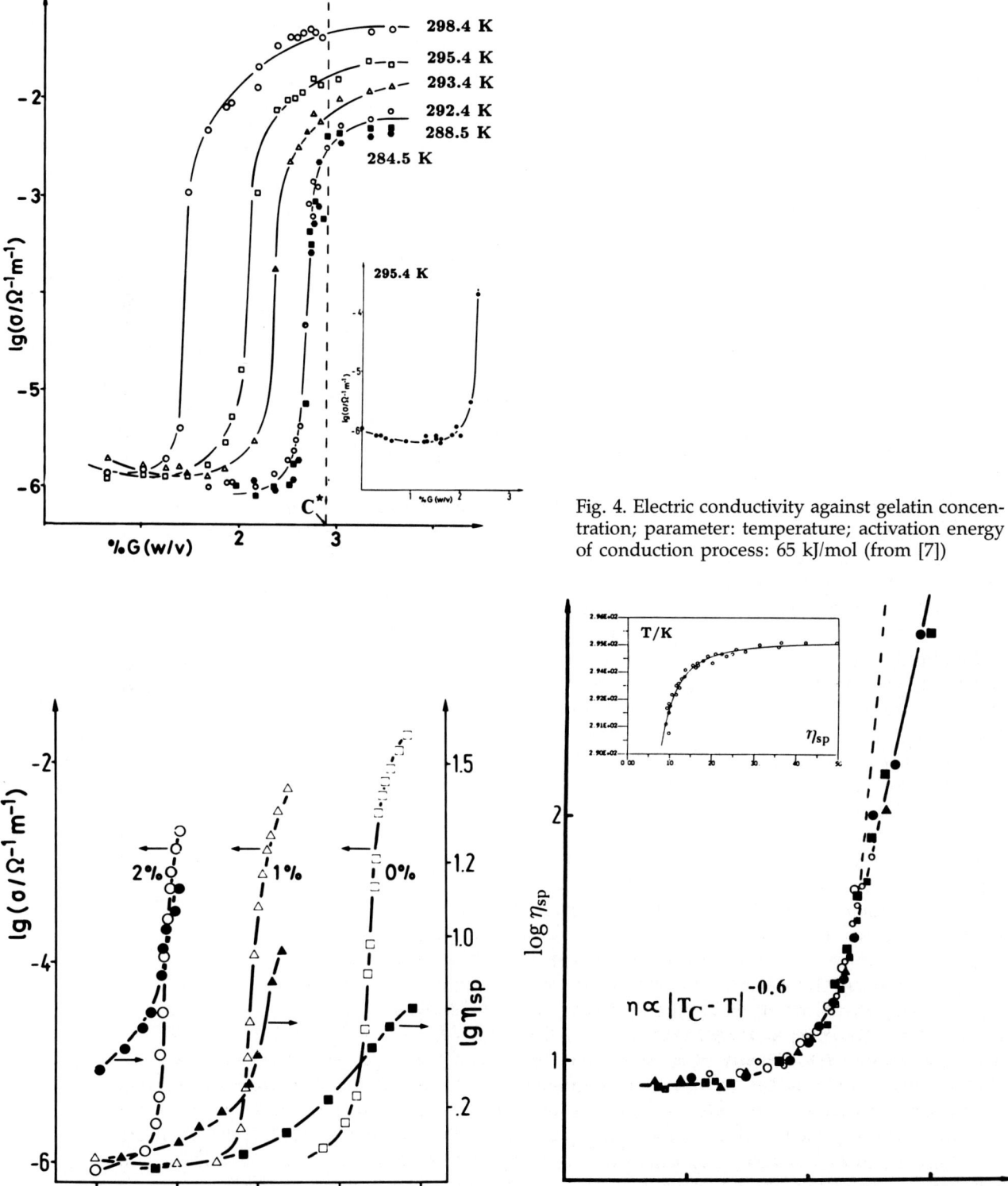

Fig. 4. Electric conductivity against gelatin concentration; parameter: temperature; activation energy of conduction process: 65 kJ/mol (from [7])

Fig. 5. Electric conductivity and specific viscosity against temperature (left); parameter: gelatin concentration, microemulsion (see Experimental and Fig. 3); master-plot of η_{sp} (right) (from [7]); broken line refers to exponent 0.6

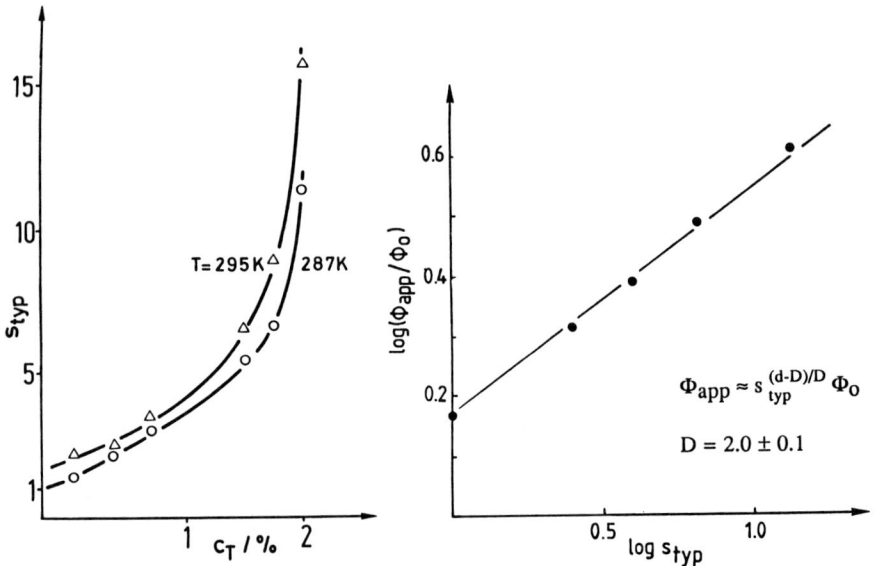

Fig. 6. Typical cluster size s_{typ} against gelatin concentration and apparent nanogel cluster volume fraction against s_{typ}; same microemulsion

tions does the system adopt a (rigid) gel state. The reason for this different behavior is due to the interactions of the hydrophilic polymer moieties within the droplets. The gelatin helices are mutually bound via hydrogen bonds. This results in a drastic increase of the lifetime of the entanglement points and, hence, leads to a stable gel. The hydrated polyoxyethylene chains are, however, sterically entangled with energetic interactions in the order of $k_B T$. This entanglement resists sudden (~1 ms) changes of configuration corresponding to activation energies of about 250 kJ/mol in the present example, while weak (gravitational) shearing energy on the order of $k_B T$ disentangles such knots, but during a much longer time, i.e., in the orders of seconds. (The reptation model predicts a strong molecular weight dependence of the disentanglement time [16].)

Oscillatory shear experiments show interesting results if applied to these systems (Fig. 7). The master plot constructed by utilizing the time-temperature superposition principle displays loss and storage moduli with at least one relaxation phenomenon. Both wings of the high-frequency relaxation process show a pronounced linearity in the $\log G'' - \log \omega$ plot. The relaxation time corresponding to the peak frequency is about 10 ms for the particular polyoxyethylene concentration. It is concentration dependent, i.e., it shifts with decreasing POE concentration to higher frequencies, while the loss modulus decreases simultaneously. Also, this relaxation phenomenon possesses a pronounced temperature dependence and an Arrhenius plot (Fig. 8), showing that activated processes are involved. It appears possible to construct a master spectrum of $G''(\omega,T)/G''(\omega_{max}(T))$ as a function of $\omega(T)/\omega_{max}(T)$ of this relaxation process (see Fig. 9). This may be interpreted as a further hint for identifying this phenomenon with the β-relaxation in the frame of the mode coupling theory.

Dynamic light-scattering measurements on the same systems display three characteristic decay times (see Fig. 10). The data points were fitted with a mono-exponential and two stretched exponential

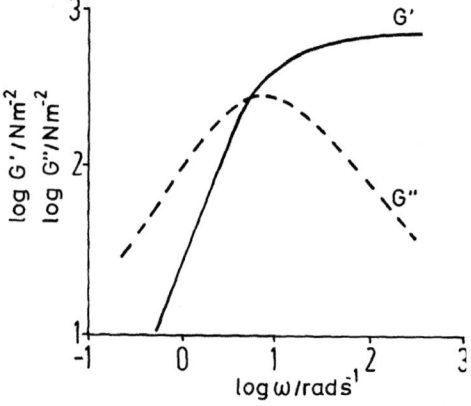

Fig. 7. Loss ($G'(\omega)$) and storage ($G''(\omega)$) moduli against frequency. Master plot. Measurements at constant strain of 0.4%; $T = 293$ K; w/o microemulsion ($w_0 = 60$), POE-b-PI-b-POE, $R = 8$

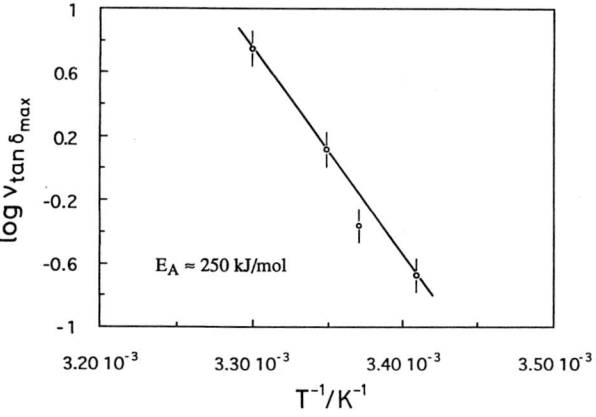

Fig. 8. Arrhenius plot of $\tan(G''/G')_{max}$ against reciprocal temperature; same system as in Fig. 7. Activation energy of relaxation process: 250 kJ/mol

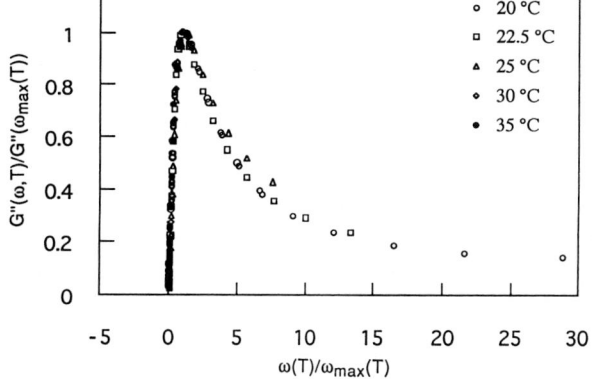

Fig. 9. Master spectrum of normalized loss moduli, $G''(\omega, T)/G''(\omega_{max}(T))$, against normalized frequency $\omega(T)/\omega_{max}(T)$; w/o microemulsion $w_0 (= [H_2O]/[AOT]) = 60$; $c_{AOT} = 0.1$ mol dm^{-3}, POE-b-PI-b-POE, R = 8

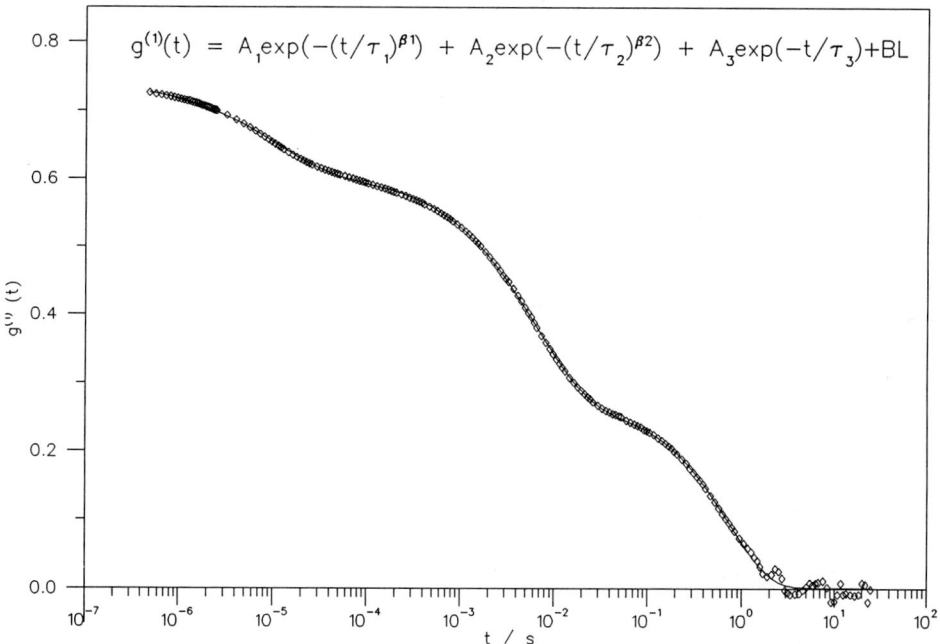

Fig. 10. Normalized field autocorrelation function against delay time. w/o microemulsion w_0 $(= [H_2O]/[AOT]) = 62$; $C_{AOT} = 0.1$ mol dm^{-3}. POE-b-PI-b-POE; R = 16.5; $\theta = 150°$; $T = 310.7$ K

functions, i.e., KWW (Kohlrausch-Williams-Watts) fits, which obtain their theoretical justification by the mode coupling theory [5]. A plot of the decay rates against q^2 (Fig. 11) yields three straight lines which we identify with diffusion processes. The hydrodynamic correlation length corresponding to the fastest decay is found to be 9 nm, which might be related to the mesh-size of the network. It ap-

pears worthwhile to mention in this context that the stretched decay, as seen in the quasi-elastic light-scattering experiment, is neither observed in strong gels nor in sols, but typically only in the glassy state of liquids [17].

All measurements have been reproduced, but are still of a preliminary character, particularly with respect to their predictive power. In view of some

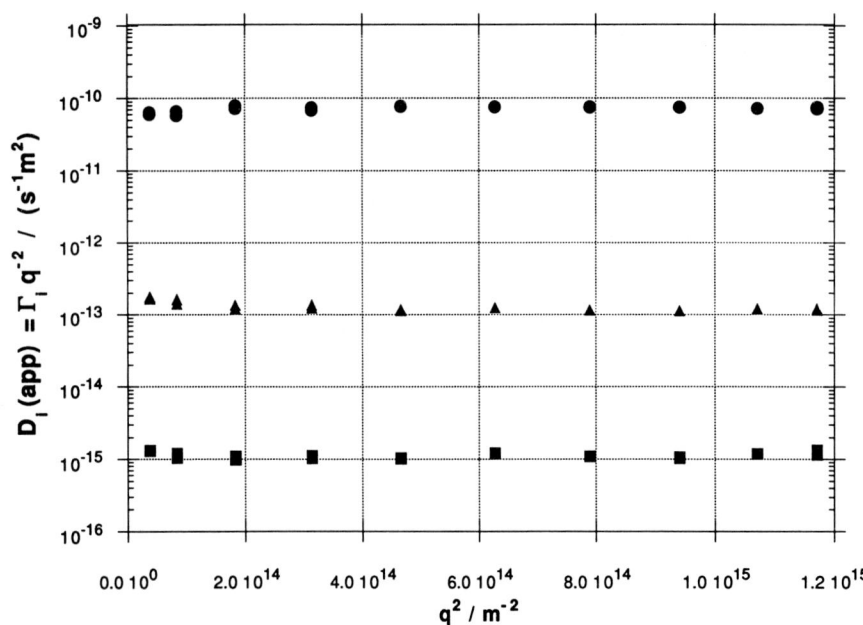

Fig. 11. Apparent diffusion coefficients of the three processes derived from the KWW fits plotted against q^2; ●: fast, ▲: intermediate, ■: slow processes

striking similarities to network-forming polymeric systems, an important aim of these investigations is to elucidate the relationship with the so-called β-relaxation process found in the above systems and its possible interpretation in the frame of the mode coupling theory.

The last example of a polymer which forms network structures in microemulsions is a α, ω telechelic ionomer [18]. This case has been selected because particular weak polymer-nanodroplet interactions are anticipated [19]: the terminal ionic groups are almost identical to those of the surfactant (AOT), thus, one expects a lifetime of the ionomer in the adsorbed state at the nanodroplet's interface similar to that of a surfactant molecule, i.e., of about 1—10 μs. From this lifetime one would conclude that no storage modulus can be detected, if such a sample is compared with a triblock copolymer viscoelastic fluid of the same nanodroplet concentration, nanodroplet size, and R-value. This is, indeed, found. Detailed investigations are in progress and will be published elsewhere.

Acknowledgment

The authors wish to thank Professor R. Schilling, Institut für Physik, Universität Mainz, FRG, for stimulating discussions. They are greatly indebted to the Swiss National Science Foundation and to the Ciba Stiftung for financial support of this work.

References

1. Jouffroy J, Levinson P, de Gennes PG (1982) J Phys (Les Ulis Fr.) 43:1241
2. Stauffer D (1979) Phys Rept 54:1
3. Borkovec M, Eicke HF, Hammerich H, DasGupta B (1988) J Phys Chem 92:206
4. Eicke H-F, Hilfiker R, Thomas H (1985) Chem Phys Lett 120:272, ibid (1986) 125:295
5. Götze W, Sjögren L (1991) Rep Prog Phys (in press)
6. Quellet C, Eicke H-F, Xu G, Hauger Y (1990) Macromolecules 23:3347
7. Quellet C, Eicke H-F, Sager W (1991) J Phys Chem 95:5642; Quellet C, Eicke H-F, Gehrke R, Sager W (1989) Europhys Lett 9:293
8. Eicke HF, Quellet C, Xu G (1989) Colloids Surf 36:97
9. Struis RPWJ, Eicke H-F (1991) J Phys Chem 95:5989
10. Hilfiker R, Eicke H-F, Steeb C, Hofmeier U (1991) J Phys Chem 95:1478
11. Allain C, Salomé L (1987) Macromolecules 20:2957
12. de Gennes PG (1976) J Phys Lett 37L:61
13. Cichocki BC, Felderhof BV (1988) J Chem Phys 89:1099
14. Hilfiker R, Eicke H-F (1987) J Chem Soc Frad Trans I 83:1621

15. Eicke H-F, Hilfiker R, Xu G (1989) Helv Chim Acta 73:213
16. de Gennes PG (1983) Physics Today 33; idem (1979) Scaling Concepts in Polymer Physics, Cornell Univ Press
17. Borsali R, Durand D, Fischer EW, Giebel L, Busnel JP (1991) Polymer Networks Blends 1:11
18. Möller M, Mühleisen E, Omeis J (1990) In: Burchard W, Ross-Murphy SB (eds) Physical Networks, Polymers and Gels, Elsevier, London, New York, p 45—64
19. Eicke H-F, Gauthier M, Hilfiker R, Struis RPWJ, Xu G (1992) J Phys Chem 96:5175

Author's address:

Prof. H.-F. Eicke
Institut für Physikal. Chemie
Klingelbergstr. 80
CH-4056 Basel, Switzerland

Theory of the mechanical and swelling properties of elastomers with chemical and physical networks

M. E. Solovjev*), A. B. Raukhvarger*), T. K. Ivashkovskaya, and V. I. Irzhak**)

*) Polytechnical Institute, Yaroslavl, Russia
**) Institute of Chemical Physics of Academy of Sciencies, Russia

Abstract: The thoeretical model of a polymer network with chemical and physical junctions is proposed on the basis of the statistical thermodynamic methods. Expressions for the average physical junction number and chemical potential of solvent in the swelling gel are obtained. Some interesting corollaries of these expressions are the non-linear character of the stress-strain dependence and the phase separation process in swelling gel.

Key words: Physical networks; mechanical properties; swelling

The concept of physical networks has had a broad acceptance in recent times. The properties of physical network are shown by many kinds of polymers: co-polymers with strong interaction of functional groups such as hydrogen bonds, many kinds of polyurethanes, block- and microcrystalline polymers.

The purpose of our work was to create a theoretical approach that allowed to describe some characteristic features of polymers with physical networks and also to outline the field of molecular parameters and testing conditions in which physical network properties are observed.

We consider the model of elastomer with molecular chains able to form strong local intermolecular bonds. Because the energy of these junctions is higher than the energy of ordinary interchain interactions, one can distinguish two levels of fluctuations in this systems: the first one is the fluctuations associated with the establishment of conformation equilibrium in the interjunctions chains, and the second one is the fluctuations associated with the establishment of the equilibrium physical network junctions number.

This assumption allows us to consider the statistical distribution of states of this system [1] on the basis of the maximum entropy condition for the model with two levels of fluctuations. This distribution was used to obtain the expressions for average physical junctions number (1) and chemical potential of solvent in the swelling gel (2):

$$\bar{N} = -\frac{N_0 + 1}{V_2^{(N_0+1)} e^{(N_0+1)x} - 1} + \frac{1}{V_2 e^x - 1}, \quad (1)$$

where:

N_0 is the highest possible number of physical junctions determined by chemical structure of the network;

V_2 is the volume fraction of polymer in the swelled gel;

$$x = \frac{\lambda^2}{V_2^{2/3}} + \frac{2}{\lambda V_2^{2/3}} - 3 - \frac{E_0 - TS_0}{kT};$$

E_0 is the energy of physical junctions;
S_0 is the entropy of physical junctions;
T is the temperature;
λ is the deformation of the network

$$\mu = \ln(1 - V_2) + V_2 + \chi V_2^2$$
$$+ \frac{\bar{N} V_1}{N_A V_0} \left[\frac{2}{3} \left(\lambda^2 + \frac{2}{\lambda} \right) V_2^{1/3} - V_2 \right], \quad (2)$$

where:

V_0 is the volume of polymer before swelling;
V_1 is the molar volume of solvent;
χ is the Flory-Huggins parameter.

Let us consider some corollaries of this distribution. In Fig. 1, we depict the average physical junctions number versus temperature for the polymer with two kinds of physical junctions. As we can see, there are two transition points and each of these is connected with the destruction of the conformable physical junctions type. This result is in agreement with the experimental observations [2, 3]. If we determine the bonds number in the stretched state, the corresponding transition points will shift to the low-temperature region with increasing of stress or strain. In the swelling state the transition points also shift to the low-temperature region. There is good empirical evidence for the validity of this result.

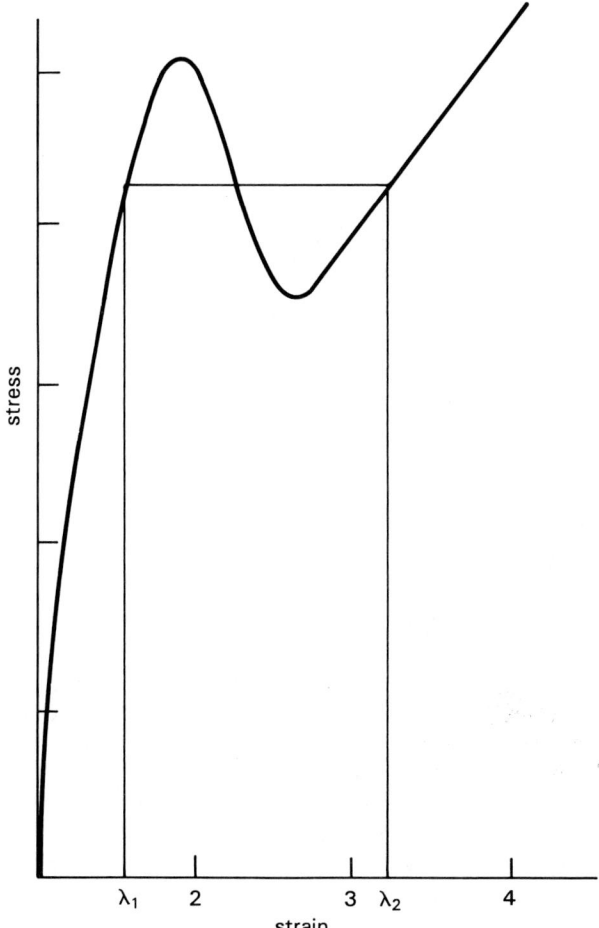

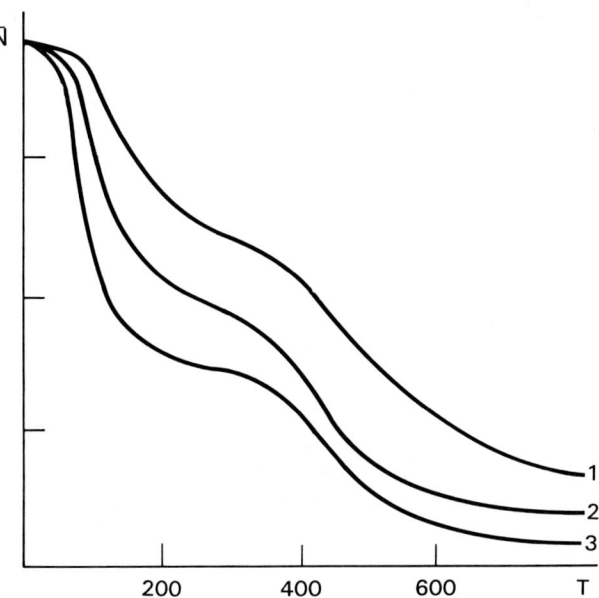

Fig. 1. The average physical junctions number versus temperature for the polymer with two kinds of physical junctions. 1 — $\lambda = 1.2$, $V_p = 1$; 2 — $\lambda = 2$, $V_p = 1$; 3 — $\lambda = 1.2$, $V_p = 0.3$, where V_p is the volume fraction of polymer in swollen gel

Fig. 2. Theoretical tensile-stress dependence on strain of polymer with the physical and chemical junctions

We can obtain some interesting consequences if we substitute the average junctions number expression into the stress-strain equation. The tensile-stress dependence on strain of polymer with physical and chemical junctions is depicted in Fig. 2. When the critical value of deformation is reached the physical junctions begin to fail intensively. In this region Van-der-Waals loop takes place. This corresponds to the first-order phase transition. The phases are regions with different bond concentrations and different deformation ratios λ_1 and λ_2 conformably. This result coincides with the experimental observations. Indeed, many kinds of polymers with a physical network possess the obvious non-linear mechanical properties, as shown, for example, in Fig. 3, where stress-strain dependence of nonilacrylate-co-acrylamide copolymer is depicted. As we can see, the strain falls dramatically when the deformation ratio reaches a critical value. In this region the sample is mudding and the Poisson parameter is decreasing.

The next important property of the physical network is the ability to form the physical gels. In Figs. 4 and 5 the dependence of chemical potential of solvent on volume fraction of polymer and phase diagram are shown. There are many ways to form gels and, in the case of a physical network, one of

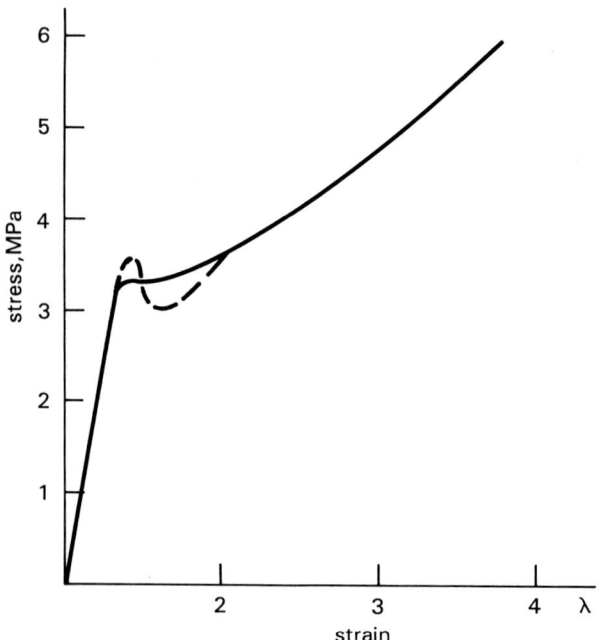

Fig. 3. Tensile-stress versus strain for nonilacrylate-coacrylamide copolymer (15% acrylamide), $T = 293$ K. The dashed curve is the theoretical dependence for conformable molecular parameters

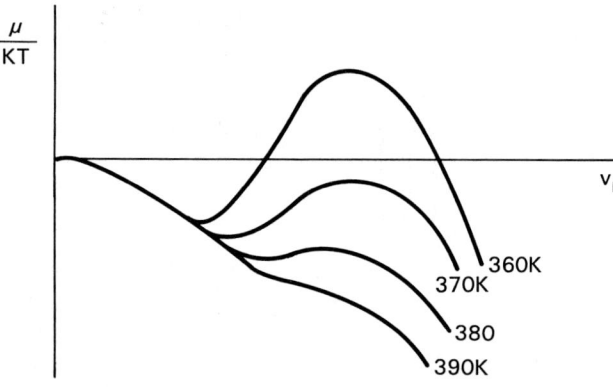

Fig. 4. Chemical potential of solvent versus volume fraction of polymer with the physical junctions

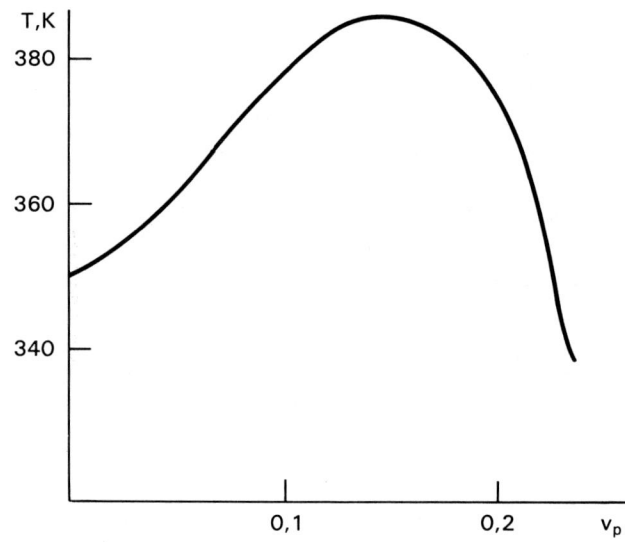

Fig. 5. Phase diagram of polymer with the physical junctions. Phases are regions with different physical crosslinks density

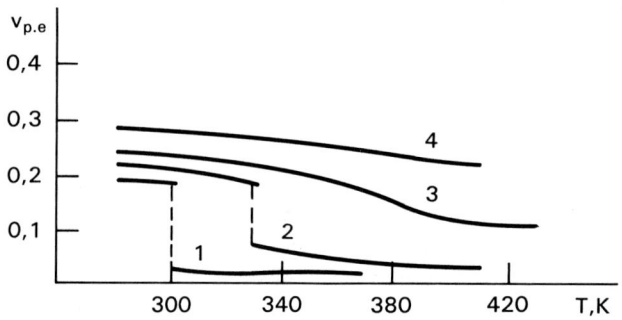

Fig. 6. Equilibrium volume fraction of polymer in swollen gel with the physical junctions versus temperature. Chemical crosslink density: $1 - 0.1 \cdot 10^{18}$ cm^{-3}, $2 - 10^{19}$ cm^{-3}, $3 - 2 \cdot 10^{19}$ cm^{-3}, $4 - 5 \cdot 10^{19}$ cm^{-3}

them is the phase separation. An important point to be made here is that the collapse of the gel depends on the chemical crosslink density of the rubber, as shown in Fig. 6. At high chemical crosslink density the phase separation process is not realized.

Thus, as we can see, the theoretical proposal of two levels of fluctuations in elastomers with the physical network may lead to some interesting results.

References

1. Raukhvarger AB, Solovjev ME, Irzhak VI (1989) Chemical Physics Letters 155 N 4,5:455—458
2. Hilger C, Stadler R (1990) Macromol Chem 191:1347
3. Birshtein VA et al. (1978) Vysokomolek Soed 20A, N 8:1885—1892

Authors' address:

Prof. M. Solovjev
Polytechnical Institute
Moscow avenue, 88
150053 Yaroslavl, Russia

The effect of free branches on the collapse of polyelectrolyte networks

O. V. Borisov, T. M. Birshtein, and E. B. Zhulina

Institute of Macromolecular Compounds of the Academy of Sciences of Russia, St. Petersburg, Russia

Abstract: The network formed by arm-linked charged polymer stars was considered as a model of a polyelectrolyte network with a large number of free branches. It is shown that the relation between the number of free branches and the number of charges on each branch determines the character of swelling/collapse transition caused by the decrease in the solvent strength: if the number of branches is small and the charge density is high, the collapse occurs as the first-order phase transition below the Θ-point, just as in ordinary polyelectrolyte networks. In the opposite case the collapse of the network occurs smoothly due to strong interchain interactions, the width of the transition range increases with degree of branching. In contrast to previous results, it is shown that under the conditions of good solvent the dependence of the network swelling coefficient on the solvent strength is always of the power character. This effect is independent of the number of free ends, but has different physical origin (intra- or interbranch interactions) in the cases of weak and strong branching, respectively.

Key words: Polyelectrolyte networks; structural defects; phase transitions; grafted polymer layers

Introduction

The theory of swelling and collapse of weakly charged polyelectrolyte networks caused by the variation in the solvent strength has been well developed in the works by Tanaka et al. [1], Khokhlov et al. [2].

In these theories the networks are suggested to be formed by linear chain molecules grafted to each other (more or less regularly) by chemical bonds. The problem of the uniform swelling (or collapse) of the polyelectrolyte network was reduced to that of the coil-globule transition in a single polyelectrolyte molecule under the condition that all its counterions are retained in the volume occupied by the chain, i.e., the condition of local electroneutrality.

In recent years a sufficient progress has been achieved in the synthesis of regularly branched, especially star-branched polymers and polyelectrolytes. The chemical linking of such star-branched polymers may result in the formation of a network with a large number of free branches. These networks should have special thermodynamic and elastic properties.

The aim of the present paper is to analyze swelling and collapse of the network formed by linking together weakly charged polyelectrolyte stars. Special attention will be paid to the influence of the degree of branching (the number of arms in the star) on this transition.

Note that this problem of uniform swelling and collapse of the network is equivalent to that of the coil-globule transition in a single star-branched polyelectrolyte containing all its counterions inside the star. The fulfillment of this local electroneutrality condition simplifies the problem sufficiently compared to the situation for an isolated star-branched polyelectrolyte in a salt-free solution where counterions are distributed, not only inside the molecule, but also in the whole volume of the solution.

Model

Let us consider the network formed by weakly charged polyelectrolyte stars (Fig. 1). Let f be the total number of arms in the star and q be the number of arms, by which each star is connected

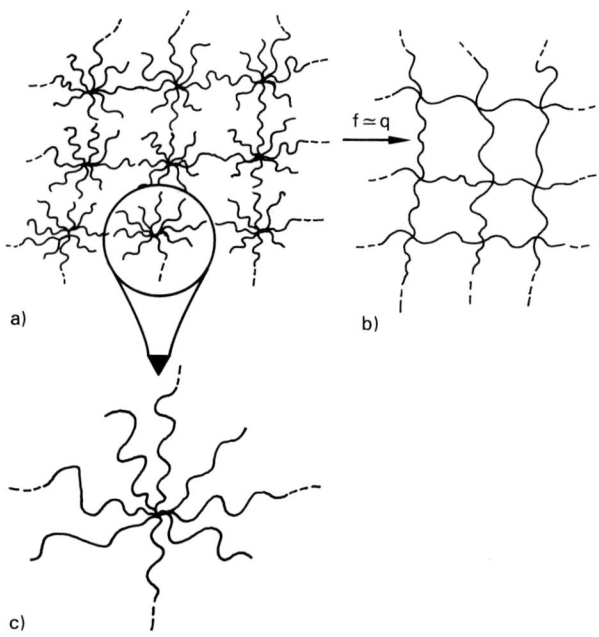

Fig. 1. The network formed by linked polyelectrolyte stars at strong branching a) and weak branching b); the star-branched fragment of the network c)

with neighboring stars. The limiting case $f \gg q$ corresponds to strong branching. In the opposite case, $f \cong q$, we return to the usual network formed by linear chains. We shall restrict ourselves by the simplest topological structure neglecting loops and other structural defects. Every star arm is a flexible chain molecule (its Kuhn segment length A is equal to the chain thickness, a) containing N units (a chain part of length a is chosen as a unit). Let every m-th unit of each branch carry an elementary charging group, so that the total number of charges per branch is $S = N/m \gg 1$. We will restrict our consideration by weakly charged polyelectrolytes, for which

$$m^{1/2} \gg u = e^2/(\varepsilon a T),$$

where ε is the dielectric constant of the solvent.

The condition of electroneutrality leads to the presence of fS mobile counterions per every star-branched fragment in the network. We shall first consider the situation when no salt is added into the solution and the osmotic effects in the network are determined by these counterions only.

The non-electrostatic volume interactions between the monomer units of the branches will be described by the second, v, and the third, w, virial coefficients. The former depends on temperature (solvent strength) near the Θ-point as $v \cong v_0 \tau$, where $\tau = (T - \Theta)/T$, and we will consider the changes in the network characteristics accompanying the variations of τ, whereas the latter is approximately independent of temperature (solvent strength), $w \cong \text{const}(\tau)$.

As the uniform swelling and collapse of the network leads to its affine deformation, we can use as the characteristic of the network (and every star) equilibrium dimensions the swelling coefficient of the star fragment (with respect to its Gaussian dimensions) $\alpha = R_{\text{star}}/(N^{1/2}a)$, or the mean monomer density $\varphi \cong N^{-1/2}\alpha^{-3}$.

Salt-free solution

General formalism

The equilibrium dimensions of the network are determined from the conditions of minimum of its free energy, which includes the contribution of different physical origin:

$$\Delta F = \Delta F_{\text{conf}} + \Delta F_{\text{conc}} + \Delta F_{\text{trans}} + \Delta F_{\text{corr}}. \quad (1)$$

In the first approximation (if we are interested in the power asymptotic dependences for α), we can neglect the inhomogeneities in the local monomer density and local branch stretching and express all contributions into ΔF as the functions of α (or φ) only:

— elastic (conformational) term [3]

$$\Delta F_{\text{conf}} \cong \begin{cases} \alpha^2 - \ln \alpha^2, & \alpha \geq 1 \\ \alpha^{-2} + \ln \alpha^2, & \alpha \leq 1 \end{cases} \quad (2)$$

— free energy of volume interactions

$$\Delta F_{\text{conc}} \cong Nv\varphi + Nw\varphi^2 \quad (3)$$

— translational entropy of mobile counterions

$$\Delta F_{\text{trans}} \cong S \ln(\varphi/m) \quad (4)$$

— electrostatic (correlation) term

$$\Delta F_{\text{corr}} \cong -N(u/m)^{3/2}\varphi^{1/2} \quad (5)$$

per one branch. Here and below all numerical coefficients are omitted and all energetic values are expressed in kT units.

The correlation (electrostatic) contribution ΔF_{corr}, Eq. (5), is written in the Debye-Hückel approximation [4]. In the most typical cases $u \cong 1$ and this approximation is valid for the solutions of weakly charged ($m \gg 1$) polyelectrolytes in a wide range of concentrations $\varphi \leq 1$.

The minimization of the free energy, Eqs. (1)—(5), with respect to a leads to the following equations for the equilibrium dimensions of the network:

$$\begin{cases} a^8 - Sa^6 + Ga^{9/2} = B\tau a^3 + C, & a \geq 1 \quad (6.1) \\ -a^4 - Sa^6 + Ga^{9/2} = B\tau a^3 + C, & a \leq 1, \quad (6.2) \end{cases}$$

where

$$B = v_0 f N^{1/2} \quad (7)$$

$$C = wf^2 \quad (8)$$

are the renormalized parameters of binary and ternary nonelectrostatic interactions and

$$G = f^{1/2}(u/m)^{3/2} N^{3/4} \quad (9)$$

is the renormalized parameter of electrostatic interactions (the latter are not significant in the case of pure polyelectrolyte network, where the chains carry the charges of one sign).

Results

The analysis of Eqs. (6) shows that, depending on the relation between the number of branches, f, and the number of charges per branch, S, two different regimes of swelling and collapse of the network exist:

Polyelectrolyte regime

$f \ll S^2 w^{-1/2}$ (weak branching, strong charging). In this regime the counterion osmotic pressure predominates over the repulsive volume interchain interactions and causes the strong swelling of the network, not only above, but also below the Θ-point. The collapse of the network occurs with a jump in its dimensions below the Θ-point, at $\tau = -\tau_e \cong -m^{-1/2} = \text{const}(f)$ just as in the usual polyelectrolyte networks (the first-order phase transition, Fig. 2a). The asymptotic dependences for a in this regime are:

$$a \cong \begin{cases} [wf/(v_0|\tau|)]^{1/3} N^{-1/6}, & \tau \ll -\tau_e \\ (N/m)^{1/2}, & -\tau_e \ll \tau \ll \tau_e \quad (10) \\ N^{1/2} \tau^{1/5} m^{-2/5}, & \tau_e \ll \tau. \end{cases}$$

In the range of τ: $-\tau_e \ll \tau \ll \tau_e$ around the Θ-point the swelling coefficient a is approximately independent of τ and f and scales with the charge density as $m^{-1/2}$.

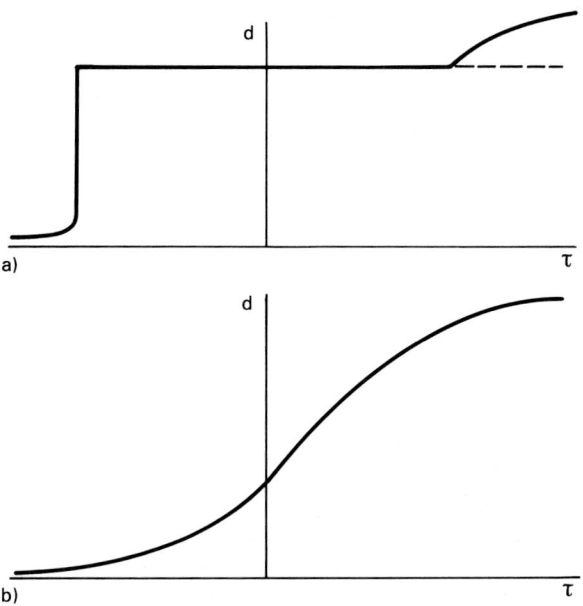

Fig. 2. The schematic dependencies of the network swelling coefficient on the solvent strength in the polyelectrolyte a) and quasineutral b) regimes

The appearance of power dependence of a on τ in the range of $\tau \gg \tau_e$ is caused by the decrease in the branches eleasticity due to the intrachain volume interactions. Actually, as is well-known, the elasticity of the chain with excluded volume interactions stretched by an external force is described, not by Gaussian, Eq. (2), but by the Pincus [5] law:

$$\Delta F_{el}(a) \cong a^{5/2} N^{-1/4} \tau^{1/2}. \quad (11)$$

In our case the role of the external force is played by the osmotic pressure of counterions which tends to swell the network. The combination of Eqs. (4), (11) describing the competition between the counterion osmotic pressure and branch elasticity leads to the last expression in (10).

The latter expression (10) does not include the degree of branching and is valid for usual as well as for branched polyelectrolyte networks. However, as far as we know, it has not been obtained previously.

Quasineutral regime

$f \gg S^2 w^{-1/2}$ (strong branching, weak charging). In this regime the volume interactions predominate over the counterion osmotic pressure in the whole range of the solvent strength. The collapse of the network occurs smoothly (non-phase cooperative transition) near the Θ-point (Fig. 2b), the width of the transition range increases with f as $\tau^* \cong f^{1/4}N^{-1/2}$. The asymptotic power dependences for the swelling coefficient are:

$$a \cong \begin{cases} [wf/(v_0|\tau|)]^{1/3}N^{-1/6}, & \tau \ll -\tau^* \\ w^{1/8}f^{1/4}, & -\tau^* \ll \tau \ll \tau^* \\ (v_0 f)^{1/5}\tau^{1/5}N^{1/10}, & \tau^* \ll \tau . \end{cases} \quad (12)$$

As it follows from Eq. (12), the network dimensions in the quasineutral regime increase monotonically with the degree of branching, f, and the solvent strength, τ. At high τ the value of a scales as $\tau^{1/5}$.

Note that the power dependences of a on τ under the conditions of a strong solvent coincides in the quasineutral and polyelectrolyte regimes although they have different physical origin: interbranch volume interactions in the former case and intrabranch interactions in the latter.

Salt-added case

Free energy

The addition of low molecular weight salt into the solution (an increase in the ionic strength) leads to an increase in the osmotic pressure in the bulk of the solution thus preventing the swelling of the network in the polyelectrolyte regime. Obviously, it does not affect the conformations of star-branched fragments in the "quasineutral" networks swollen by volume interactions.

In order to consider the dependence of the dimensions of the swollen star-branched polyelectrolyte network on the concentration of added salt, we suppose that simple 1:1 electrolyte is added so that its dimensionless concentration in the bulk of the solution is equal to ϕ_s. As the network contains its own counterions the concentration of the added salt is different inside and outside the network. The equilibrium swelling coefficient and salt concentration φ_s inside the network are determined by the following conditions:

$$-a^{-2}(\partial \Delta F/\partial a)_{\varphi_s} \cong \pi^{bulk}N^{3/2}a^3 f^{-1} \quad (13)$$

$$(\partial \Delta F/\partial \varphi_s)_a \cong \mu_s^{bulk} a^3 N^{3/2} a^3 f^{-1}, \quad (14)$$

where

$$\mu_s^{bulk}(\phi_s) = \partial[2\phi_s \ln \phi_s]/\partial \phi_s \quad (15)$$

is the chemical potential and $\pi^{bulk}(\phi_s) = 2\phi_s$ is the osmotic pressure of salt in the bulk of the solution.

Note that the free energy, ΔF, in Eqs. (13), (14) should include the contribution describing the translational entropy of all mobile ions (co- and counterions) which contains the network.

The analysis of Eqs. (13), (14) shows that the increase in the concentration of salt in the bulk of the solution leads to the monotonic decrease in the dimensions of the network.

When the concentration of salt in the bulk of solution exceeds the concentration of counterions in the swollen polyelectrolyte network greatly, i.e.,

$$\phi_s \gg \begin{cases} fm^{1/2}N^{-2}, & \tau \ll \tau_e \\ fm^{1/5}N^{-2}\tau^{-3/5}, & \tau \gg \tau_e \end{cases}$$

the situation of salt dominance takes places. Under the condition of salt dominance summary contribution of the osmotic pressures of counterions and salt ions in the free energy of the network may be described by the effective binary interunit interaction with an effective second virial coefficient $v_{eff} = (m^2 \phi_s)^{-1}$. This leads to the simple power dependence of the swelling coefficient on the salt concentration:

$$a \cong \begin{cases} f^{1/5}(m^2\phi_s)^{-1/5}N^{1/10}, & \tau \ll \hat{\tau} \\ f^{2/11}(m^2\phi_s)^{2/11}\tau^{1/11}N^{3/22}, & \tau \gg \hat{\tau}, \end{cases} \quad (16)$$

where

$$\hat{\tau} \cong (fm^{-2}\phi_s^{-1})^{1/5}N^{-2/5}. \quad (17)$$

As it follows from Eqs. (16), (17) the presence of the large number of free branches, which does not effect the swelling of the network in the polyelectrolyte regime in the salt-free solution, prevents the collapse of the network caused by the increase in the ionic strength of the solution.

At very high salt concentration, $\phi_s \gg N^{1/2}f^{-1/4}m^{-2}$, the osmotic effect become irrelevant and the network passes into quasineutral regime.

In the intermediate range of salt concentration corresponding to the salt dominance conditions, $N^{-2}fm^{1/2} \ll \phi_s \ll N^{1/2}f^{-1/4}m^{-2}$ the network remains partially swollen by the osmotic force. The transition into the collapsed state caused by the decrease in τ retains the jump-wise character at $N^{-2}fm^{1/2} \ll \phi_s \ll m^{-3/2}$, but becomes smooth at $m^{-3/2} \ll \phi_s$.

Note that the main relationships of the collapse of branched polyelectrolyte networks caused by the decrease in the solvent strength or by the increase in the ionic strength of the solution coincide with that obtained previously for grafted polyelectrolyte layers [6—8].

References

1. Tanaka T, Fillmore D, Sun S-T, Nishio I, Swislow G, Shan A (1980) Phys Rev Lett 45:1636—1639
2. Vasilevskaya VA, Khokhlov AR (1982) In: Mathematical methods in polymers. USSR Acad Sci Biological Research Center, Poustchino, pp 45—52
3. Birshtein TM, Priamitsyn VA (1991) Macromolecules 24:1554—1560
4. Landau LD, Lifshitz EM, Statistical Physics, Part 1, Nauka: Moscow, 1976
5. Pincus P (1976) Macromolecules 9:386—391
6. Borisov OV, Birshtein TM, Zhulina EB (1991) J Phys II 1:521—526
7. Birshtein TM, Borisov OV, Mercuryeva AA, Zhulina EB (1991) Prog Colloid Polym Sci 85:38—45
8. Zhulina EB, Borisov OV, Birshtein TM (1991) J Phys II, 2:63—74

Authors' address:

Dr. O. V. Borisov
Institute of Macromolecular Compounds
of the Academy of Sciences of Russia
199004, St. Petersburg, Russia

Local order and statistics of a polymer chain in an external electric field

J. Walasek and S. Grela

Technical University of Radom, Department of Physics, Radom, Poland

Abstract: A solution of polymer macromolecules in the presence of an external electric field is considered. Dipole-like interactions between polymer chain segment-vectors and the electric field are assumed to be proportional to $\cos v$, where v is the angle between the segment-vector and the direction of the electric field. Parameters describing the local order at the segmental and chain level, i.e., moments of the first and second Legendre polynomials $\langle P_1(v) \rangle$ and $\langle P_2(v) \rangle$ are expressed as power series on parameter q, where q is the intensity of the external electric field in kT units. The optical anisotropy induced by the external electric field and the possibility of application of the Kerr equation for description of this effect are discussed.

Key words: Polymer solution; electric field; interactions; local order; optical anisotropy; Kerr equation

Introduction

The Kerr effect stems from a local order in the system in the presence of the external electric field, and is observed in a wide variety of solutions of flexible macromolecules and in liquid crystalline polymers [1].

In this work, the hypothetical semiflexible chain model consisting of rigid rods connected consecutively by completely flexible joints, proposed by Flory et al. [2—5], is applied. Results on the lyotropic behavior of various polymer solutions are in good agreement with predictions of this model [6—9].

The orientation-dependent intermolecular interactions within the system and (or) with an external orienting field are responsible for the behavior of polymer systems [10—13].

In [13] it was shown that the local order parameters $\langle P_1(v) \rangle$ and $\langle P_2(v) \rangle$ are functions of the external electric field intensity q. The general results do not have a simple mathematical form. For weak fields, the local order parameters $\langle P_1(v) \rangle$ and $\langle P_2(v) \rangle$ can be expressed by simple functions of q. The main aim of this work is to estimate the range of values q, where these simple functions can be applied.

General considerations

The chain model consists of N segment-vectors $\mathbf{l}$, of constant length l, connected consecutively by completely flexible joints. The permanent dipole moment p of each segment-vector is assumed to be directed along the vector $\mathbf{l}$.

The orientation-dependent potential is assumed as:

$$V(l) = -kT\mathbf{q} \cdot \mathbf{l} + \text{const.} \quad (1)$$

Here, the vector $\mathbf{q}$ reads:

$$\mathbf{q} = \varkappa \mathbf{E}/kT, \quad (2)$$

where $\mathbf{E}$ is the vector of field strength, $\varkappa$ is a positive potential constant, and $\mathbf{q} \cdot \mathbf{l} = ql\cos v$. The effects of an induced molecular field are of small order and can be neglected [11—13].

In [13] it was shown that, for the system considered here, the distribution function of segmental orientations is:

$$w(l,r;q) = C \exp\left\{-N\left[\int_0^r L^*(\tau)d\tau - \mathbf{r} \cdot \mathbf{q}\right] + L^*(r)\mathbf{r} \cdot \mathbf{l}/rl\right\}, \quad (3)$$

where $r = h/Nl$, h is the polymer chain end-to-end vector, r is modulus of the vector r and $L^*(\tau)$ is the inverse Langevin function. C is a constant normalizing $w(l,r;q)$ to unity.

The moments $\langle P_1(v)\rangle$ and $\langle P_2(v)\rangle$ are calculated with the distribution function given by Eq. (3) and can be expressed as the power series [13]:

$$\langle P_1(v)\rangle = \sum_{k=1}^{\infty} A_{2k-1} q^{2k-1} \quad (4)$$

$$\langle P_2(v)\rangle = \sum_{k=1}^{\infty} A_{2k} q^{2k}, \quad (5)$$

where A_{2k-1} and A_{2k} are dependent on N. This dependence will be discussed in the next section.

The total birefringence Δn is given by the equation:

$$\Delta n/\Delta n_s = \langle P_2\rangle, \quad (6)$$

where Δn_s is birefringence for a segment [14]. Hence, for small q, the birefringence Δn is:

$$\Delta n = KE^2, \quad (7)$$

where, for q in terms of Eq. (2), the Kerr constant K is of the form:

$$K = A_2(\varkappa/kT)^2 \Delta n_s. \quad (8)$$

In the next section we discuss the range of q (or E) where the Kerr equation can be applied.

Results and discussion

Figure 1 shows a plot of exact values of the moments $\langle P_1\rangle = \langle P_1(v)\rangle$ and $\langle P_2\rangle = \langle P_2(v)\rangle$, calculated in [13], vs q. A simple theoretical derivation show that the moments $\langle P_1\rangle$ and $\langle P_2\rangle$ increase to unity as q tends to infinity [13]. The increase is steeper for large N than for small N. The discrepancy vanishes with increase of N. For example, for $N = 100$ and $N = 1000$ the plots of $\langle P_1\rangle$ and $\langle P_2\rangle$ vs q, are practically the same. The relative error d_1 and d_2 of the approximation $\langle P_1\rangle$ by $A_1 q$ and $\langle P_2\rangle$ by $A_2 q^2$ are defined, respectively, by the equations:

$$d_1 = [A_1 q - \langle P_1\rangle]/\langle P_1\rangle \quad (9)$$

$$d_2 = [A_2 q^2 - \langle P_2\rangle]/\langle P_2\rangle. \quad (10)$$

Plots of d_1 and d_2 vs q are shown in Fig. 2. Hence, we can estimate the range of values q, where the dependence between Δn and q (see Eq. (6)) is in good accordance with the Kerr equation (7). The behavior with respect to N, of plots d_1 and d_2 vs q, is analogical, as in Fig. 1.

Figure 3 shows a plot of coefficient A_1 and A_2 vs N. The coefficient A_1 increases to 1/3 and A_2 increase to 2/30 as N tends to infinity. For example, for N equal to 10, 100, 1000, respectively, the coefficient

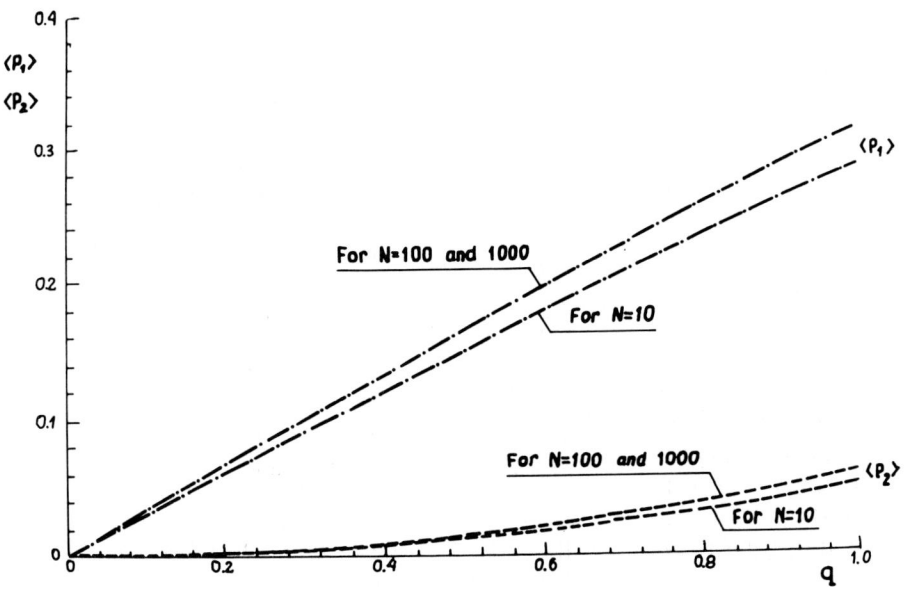

Fig. 1. Plot of $\langle P_1\rangle$ and $\langle P_2\rangle$ vs q for various N, where N is the number of segments per chain

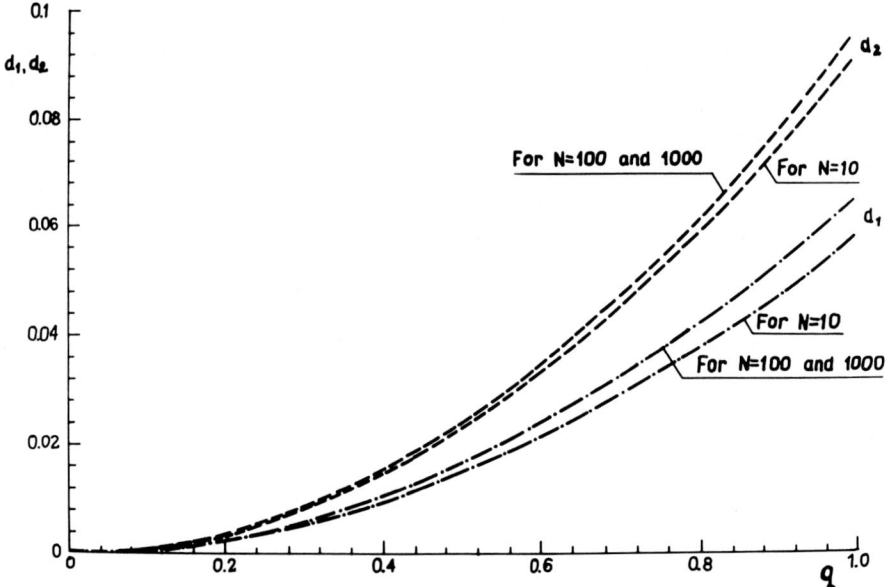

Fig. 2. Plot of d_1 and d_2 vs q, for N as in Fig. 1

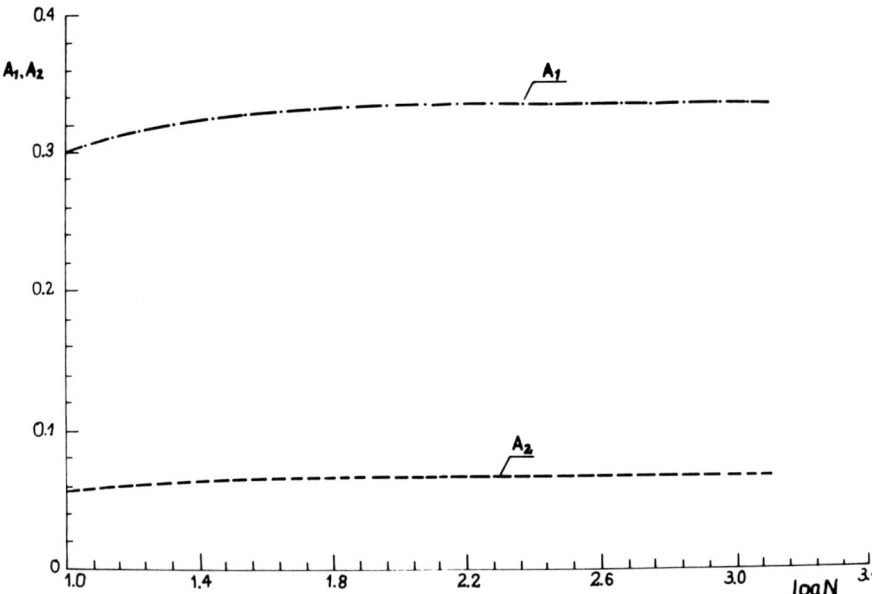

Fig. 3. Plot of coefficients A_1 and A_2 vs N

A_1 is equal to 0.300, 0.330 and 0.333, and A_2 is equal to 0.056, 0.066 and 0.067. In general, the ratio $A_{2(k-1)}/A_{2k}$ and A_{2k-3}/A_{2k-1} is of the order 1/10. As inspection of Fig. 3 shows. A_1 and A_2 are, practically, constant with increasing N. Hence, the Kerr constant is, practically, independent of the number of segments per chain.

References

1. Tsvetkov VN (1986) Zhesdkocepnyie Polymernyie Molekuly (Semi-flexible Polymer Molecules). Nauka, Leningrad
2. Flory PJ (1978) Macromolecules 11:1119—1144
3. Ronca G, Yoon DY (1982) J Chem Phys 76:3295
4. Matheson RR, Flory PJ (1981) Macromolecules 14:954

5. Flory PJ (1985) In: Chapoy LL (ed) Recent Advances in Liquid Crystalline Polymers. Elsevier Science Publishers, London-New York
6. Aharoni SM (1980) Mol Cryst Liq Cryst 56:237
7. Dayan S, Maissa P, Vellutini MJ, Sixou PJ (1982) J Polym Sci, Polym Lett Ed 20:33
8. Bheda J, Fellers JF, White JL (1980) Colloid Polym Sci 258:1335
9. Conio G, Bianchi E, Cifferi A, Tealdi A, Aden MA (1983) Macromolecules 16:1264
10. Walasek J (1988) J Polym Sci Part B: Physic Vol 26:1907
11. Walasek J (1990) J Polym Sci Part B: Physic Vol 28:1075
12. Walasek J (1990) J Polym Sci Part B: Physics Vol 28:2473
13. Walasek J (1992) J Polym Sci Part B: Physics Vol 30:401
14. Volkenstein MV (1963) Configurational Statistics of Polymeric Chains. Interscience Publishers, New York—London

Authors' address:

Dr. Janusz Walasek
Technical University of Radom
Department of Physics
ul. Malczewskiego 29
26-600 Radom, Poland

Charge photogeneration in carbazole-containing compounds and valency bands of oligomers

A. Tamulis and L. Bazhan*)

Institute of Theoretical Physics and Astronomy, Vilnius, Lithuania
*) Polytechnic Institute of Nizhniii Novgorod, Nizhnii Novgorod, Russia

Abstract: Quantum chemical investigations of the electronic structure of the design carbazole (Cz)-containing supermolecules, supramolecules, oligovinylcarbazole (OVK) and oligoepoxypropylcarbazole (OEPK) were performed in the framework of the semiempirical modified neglect of differential overlap (MNDO) method. The data obtained show that donor-acceptor properties of supermolecules: Cz-σ (or π) bridge-acceptors slightly depend on: 1) the number of alkane or alkene groups in the bridges, 2) the presence of π bonds, and 3) the various conformations in the bridge. Equilibrium distances between Cz and acceptor molecules in the supramolecular complexes and the sequence of the stability are found taken into account the oriented electrostatic attraction forces. There is a small charge redistribution in the ground state of the Cz-containing supermolecules and supramolecules. These designed derivatives are potential molecular photodiodes. The MNDO calculation data obtained show that the valence bands of the OVK and OEPK are narrow, therefore there is a electron hole hopping transfer mechanism in these systems.

Key words: Carbazole-containing oligomers; supermolecules; supramolecules; self assembly; photo-induced charge separation; molecular photodiodes

Introduction

The modeling of the processes of the photogeneration and photosensibilization in the chaotic carbazolyl containing photoconductors [1—3] and the studies of electron transfer in supermolecules and supramolecules [4—18] led us to design as well as to the quantum chemical investigation of the supermolecules: carbazole-σ (or π) bridge-acceptors and supramolecules: carbazole:acceptors. These design derivatives can be used as the molecular photo-rectifiers, as the basic elements of the molecular computers and other molecular nanoelectronics devices. Langmuir-Blodgett technology and self assembly phenomenon can be used in the construction of regular layers from these supermolecules and supramolecules. We expect that the regular layers must be effective in the construction of electrophotographic films, radiation solar energy cells, optical switches, memory discs and other macroscopic molecular electronics devices.

Quantum chemical calculations of the design carbazole-containing supermolecules, supramolecules, oligovinylcarbazole (OVK), and oligoepoxypropylcarbazole (OEPK) were performed in the framework of MNDO method [19] in this work. The obtained values of intermolecular distances in the supramolecules are longer than the experimental [20] values in EDA-complexes. The detailed investigations of the applicability of MNDO, AM1, and PM3 methods to the EDA-complexes can be found in [21]. The intermolecular distances calculated by MNDO method are longer than the experimental ones, thus, we can make only a qualitative analysis concerning the geometry of the various calculated supramolecules.

The MNDO method was also used for the investigation of monomers: donors, acceptors, insulators (see also [22]).

Table 1. Calculation (I_K, A_K) and experimental (Exp.) data of the ionization potential and electron affinity of the various molecules. All values are in electron volts

Molecule	I_K	A_K	Exp.
Cz	—8.25		—7.6 [23, 24], —7.57 [25]
TNF		—3.21	—0.94 [26, 27]
TeNF		—3.70	—1.16 [26] —1.19 [27]
TCNQ		—2.82	—2.80 [24, 28]
TN9(CN)$_2$F		—3.09	
TeN9(CN)$_2$F		—3.53	
C$_2$H$_6$	—12.70	3.75	—12.1 [29], —12.69 [30]
C$_3$H$_8$	—12.41	3.57	—11.5 [29], —12.04 [30]
C$_5$H$_{12}$	—12.07	3.35	—10.9 [29]

The calculation data of ionization potential (I_K) and electron affinity as well as experimental data are presented in Table 1. The values of I_K and A_K are presented by using Koopman's approximation (the index K it means). We use the following abbreviations of the electron acceptor molecules: 2,4,7-trinitro-9-fluorenone (TNF), 2,4,5,7-tetranitro-9-fluorenone (TeNF), 2,4,7-trinitro-9-dicyanofluorenene (TN9(CN)$_2$F), 2,4,5,7-tetranitro-9-dicyanofluorene (TeN9(CN)$_2$F), 7,7,8,8-tetracyanoquinodimethane (TCNQ).

Electronic structure of the supermolecules

Two types of Cz-containing supermolecules were designed and calculated: 1) carbazole-(CH$_{1,2}$)$_n$-TCNQ and 2) carbazole-(CH)$_4$-TNF. It was assumed that the alkane or alkene groups most probably join with the N atom in Cz and with the C atom in TCNQ or TNF ring (see Figs. 1 and 2). All the values in Figs. 1—3, and 5—8 must be multiplied by 10^{-3}.

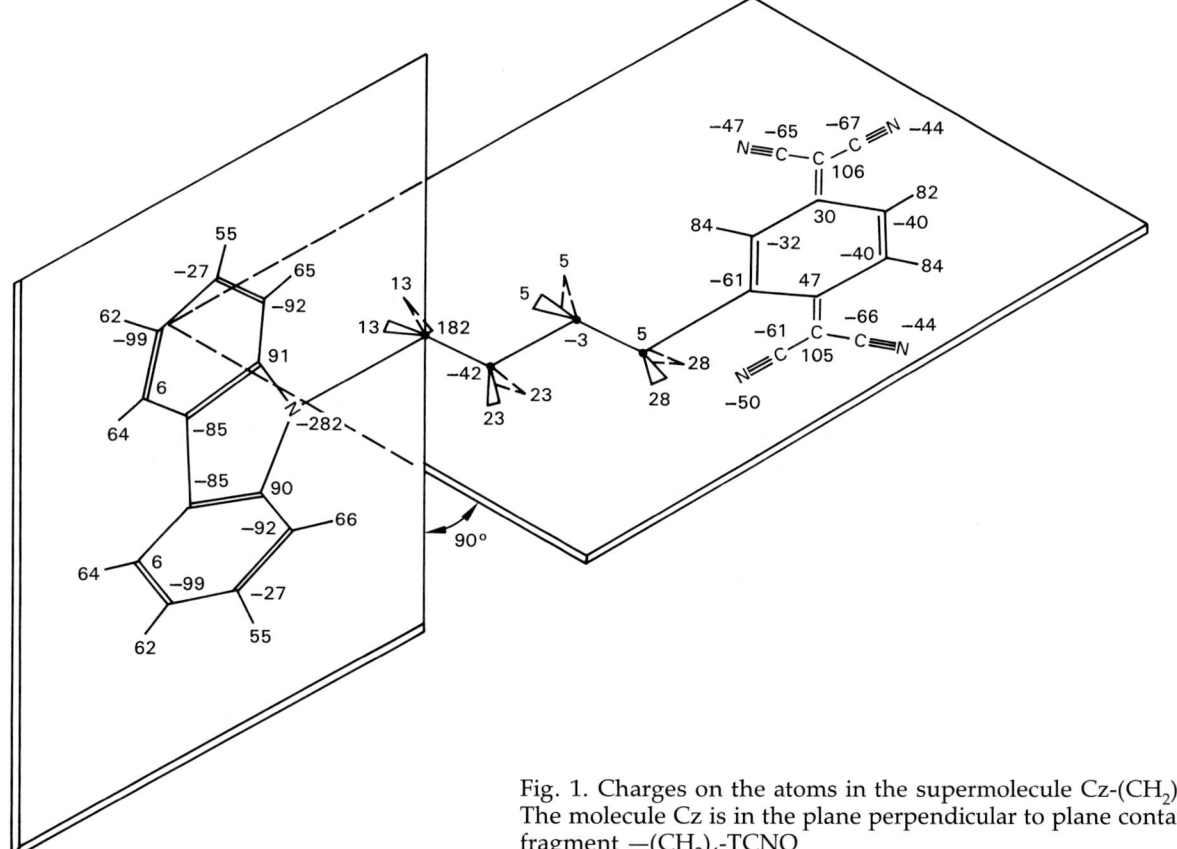

Fig. 1. Charges on the atoms in the supermolecule Cz-(CH$_2$)$_4$-TNCQ. The molecule Cz is in the plane perpendicular to plane containing the fragment —(CH$_2$)$_4$-TCNQ

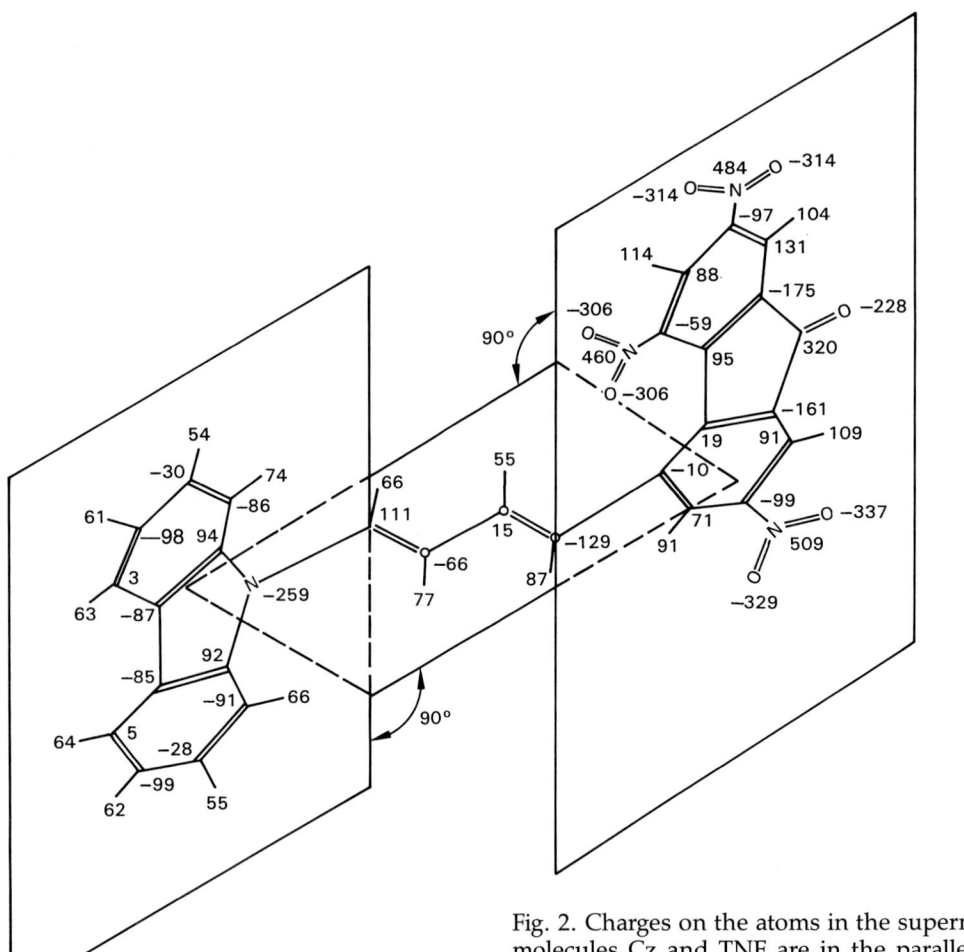

Fig. 2. Charges on the atoms in the supermolecule Cz-(CH)$_4$-TNF. The molecules Cz and TNF are in the parallel planes which are perpendicular to the plane containing the fragment —(CH)$_4$—

There is a small charge redistribution in the ground state of the supermolecules constructed of Cz, various alkane or alkene chains and TNF or TCNQ molecules. Several conformations of the supermolecules with the number of alkane or alkene groups n = 2, 3, 4 were investigated. The data obtained show that ionization potential of 1 supermolecule changes in the interval: —8.35 eV ÷ —8.21 eV, electron affinity: —2.89 eV ÷ —2.85 eV, and dipole moment: 0.58 Deb ÷ 1.18 Deb. These data show that the donor-acceptor properties of supermolecules slightly depend on: 1) the number of alkane or alkene groups in the bridges, 2) the presence of π bonds, and 3) the various conformations in the bridges.

Ionization potential of 2 supermolecule I_K = —8.25 eV, electron affinity A_K = 2.84 eV and dipole moment d = 0.64 Deb.

The calculation data show that Cz-containing supermolecules can be considered as systems with isolated electron donor and electron acceptor parts. Therefore, these systems are potential molecular photodiodes, because we can expect only photo-induced charge separation.

The modeling of the photoinduced charge separation in the supramolecules

The supramolecules' self assembly of Cz and TNF or TeNF, TN9(CN)$_2$F, TeN9(CN)$_2$F, TCNQ molecules was investigated.

First, we investigated the complex of Cz and TCNQ molecules (supramolecule Cz:TCNQ). Cz and TCNQ were placed in the parallel planes and

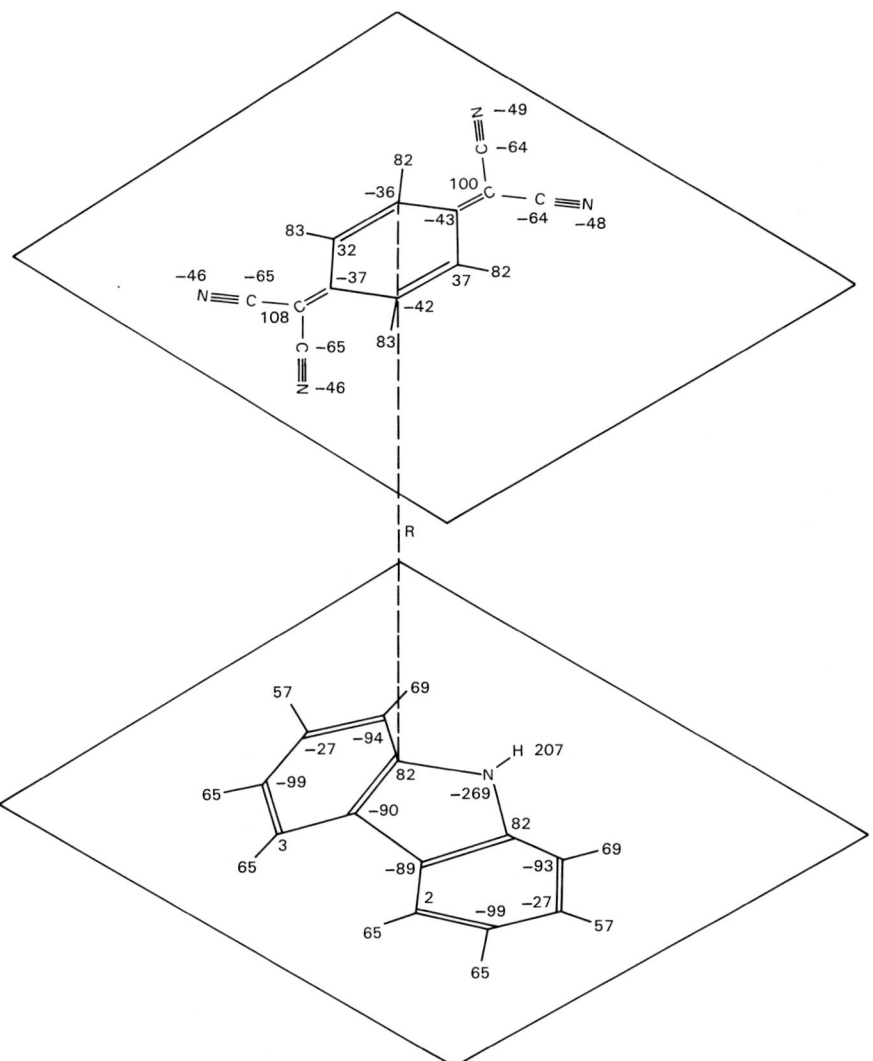

Fig. 3. Charges on the atoms in the supramolecule Cz:TCNQ

were rotated until the most probable energy conformation was obtained (see Fig. 3). The distance (R) between molecules was changed in the range of 3.0 Å—6.0 Å (see Fig. 4). It was found that the maximum of attraction energy of oriented electrostatic forces is $E = 0.517$ kcal/mol when the distance is 5.2 Å. The ionization potential in this distance is $I_K = -8.46$ eV and electron affinity $A_K = -2.77$ eV.

The molecules Cz, TNF, TeNF, and TN9(CN)$_2$ in the supramolecules Cz:TNF, Cz:TeNF and Cz:TN9(CN)$_2$F) were oriented in the parallel planes and with the maximum coincidence of the ring projection. The maximum of the attraction energy (E) in the equilibrium distance (R), ionization potentials (I_K), and electron affinities (A_K) of the above-named supramolecules are presented in Table 2.

Thus, we can say that the supramolecule Cz:TeNF is more stable than Cz:TNF, Cz:TN9(CN)$_2$F and Cz:TCNQ supramolecules. This calculated result correlates with the experimental result [31] that disc-like pentaynes and TNF EDA-complexes are more stable as disc-like pentaynes and other acceptor molecules EDA-complexes.

We established only a qualitative row of stability of the supramolecules: Cz:TeNF > Cz:TNF > Cz:TN9(CN)$_2$F > Cz:TCNQ. We did not evaluate the basis set superposition error [32], so all the

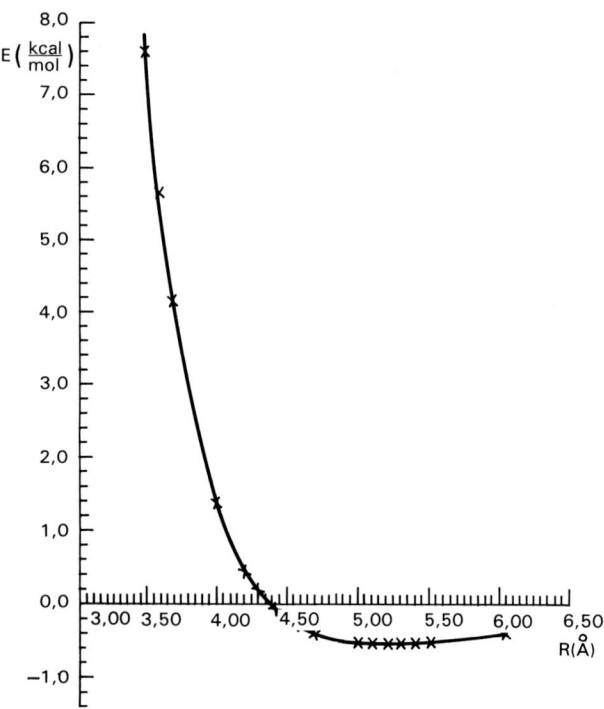

Fig. 4. The interaction energy dependence upon the intermolecular distance in the supramolecule Cz:TCNQ

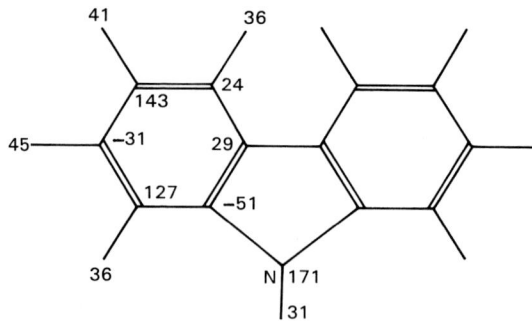

Fig. 5. The additional electron hole distribution on the Cz molecule

negative charged atoms of the neutral Cz molecule (see Fig. 5) while the additional electron is localized on the positive charged and oxygen atoms of the neutral TNF (Fig. 6), TeNF (Fig. 7) molecules, and on the $-C(CN)_2$ fragments of the neutral TCNQ

Table 2. Calculated equilibrium distance (R), ionization potential (I_K), electron affinity (A_K), and attraction energy (E) of the various supramolecules

Supramolecule	R in Å	I_K in eV	A_K in eV	E in kcal/mol
Cz:TCNQ	5.2	−8.46	−2.77	0.517
Cz:TNF	5.1	−8.58	−2.79	0.97
Cz:TeNF	5.1	−8.62	−3.23	1.056
Cz:TN9(CN)$_2$F	5.1	−8.64	−3.04	0.526

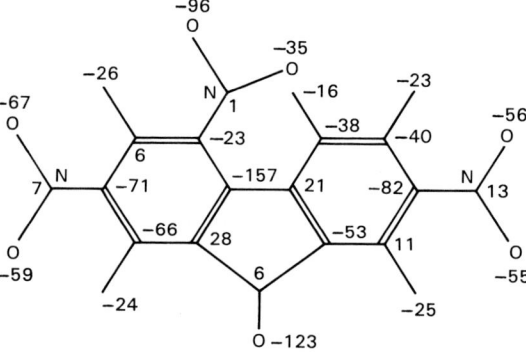

Fig. 6. The additional electron distribution on the TNF molecule

values of attraction energy (E) in Table 2 are approximate and could be useful only for establishing the row of stability of supramolecules.

The destabilization energy at room temperature is more than oriented electrostatic attraction forces for Cz:TCNQ and Cz:TN9(CN)$_2$F, but we did not calculate the dispersion forces, and supramolecules can be stable.

The calculations of positive charged Cz and negative charged accepter molecules were performed. The additional electron hole is localized on the

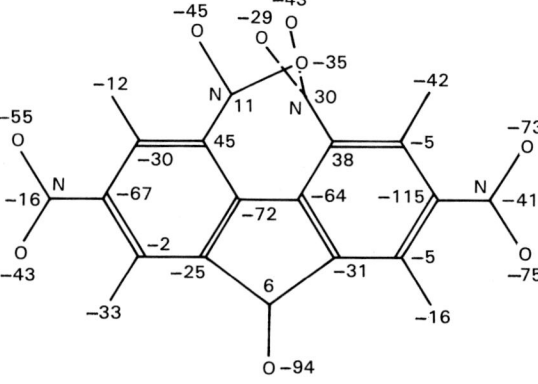

Fig. 7. The additional electron distribution on the TeNF molecule

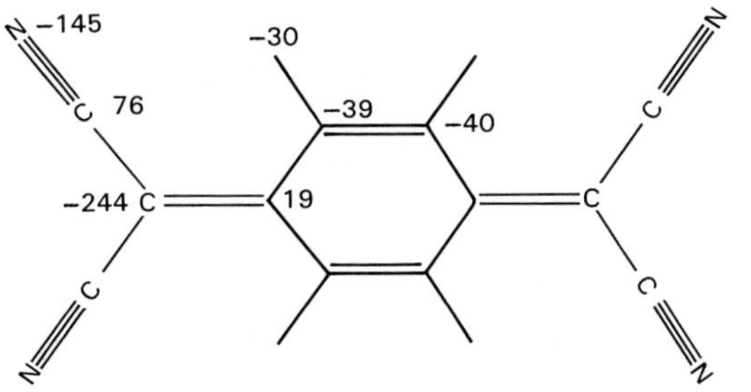

Fig. 8. The additional electron distribution on the TCNQ molecule

molecule (see Fig. 8). We expect that these local charge distribution regions are present during the photoinduced charge separation in the Cz-containing supermolecules and supramolecules.

Electronic structure of the carbazole-containing oligomers of the PVK and PEPK

The electronic structure of the oligovinylcarbazole (see Fig. 9) and oligoepoxypropylcarbazole (Fig. 10) were calculated. The geometry of the chain of 13 alkane groups and two carbazolyl fragments was taken as in the polyvinylcarbazole (PVK) crystals [33]. There were calculated three different cases when carbazolyl fragments were joined: 1) to the fourth and sixth carbon atoms in the alkane chain, 2) to the fourth and eight atoms, and 3) to the fourth and tenth atoms.

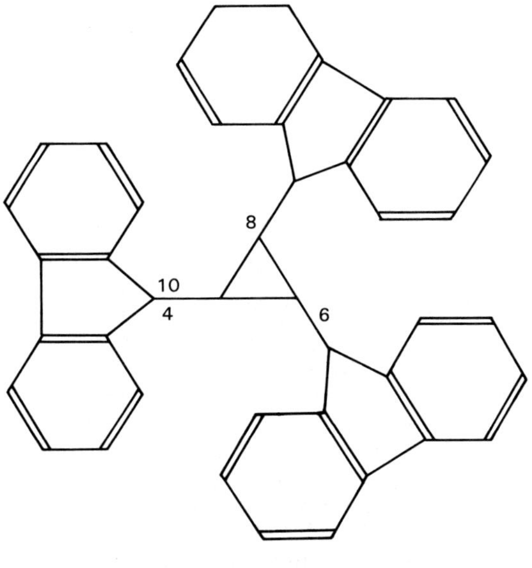

Fig. 9. Schematic image of the geometrical structure of the PVK polymer

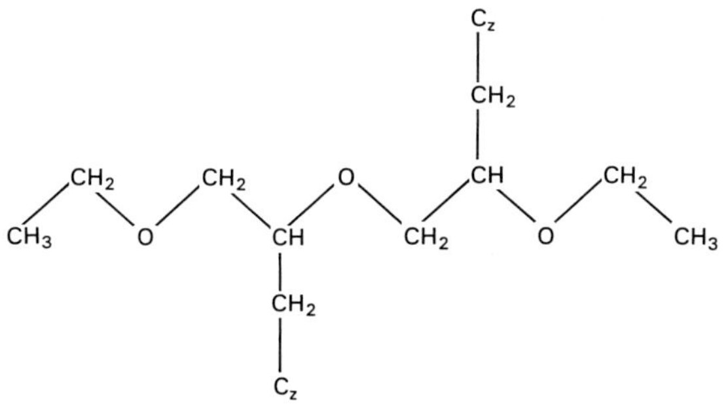

Fig. 10. Schematic image of the geometrical structure of the OEPK. Cz-fragment of the carbazolyl

The calculations and experimental [34] data obtained show that the valence bands of the OVK and OEPK are narrow (see Fig. 11). This fact confirms the presence of an electron hole hopping transfer mechanism in carbazolyl containing polymers. We

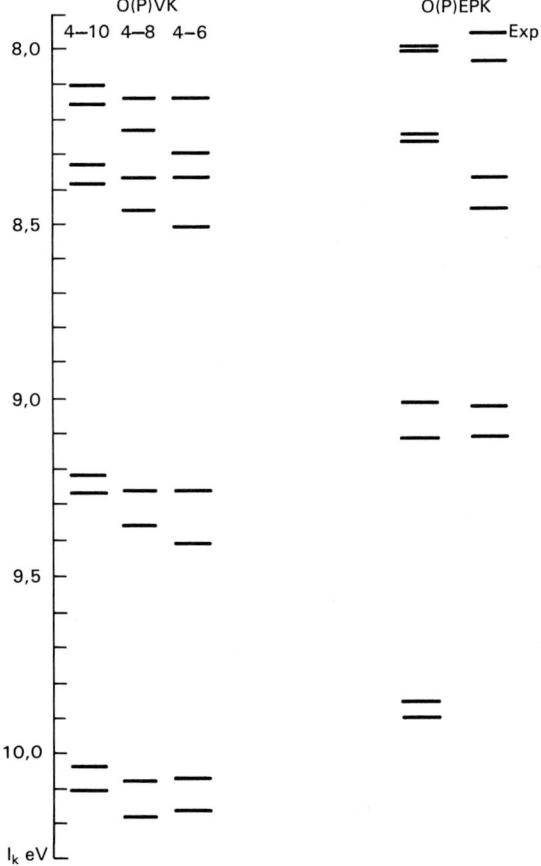

Fig. 11. The valence bands structure of the calculated OVK, OEPK and experimental results of the PEPK [34]

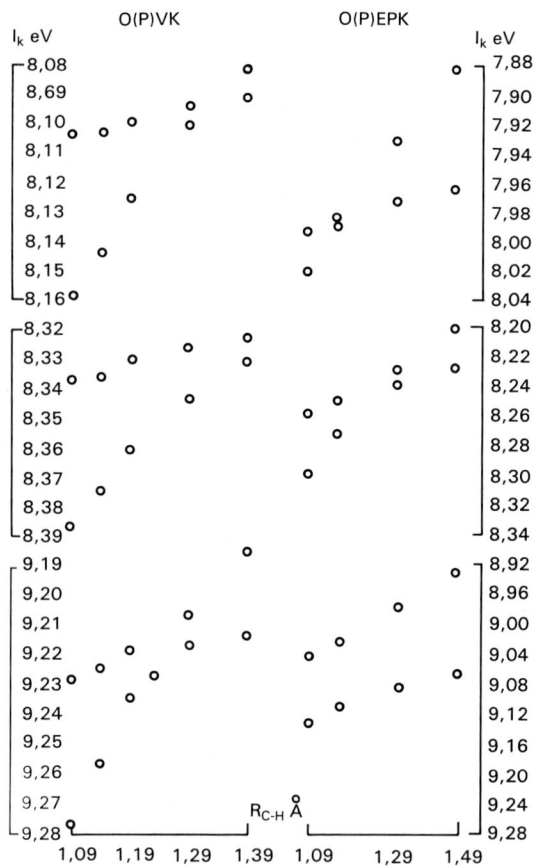

Fig. 12. The valence bands' width dependence upon the valence bond distance (R_{C-H}) in the carbazolyl fragment. The right column represents three experimental valence bands [34].

also investigated the dependence of the value of OVK and OEPK valence band splitting on the parameter R_{C-H}. R_{C-H} is the valence bond distance in the carbazolyl fragment. The value of the minimal valence splitting was obtained (see Fig. 12). This value is correlate with the value of the resonance interaction due to the electron hole tunneling from the carbazolyl fragment joined to the fourth atom in the alcane chain into the carbazolyl fragment joined to the 10th atom in the alkane chain [2].

Conclusions

Donor-acceptor properties of supermolecules slightly depend on:
1) the number of alkane or alkene groups in the bridges;
2) presence of π bonds, and
3) various conformations in the bridges.

There is a small charge redistribution in the ground state of Cz containing supermolecules. Cz containing supermolecules and supramolecules can be considered as systems with isolated electron donor and electron acceptor parts. Therefore, these systems are potential molecular photodiodes, because we can expect only photoinduced charge separation.

The row of the supramolecules' stability given from the attraction energy calculations is: Cz:TeNF > Cz:TNF > Cz:TN9(CN)$_2$F > Cz:TCNQ.

The additional electron hole is localized on the negative charged atoms of neutral Cz molecule while the additional electron is localized on the —C(CN)$_2$ and on the —NO$_2$ fragments of the acceptor molecules.

Valence bands of the OVK and OEPK are narrow, thus, we expect the electron hole hopping transfer mechanism in these systems.

Acknowledgements

The authors would like to thank Dr. Š. Kudžmauskas for stimulating discussion. Financial support by Dr. Sci. S. Janušonis is gratefully acknowledged.

References

1. Tamulis A, Kudžmauskas Š (1987) Investigations of Electronic Structure of Carbazole, 2,4,7-Trinitro-9-Fluorenone, 2,4,5,7-Tetranitro-9-Fluorenone Molecules and its Single Charge Ions by Using the NDO Methods. Deposited in the Lithuanian Research Institute for Scientific and Technical Information. 232659, Vilnius, Kalvariju 3, The Repulic of Lithuania, No 1908 LI (in Russian)
2. Kudžmauskas Š, Vektaris G, Tamulis A, Liuolia V, Gaidelis V, Undzėnas A, The Modeling of the Processes of the Photogeneration, Photosensibilization and Charge Carrier Transport in the Carbazolyl Containing Photoconductors, Deposited the Soviet Union Research Institute for Scientific and Technical Information, Moscow, Inventor No 02890037345, 1988.12.12. (in Russian)
3. Tamulis A, Undzėnas A, Bažan L The Electronic Processes in Carbazolyl Containing Polymers Sensitized by Pyrylium and Rhodamine 6G, Proc. of the 2nd International Symposium "Organic Materials for Non-Linear Optics", University of Oxford, UK, 4—6 September, 1990. Hann RA, Bloor D (eds) Published by Royal Society of Chemistry, Thomas Graham House, Science Park, Cambridge CB4 4WF (1991) pp 267—272

4. Nishitani S, Kurata N, Sakata Y, Misumi S, Karen A, Okada T, Mataga N (1983) J Amer Chem Soc 105:7771—7772
5. Heitele H, Michel-Beyerle ME, Finckh P, Rettig W (1986) In: Jortner J, Pullman B (eds) Tunelling. D. Reidel Publishing Company, pp 333—344
6. Lehn J-M (1988) Supramolecular chemistry-scope and perspectives. Molecules, supermolecules, and molecular devices (Nobel lecture) Angewandte Chemie 27:89—112
7. Popovic ZD, Hor Ah-M, Loutfy RO (1988) Chem Phys 127:451—457
8. Fujihara M (1989) Proc of the 11th Annual International Conference of "IEEE Engineering in Medicine and Biology Society (EMBS)" — Molecule electronics technical sessions and workshop, Seattle, Washington, U.S.A., pp 1318—1319
9. Birge RR, Ware BR, Dowben PA, Lawerence AP (1989) Proceedings of the Engineering Foundation Conference on Molecular Electronics — Sciences and Technology. Keauhan-Kona, Havaii, February, 19—24
10. Hopfield JJ, Onuchic JN, Baratan DN (1989) J Phys Chem 93:6350—6357
11. Bakker NAC, Wiering PG, Brouwer AM, Warman JM, Verhoeven JW (1990) Mol Cryst Liq Cryst 183:31—39
12. Nazeeruddin MK, Liska P, Moser J, Vlachopoulos N, Gratzel M (1990) Helvetica Chimica Acta 73:1788—1803
13. Gust D, Moore TA, Moore AL, Lee S-J, Bittersmann E, Lutrull DK, Rehms AA, DeGraziano JM, Ma KC, Gao F, Belford RE, Trier TT (1990) Science 248:199—201
14. Waite J, Papadopoulos MG (1990) J Phys Chem 94:6244—6249
15. Zelent B, Messier P, Gauthier S, Gravel D, Durocher G (1990) J Photochem Photobiol 52:165—178
16. Deronzier A, Essakalli M (1991) J Phys Chem 95:1737—1742
17. Tanaka M, Yoshida H, Ogasavara M (1991) J Phys Chem 95:955—960
18. Harriman A, Magda DJ, Sessler JL (1991) J Phys Chem 95:1530—1532
19. Dewar MJS, Thiel W (1977) J Amer Chem Soc 99:4899—4907
20. Barltrop JA, Coyle JD (1975) Excited states in organic chemistry. John Wiley and Sons, London, New York, Sidney, Toronto
21. Messinger J, Buss V (1990) Tetrahedron 46:423—433
22. Tamulis A, Kudžmauskas Š (1989) Soviet Physics Collection (Litovskii Fizicheskii Sbornik/Lietuvos Fizikos Rinkinys), Allerton Press Inc 29:15—21
23. Basak B, Lahiri B (1969) Ind J Pure Apl Phys 7:234—235
24. Neiland OJ (1990) Organicheskaja khimija, Visshaja shkola, Moscow (in Russian)
25. Pozharskii AS (1985) Teoreticheskie osnovy khimii heterocklov, Khimija, Moscow (in Russian)
26. Mukkerjee TK (1968) Tetrahedron 24:721—728
27. Wagner WJ, Gasner EL (1970) Photographic science and engineering 14:205—209
28. Metzger RM, Panetta ChA (1989) J Molec Electron 5:1—17
29. Bieri G, Burger F, Heilbronner E, Maier JP (1977) Helvetica Chimica Acta 60:2213—2233
30. Pireaux JJ, Svensson S, Basilier E, Malmgvist P-Å, Gelius U, Caudano R, Siegbahn K (1976) Physical Rev A 14:2133—2145
31. Praefcke K, Singer D, Langner M, Kohne B, Ebert M, Liebmann A, Wendorff JH (in press since 11/91) Molecular Crystals and Liquid Crystals
32. Clementi E, Corongiu G, Chakravorty S (1990) in Clementi E (ed) Modern Techniques in Computational Chemistry: MOTECC-90. ESCOM, The Netherlands, pp 359—360, McLean AD, Yoshimine M, Lengsfield BH, Bagus PS, Liu B, ibid, pp 593—614
33. Penwell RC, Gangulary BN, Smith TW (1978) J Polym Sci: Macromolecular Rev 13:63—160
34. Davidenko NA, Kuvshinskii NG, Stirka VCh (1988) Khimicheskaja Fizika 7:1245—1249 (in Russian)

Authors' address:

Dr. Arvydas Tamulis
Didlaukio 27—40
Vilnius, 2057, The Republic of Lithuania

The peculiarities and nature of large-scale motion of highly crosslinked polymers

B. A. Rozenberg and V. I. Irzhak

Department of Polymer and Composite Materials of the Institute of Chemical Physics in Chernogolovka, Russian Academy of Sciences, Moscow, Russia

Abstract: The peculiarities of highly crosslinked epoxy-amine polymer relaxation studied by pulse NMR method have been analyzed. A qualitative model of large-scale motion of highly crosslinked polymers based on the supposition of an anisotropic character of the interknot chain segmental motion and its dependence on the chain rigidity and connectivity of knots that adjoined to it has been developed. On the basis of this model the probable molecular and topological structure of epoxy-amine network fragments responsible for the α-relaxation has been given. The undissolved problems of highly crosslinked polymer α-relaxation have been discussed.

Key words: Large-scale motion in networks; glass transition temperature; anisotropy of segmental motion; connectivity of network; network crosslinks

Introduction

Large-scale motion is the most important relaxation phenomenon since it is responsible for physical-mechanical properties of polymeric materials. There are numerous data in the literature, generalized in [1—2], about the relaxation properties of various network polymers obtained by different methods. Most of them are difficult to interpret due to incomplete characterization of chemical and topological structure of the investigated samples. Sometimes, the difficulties of polymer characterization arise from the industrial origin of resins and curing agents. To obtain meaningful results, good model systems which can be quantitatively characterized on each step of the experimental study are necessary. Such approach was developed for the solution of the problem of the nature of the network polymer large-scale motion mainly in the authors' laboratory at the Institute of Chemical Physics Russian Academy of Sciences in Chernogolovka. This paper is devoted to a review of the results in the field under consideration obtained in the last decade using mainly pulse NMR as a method and epoxy polymers as objects of investigation. Experimental details are described in the [2] and in the original authors' papers.

Model of large-scale motion of highly crosslinked polymers

The crosslinking of polymers cannot be simply reduced to an effective increase of its molecular weight, taking into account not only equilibrium properties of the polymer, but also its relaxation properties. Thus, over some value the molecular weight of polymer does not affect T_g, while for network polymer it increases approximately linearly with the network knots (crosslinks) concentration [1].

The kinetic unit responsible for the large-scale motion (α-process) is the segment, the value of which is determined in the linear polymers by minimal chain size when the correlation of motion is absent. It can be easily evaluated by studying the α-process kinetic parameters as a functionof the polymer molecular weight. When we come to the network polymers the situation becomes much more complicated as far as there are gel- and sol-fractions and both contain three different types of fragments: knots, interknot chains and tails (dangling chains); thus, several types of segments can exist simultaneously. When the crosslinking density is increased, the role of the former type of segments has to grow and, as a limit, the segmental motion has to be determined by the knot mobility. However, even in this case we can expect an inhomogeneous segmental motion since new possibilities appear, namely, one, two or more knots can take part in a segmental motion simultaneously. The realization of these opportunities must depend on the rigidity of interknot chains. The knot motion must be correlated for rigid interknot chains, and the degree of the correlation of the knot motion must decrease with increase of interknot chain flexibility.

It is obvious from these considerations that any physically reasonable model of large-scale motion in network polymers has to take into consideration the fact of the existence of network knots. The role of the latter in the determination of relaxation properties of network polymers was taken into account in an evident way in the model of Gotlib and coworkers [3]. According to this model, the network knots fixed in space restrict the isotropy of segment movements, knots prevent stochastic rotational motions of interknot chains and transform the isotropic motion of chains to anisotropic one. The more the crosslinking density the higher the anisotropy of the interknot chain motion. This model was developed further in Provotorov and coworkers' theory [4] that describes the behavior of flexible Gaussian internot chains at high temperatures ($T > T_g$). It was shown that, on the basis of the theory, the molecular weight distribution of elastomers interknot chains can be evaluated from the NMR data.

We used Gotlib and coworkers' idea as the basis for an improved model of network polymer large-scale motion which displays the role of network knot mobility and connectivity. It is reasonable to assume that at sufficiently high temperature (in the rubbery state), the network knots themselves can move and the intensity of their movement at the particular temperature depends on their connectivity [5—6]. Thus, if the knot is not connected with the network (knot with zero connectivity), its motion is reduced to free rotational diffusion and such a fragment must be the most mobile element of the network. A network knot connected with one chain is less mobile, a double-connected knot is still less mobile, and so on. Thus, the model of large-scale motion of highly crosslinked polymers must describe the network chain motion limited by the network knots. The degree of anisotropy of chain motion is decreased and limited by slow network knots motion. The degree of anisotropy increases with the increase of the degree of the knot connectivity.

This model leads also to the conclusion that if knots have the same connectivity (functionality) the motion of the network must be characterized by single relaxation time, but, the value of the latter depends on the crosslinking density. If the network is characterized by some distribution of knot connectivity the spectrum of relaxation time has to be observed.

These conclusions can be checked experimentally since the inhomogeneity of motions in the region of α-relaxation can be determined by some relaxation experiments. Especially useful information can be obtained by the pulse NMR method. In this case, every kinetic phase can be characterized, not only by the relaxation time and its temperature dependence, but also by its proton population, i.e., fraction of protons associated with the particular kinetic phase. The value of the latter can be evaluated by statistical calculation for the fragments of the network polymer whose chemical and supermolecular structure is well characterized. The coincidence of the experimental value of the kinetic phase proton population with the calculated one gives the possibility to identify the chemical structure of the relaxing element in the region of α-relaxation, i.e., to identify the segment nature responsible for the large-scale motion in the highly crosslinked networks.

Below, we will demonstrate these conclusions by some experimental data.

The influence of the crosslinking density on the parameters of the α-process

The objective of this part is an attempt, using a highly crosslinked network polymer system (epoxyamine networks), to answer whether either the knots are involved in the segmental motion or not. If the answer is positive, the next question arises. How many knots are included in the segment? The solution of these problems can be achieved by the investigation of the α-process relaxation parameters (T_g, activation energy E_a and preexponential factor τ_0) as a function of the crosslinking density. The latter can be changed in different ways: by changing the conversion at network formation, or the interknot chains length, or the reagent's stoichiometry, or by cutting the interknot chains — which can be done by adding a monofunctional reagent instead of the bifunctional one. The latter approach is the most convenient because, in this case, the interpretation of experimental results is much easier in comparison with the first two.

We used well purified diglycidyl ether of resorcinol (DGER) and phenyl glycidyl ether (PhGE). 2,6-diaminopyridine (DAP), m-phenylenediamine (m-PhDA) or their mixtures with aniline (A) were used as curing agents. The equivalent ratio of functional groups was preserved in all cases. All systems were studied in a wide temperature range (260—473 K). In Table 1 results obtained by the multi-pulse NMR method are presented [7].

Table 1. Relaxation parameters of the system DGER-PhGE-DAP

PhGE content, %	T_g, K	E_a, kJ · mol^{-1}	τ_0, s
0	440	—	—
5	421	96 ± 6	$2 \cdot 10^{-17}$
10	407	96 ± 4	$6 \cdot 10^{-18}$
20	383	96 ± 4	$3 \cdot 10^{-18}$
40	362	96 ± 4	$7 \cdot 10^{-19}$
100	316	96 ± 4	$1 \cdot 10^{-20}$

It should be noted that the correlation time in the minimum region on the curves of relaxation time T_e vs. temperature is quite satisfactorily described in term of an Arrhenius plot.

As one can see from Table 1, the glass transition temperature and the preexponential factor regularly decrease with decreasing knots concentration, but the activation energy does not depend on the knot concentration. Even for a low molecular compound (the last line in the table), E_a is the same as for all measured samples with high concentration of knots.

These results mean that, in the case under consideration, the segment is the knot with adjoined chains, i.e., the segment includes one knot only and there is no correlation in the network knot motion. Such a situation is probably due to the molecular structure of the epoxy-amine network containing rather flexible interknot chains with oxyether fragments [7].

In the following, we will discuss the problem of inhomogeneity of segment types in the epoxy-amine networks that can be revealed as the existence of kinetic phases from the analysis of curves of free induction decay.

Kinetic phases in network polymers

Data on the relaxation behavior of the DGER-PhGE-DAP systems are presented on Fig. 1 [6, 8]. The addition of PhGE leads to the complex nonexponential curves of free induction decay and at least two kinetic phases appear. Figure 1 shows the temperature dependence of the spin-spin relaxation time (T_2) of each kinetic phase. Analogous results were obtained for various diepoxy-diamine systems with the addition of monofunctional epoxide [8].

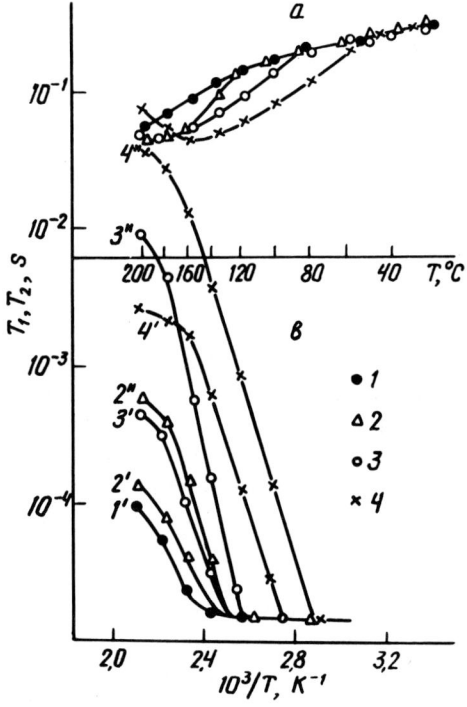

Fig. 1. Temperature dependence of spin-lattice T_1 (a) and spin-spin T_2 (b) relaxation time for DGER-PhGE-DAP systems. DGER/PhGE molar ratio: 1:0 (1); 9:2 (2); 8:4 (3); 6:8 (4). 2',3',4' are the less mobile phase, 2'',3'',4'' are the more mobile phase

The stoichiometric epoxy-amine networks with full conversion of the functional groups are characterized by single values of T_2 at given temperature [8], although the relaxation properties of the same incompletely reacted polymers are characterized by at least two kinetic phases. The lengthening of interknot chains by the addition of aniline to the diepoxy-diamine systems does not change the character of the dependencies of T_2 vs. T. According to the expectancy, the lower the crosslinking density the lower the anisotropy of large-scale motion, i.e., the limit value of T_2 increases (Figs. 1 and 2) and the latter is absent for the linear polymer (curve 7 in Fig. 2).

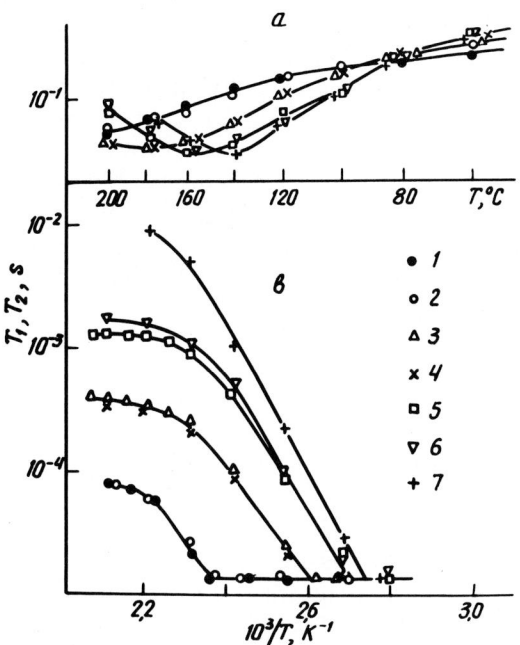

Fig. 2. Experimental pulse NMR data for systems DGER-DAP-A (curves 1, 3 and 5) and DGER-m-PhDA-A (curves 2, 4, 6). Diamine-aniline molar ratio: 1:0 (1, 2); 1:2 (3, 4); 1:18 (5, 6); 0:1 (7)

All experimental results considered above show that the pulse NMR method is rather sensitive to topological structure peculiarities of networks. These phenomenological qualitative results unambiguously show that the presence of various types of network knots essentially widens the relaxation time spectrum while, in the case of one type of knots, the pulse NMR method detects the presence of only one kinetic phase. However, it is not possible to differentiate between the contribution of knots and network chains in the segmental motion-nor to ascertain the nature of segments. We can approach solving the problems indicated by comparing the measured proton population and the calculated value of proton concentration by selecting the suitable fragments for each kinetic phase. However, before we start to discuss this problem for the epoxy networks, it is necessary to make some remarks about the used definition and applicability of the statistical description of the epoxy network topology.

Some remarks on the definitions used

Describing the dynamical behavior of a network or its equilibrium properties (modulus of elasticity) such terms as chains, knots, and tails that denote the elements of its molecular and topological structure are usually used. However, when we take into consideration the variety of knots due to their differences in connectivity or, in other words, in their already realized functionality (or simply functionality) misunderstanding can occur, since the knot connectivity means not only the number of reacted functional groups, but also the type of connected chains. The knot will have 0-functionality if it connects tails only; the higher the knot functionality, the lower the number of connected tails. On the other hand, the type of network chain will depend on the functionality of knots joining the chain ends. Thus, a free chain connects two one-functional knots; a tail has a one-functional (a free end should also be considered as a one-functional knot) and a multi-functional knot; a chain joining two two-functional knots is a fragment of a chain with more length; a network chain has knots on its ends and its functionality is more than 2.

Applicability of the statistical theory for the epoxy-amine networks structure description

This problem has been the subject of various publications [1, 9—10]. Here, we give the direct evidence for the applicability of the statistical theory for the solution of the problem under consideration. This approach is based on the statistical calculation of the concentration of corresponding structural elements formed during the polycondensation of diepoxide with diamine using the cascade theory [1, 9—11] up to the gel-point and direct measurement of these element concentrations by GPC [12]. The results of such investigations are given in Fig. 3.

As seen from the data of Fig. 3, the coincidence of the experimentally measured and calculated values of oligomer concentrations is quite satisfactory. This means that the statistical method is suitable for the calculation of the structural elements formed during the particular epoxy-amine systems cure.

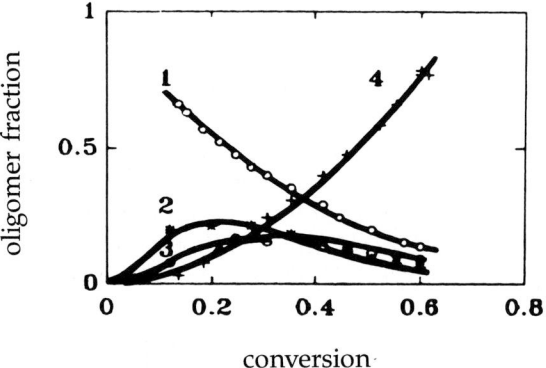

Fig. 3. Change of weight fraction of reaction system fragments during cure of diglycidyl ether of bisphenol A (E) and 3,3'-dichloro-4,4'-diaminodiphenylmethane (A) (1:1 on functionality) at 120°C. Fragments: 1 — E + A; 2 — EA; 3 — $E_2A + A_2E$; 4 — $\sum E_i A_j$, where $i + j \geq 4$

The nature of the most mobile phase

As mentioned above, because of the complicated topological structure of the networks the inhomogeneity of the relaxation behavior of the epoxy-amine network polymers appears and brightly displays itself in the pulse NMR experiments. Analyzing the free induction decay, we can usually see two kinetic phases with different mobility, more and less mobile ones, but sometimes a third kinetic phase can appear [5, 8]. However, we will ignore this case and will analyze here only curves containing two kinetic phases. Usually, fragments with the simplest topological structure, including tails in gel-fraction and sol-fraction, form the most mobile phase. More complicated structural elements, including the network chains connected by knots with connectivity equal to 2 or less, are involved in the most mobile phase when the temperature decreases or the total connectivity of knots increases in the system.

Symbols used for the presentation of the network fragments are given in Table 2. These symbols allow us to enumerate all basic fragments containing one network chain with the adjoined knots of different connectivity. With the aid of these basic fragments we can design other network fragments of any complexity.

The comparison of calculated, based on the statistical theory, and measured proton fractions, associated with the most mobile kinetic phase, allows us to make a supposition about its composition, i.e., to ascertain its molecular nature. An example of such investigation is presented in Fig. 4 [12].

It is obvious that the best description of the experimental kinetic curve is realized when the composition of the relaxing element corresponds to fragments 4 and 5 of Fig. 4 and only near the gel point is a better coincidence fulfilled for the more complex fragment 6 containing two amine molecules.

Analogous calculations were performed for some diepoxy-monoepoxydiamine systems and the results are compared with the measured proton population of the most mobile kinetic phase (Table 3). In Table 3 C_m is given by:

$$C_m = \frac{[\text{monoepoxy}]}{[\text{monoepoxy}] + 2 \cdot [\text{diepoxy}]} .$$

As one can see, the developed approach allows us to quite satisfactorily determine the structure of the kinetic unit responsible for the motion of the most mobile kinetic phase.

It is rather characteristic for the systems investigated that the fraction of protons in the most mobile kinetic phase decreases with increasing temperature, i.e., at elevated temperatures only the very simple fragments are in this phase. This is a nontrivial result since we should expect a quite opposite one. The reason for this phenomenon can be connected with the fact that the temperature dependence of T_2 at higher temperatures is not determined by segmental motions of a chain, but rather by its anisotropy which is controlled by the knots' mobility and, thus by their topological structure (connectivity). Unfortunately, the theory of the relationship between the knot connectivity and the anisotropy of the segmental motion of a chain is not yet available. Therefore, the reason for indicated regularities is not yet clear.

The proton fraction in the most mobile kinetic phase for system 2 (of the same composition as 3)

Table 2. Symbols for network fragments based on tetrafunctional knots ($\gtrless X \lessgtr$), mono- ($>-$) and bifunctional ($>-<$) oligomers*)

Schematic representation	Knot	Oligomer	
		Monofunctional	Bifunctional
−<>X<>− −<>X<>−	0k	0t	—
>−<>X<>− −<>X<>−	1k	1t	—
>−<>X<>−< −<>X<>−<	3k	3t	—
−<>X<>—<>X<>−< −<>X<>− −<>X<>−<	1k + 3k	1t + 3t	13c
>−<>X<>—<>X<>−< >−<>X<>− >−<>X<>−<	3k + 4k	3t	34c

*) Here letters k, t and c denote knot, tail, and chain, respectively. Knot connectivity is denoted by figure, structural unit — by letter. Two figures at chain index denote the connectivity of both knots connected with the particular chain.

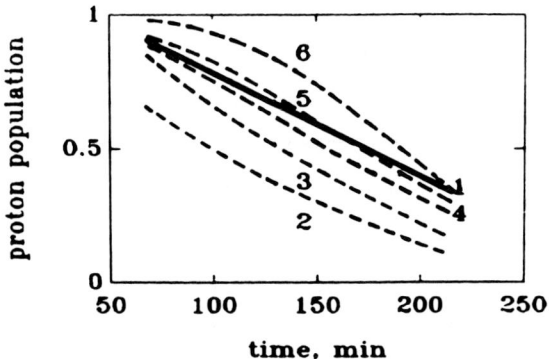

Fig. 4. Time dependence of the experimental (1) and calculated (2—6) weight fraction of mobile kinetic phase protons for the same system as in Fig. 3. The fragments:
$2 - A + E; \; 4 - A + E + \sum_{1}^{2} AE_i; \; 5 - A + E + \sum_{1}^{4} AE_i;$
$6 - A + E + \sum_{1}^{4} AE_i + \sum_{1}^{7} A_2E_i$

under the same temperature conditions differs essentially from that of system 3. It is considerably more for the latter which contains many more flexible curing agents, i.e., flexible knots. This result is quite clear: the most mobile kinetic phase with flexible knots contains more complex fragments than the kinetic phase with the rigid knots.

Thus, the most mobile kinetic phase is stipulated by the motion of the simplest network fragments that can be identified with some probability using the method discussed above.

The structure of fragments is rather sensitive to the structural elements, knots and chains, and their rigidity.

The large-scale motion of the less mobile kinetic phase is connected with the motion of the network fragments containing the knots of maximal connectivity. This conclusion will be demonstrated in the next part.

The nature of the less mobile kinetic phase

A similar approach was used for the determination of the fragment structure responsible for the motion of the less mobile kinetic phase. The experimental and calculated proton populations of the less mobile phase as the function of the stoichiometric coefficient $K = [NH]/[epoxy]$ for some epoxy-amine systems are presented in Fig. 5.

As one can see, the coincidence of the experimental and calculated data obtained from the supposition that fragments responsible for the motion contain the knots with the connectivity equal to 4 and 3 is rather good. The correlation between calculated values of the fragment molar fraction (W) containing the knots with the connectivity equal to 4 and 3 (W) and the composition of the network (K) is presented in Fig. 6. The results of T_g measurements by the thermomechanical and NMR methods for some epoxy-amine network polymers are also plotted in this figure.

Table 3. The fraction of protons (P) in the most mobile phase for diepoxy-monoepoxy-diamine systems and the probable structure of segments responsible for large-scale motion [5]*)

System	T, °C	C_m	P_{exp}	P_{calc}	Chain fragment
1	130—180	0.25	0.60	0.60	all, except 34c, 3k, 4k
1	160—180	0.33	0.50	0.49	all, except 33c, 34c, 3k, 4k
1	150—160	0.40	0.50	0.50	all, except 23c, 24c, 33c, 34c, 44c, 2k, 3k, 4k
1	105—160	0.50	0.41	0.40	0t, 1t, 2t, 3t, 0k
1	90—125	0.75	0.70	0.72	0t, 1t, 2t, 3t, 0k
1	200	0.25	0.43	0.42	all, except 33c, 34c, 44c, 3k, 4k
1	200	0.33	0.25	0.25	0t, 1t, 2t, 3t, 0k
1	200	0.40	0.24	0.25	0t, 1t, 2t, 3t, 0k
1	200	0.50	0.35	0.35	0t, 1t, 2t, 0k
1	200	0.75	0.54	0.56	0t, 1t, 0k
2	140	0.20	0.41	0.38	all, except 33c, 34c, 44c, 3k, 4k
2	100	0.40	0.42	0.43	0t, 1t, 2t, 0k
2	200	0.20	0.10	0.11	0t, 1t, 2t, 0k
2	160—180	0.40	0.34	0.34	0t, 1t, 2t, 0k
3	80—120	0.20	0.55	0.53	all, except 34c, 44c, 4k
3	45	0.40	0.58	0.58	0t, 1t, 2t, 3t, 0k, 1k, 2k, 3k, 11c, 12c
3	200	0.20	0.43	0.45	all, except 33c, 34c, 44c, 4k
3	150—200	0.40	0.10	0.14	0t, 1t, 0k

*) 1 — DGEBA-PhGE-3,3'-dichloro-4,4'diaminodiphenylmethane; 2 — DGER-PhGE-DAP; 3 — DGER-PhGE-diaminopentane.

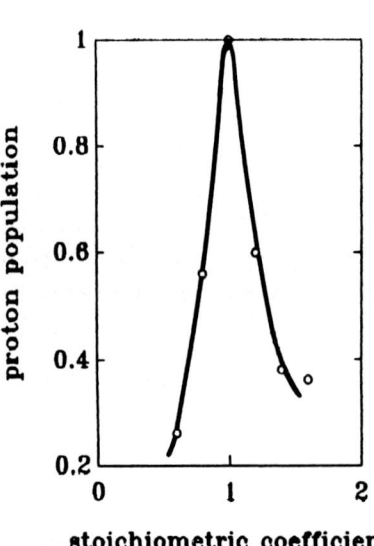

Fig. 5. The dependence of the proton fraction (P) in the less mobile kinetic phase for epoxy-amine networks (see Fig. 3) on their composition. Experimental and calculated values of P are given by dots and solid lines, respectively

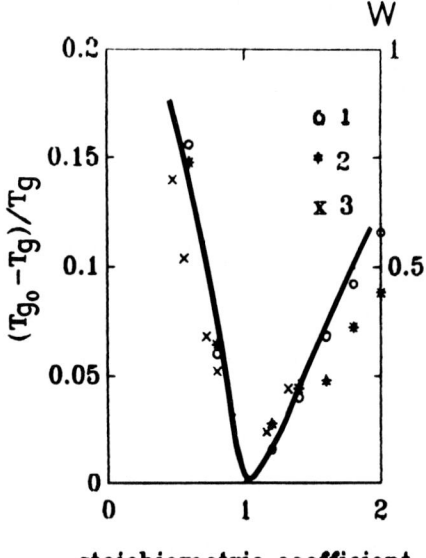

Fig. 6. The dependence of the fragment molar fraction (W) containing the knots with the connectivity equal to 4 and 3 (solid lines) and relative change of T_g (1—3) on the stoichiometric coeffient (K) for systems DGEBA-3,3'-dichloro-4,4'-diaminodiphenylmethane (1, 2) and DGER-m-PhDA (3). T_g was measured by thermomechanical (1) and NMR (2, 3) methods

The highest T_g is observed at stoichiometric composition when the concentration of the four-functional knots is the highest. There is a linear dependence between T_g and the concentration of the four-functional knots [13]. These results show without any doubt that just the network fragments containing the knots of highest connectivity are the most immobile ones and they are responsible for the α-transition.

Conclusions

The qualitative model of large-scale motion of highly cross-linked network polymers is developed. It is based on the supposition that network knots possess own mobility at high temperatures (in the rubbery state) and that the intensity of their motion depends on their connectivity.

The model gives a possibility to identify the molecular and topological structure of network fragments responsible for the rather complicated α-relaxation process of networks. The most probable nature of the segments for some epoxy-amine networks is determined.

Unfortunately, there is no theory of the relationship between the knot connectivity and anisotropy of the segmental motion of a chain. Such a theory should provide additional information on the structure of the knots involved into the segmental motion from the character of the temperature dependence of T_2 in NMR pulse experiments.

References

1. Irzhak VI, Rozenberg BA, Enicolopyan NS (1979) Network polymers: synthesis, structure, properties (in Russian). Chemistry, Moscow, pp 248
2. Andreis M, Koenig JI (1989) Adv Pol Sci 89:71—160
3. Gotlib JY, Livshits MI, Shevelev VA, Lishanskii IS, Balakina NV (1976) Vysokomol soed 18A:2299
4. Kulagina TP, Marchenkov VV, Provotorov BN (1989) Vysokomol soed 31A:381—386
5. Zakirov IN, Irzhak VI, Lantsov VM, Rozenberg BA (1988) Vysokomol Soed 30A:915—921
6. Irzhak VI, Lantsov VM, Rozenberg Ba (1987) in: Sedlachek B, Kahovec J (eds) Crosslinked Epoxy. Walter de Gruyter & Co. Berlin-New York, pp 359—372
7. Tarasov VP, Smirnov YN, Erofeev LN, Irzhak VI, Rozenberg BA (1982) Vysokomol Soed 24A:2379—2382
8. Zakirov IN, Lantsov VM, Derinovski VS, Smirnov YN, Efremova AI, Irzhak VI, Rozenberg BA (1986) Vysokomol Soed 28A:1719—1726
9. Dušek K (1986) Adv Pol Sci 78:1
10. Dušek K (1987) Makr Chem Macr Symp 7:37
11. Dušek K (1985) Brit Polym J 17:185
12. Lantsov VM, Pakter MK, Irzhak VI, Abdrakhmanova LA, Kuznetsova LM, Borisenko GV, Zaitsev YS (1987) Vysokomol Soed 29A:2292—2296
13. Oleinik EF (1986) Adv Pol Sci 80:49—99

Authors' address:

Prof. Dr. B. A. Rozenberg
Institute of Chemical Physics
Russian Academy of Sciences
142432 Chernogolovka, Moscow region, Russia

FTIR spectroscopy studies on epoxy networks

T. Scherzer[1], V. Strehmel[2], W. Tänzer[2], and S. Wartewig[1]

Departments of Physics[1] and Chemistry[2], Technical University Merseburg, FRG

Abstract: Epoxy networks on the basis of diglycidyl ether of bisphenol-A (DGEBA) cured with aliphatic diols of different chain length in the presence of magnesium perchlorate or with 4,4'-diaminodiphenylmethane (DDM) and imidazole, respectively, were investigated by FTIR spectroscopy. The influence of the accelerator on the epoxy/amine system was studied. The addition of a low amount of imidazole leads to a decrease of the epoxy absorption band at 915 cm^{-1} and to the appearance of a shoulder at 1132 cm^{-1} assigned to branched ether structures due to the reaction of epoxy groups with secondary hydroxyl groups. At higher concentrations only minimum changes of these bands are observed, but absorptions of unreacted amine functionalities near 3300 cm^{-1} remain in the spectra. — In the spectra of the epoxy/diol networks the epoxy bands have disappeared. The formation of ether structures indicated by absorption bands near 1120 cm^{-1} and the behavior of the hydroxyl band are strongly dependent on diol chain length and concentration. Only a small amount of diol is incorporated into the network by covalent bonding. The remaining diol acts as platicizer.

Key words: FTIR spectroscopy; polymer networks; epoxies; amine; diol; accelerator

Introduction

Imidazole is used as accelerator in a variety of commercial epoxy resin systems. The resulting networks exhibit excellent chemical stability, improved heat and water resistance as well as good electrical and mechanical properties. Generally, imidazole accelerates etherification reactions. However, despite the widespread commercial use of imidazoles, little is known about their role in the curing process. In amine cured systems imidazole has been described as an accelerator in the patent literature only. The aim of the first part of the present paper is to determine the influence of imidazole on networks formed from diglycidyl ether of bisphenol-A (DGEBA) and 4,4'-diaminodiphenylmethane (DDM).

Few studies have been reported concerning the reaction of DGEBA with linear aliphatic diols [1—7]. Part 2 of this paper deals with the characterization of diol modified epoxy resins. The chemical composition was varied in a systematic manner to monitor the influence of diol concentration and chain length on the network structure.

Crosslinked polymers such as epoxy networks cannot be characterized easily because of their insolubility and infusibility. FTIR spectroscopy has been proved to be a suitable nondestructive method to investigate cured epoxy resins. Both transmission and diffuse reflectance (DRIFT) spectra were recorded to monitor changes and differences in the structure of various epoxy networks.

Experimental

Sample preparation

All starting materials are commercially available products. DGEBA was recrystallized from a mixture of acetone and methanol (m.p. = 42 °C).

In amine cured systems with reactant molar ratios $r_1 = [NH]_0/[Epoxide]_0 = 1$ or 0.5, respectively, the imidazole concentration was varied from 0 to 2.5 mole-%. DGEBA, DDM, and imidazole were mixed at 70 °C for 5 min and cured at 120 °C for 5 h using a PTFE mould. For measuring DRIFT spectra the samples were powdered and mixed with potassium bromide.

Linear aliphatic diols (HO-[CH$_2$]$_n$-OH) with chain lengths from n = 2 to 12 were used to prepare modified epoxy resins. They were added in a ratio of 1, 0.5 or 0.25 mole diol to 1 mole resin (r_2 = [OH]$_0$/[Epoxide]$_0$). DGEBA, diol and 3 mole-% magnesium perchlorate were stirred at 95°C for 1 h. Then, samples were cured between steel plates covered with release films (Richmond Aircraft Products, Santa Fe Springs/California) at 120°C for 48 h. The complete synthesis scheme is described elsewhere [1, 2].

FTIR spectroscopy

FTIR spectra were obtained with a BRUKER IFS 66 spectrometer equipped with a DTGS detector. Transmission spectra were recorded coadding 32 scans at a resolution of 2 cm^{-1}. The epoxy/diol films were sufficiently thin (about 10—20 μm) to absorb in the range where the Lambert-Beer law is valid. For DRIFT spectra 100 scans were collected at 2 cm^{-1} resolution. Subsequent data analysis is based on baseline corrected absorbance intensity. The band at 1182 cm^{-1} assigned to an in-plane deformation vibration of the para-disubstituted benzene ring was used as a reference peak.

Results and discussion

DGEBA cured with DDM

Figures 1 and 2 show the FTIR spectra of epoxy resins cured with DDM in the presence of imidazole.

Imidazole accelerates the epoxide conversion and the formation of ether structures (~1120 cm^{-1}) in systems with molar ratios of both 1 and 0.5. In the epoxy-rich system a considerably higher amount of epoxide groups (915 cm^{-1}) remains in the network in the absence of an accelerator and at imidazole concentrations up to 1.0 mole-%. The addition of more imidazole improves the epoxide conversion distinctly, but the same level as in the networks with stoichiometric reactant ratio is not reached. In those systems 0.5 mole-% imidazole is sufficient for nearly complete consumption of the epoxide groups.

Etherification is indicated by a small shoulder at 1132 cm^{-1}. In the case of r_1 = 1 this shoulder increases dramatically if a very low amount of im-

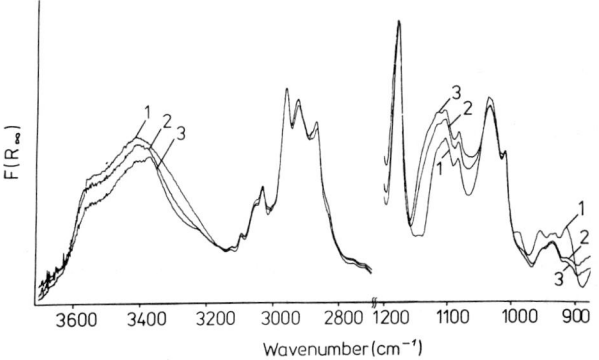

Fig. 1. FTIR-DRIFT spectra of DGEBA/DDM ([NH]$_0$/[Epoxide]$_0$ = 1), accelerated with different amounts of imidazole (1) 0 mole-%, 2) 1 mole-%, 3) 2 mole-%)

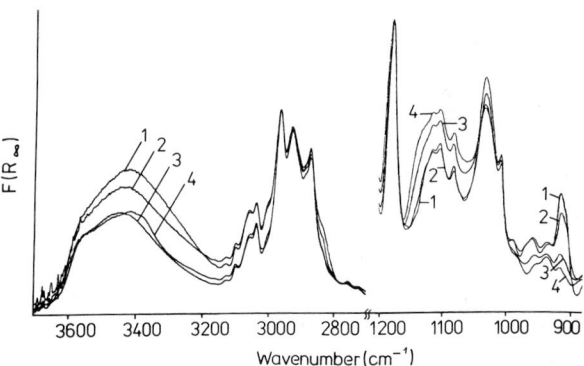

Fig. 2. FTIR-DRIFT spectra of DGEBA/DDM ([NH]$_0$/[Epoxide]$_0$ = 0.5), accelerated with different amounts of imidazole (1) 0.1 mole-%, 2) 0.2 mole-%, 3) 1 mole-%, 4) 2.5 mole-%)

idazole was added. When more imidazole is used, etherification is nearly independent of the accelerator concentration. In the cured samples with r_1 = 0.5 etherification increases continuously with imidazole concentration up to 1.5 mole-%. The glass transition temperatures T_g of the networks [8] confirm these results. T_g rises with increasing accelerator concentration, but remains constant for samples with 1.5 mole-% or more imidazole.

For low imidazole concentrations no bands of primary and secondary amine were found in the spectra. The addition of 2 mole-% imidazole or more leads to incomplete amine conversion in the networks with stoichiometric ratios. Amine bands at 3375, 3220 and 513 cm^{-1} still appear in the spectra as small shoulders. In samples with an excess of DGEBA the amine conversion is nearly independent of the amount of added accelerator and reaches completion.

The intensity of the hydroxyl bands is weakly dependent on the molar ratio r_1 and the accelerator concentration. In samples with epoxy excess the ratio between free and associated hydroxyl groups increases with imidazole content because of higher network density and decreasing concentration of unreacted glycidyl groups acting as plasticizers. Simultaneously, the maximum of the hydroxyl band shifts to higher wavenumbers and the total intensity diminishes slightly. Samples with stoichiometric ratio of reactants show no band shifting, but small intensity changes were found too. They are partially overwhelmed by the reappearing amine bands. A better separation of the amine and hydroxyl absorptions can be achieved by near infrared spectroscopy. Quantitative results of the behavior of the overtone and combination bands will be published [9].

DGEBA modified with diols

The FTIR spectra of epoxy resins modified with diols are shown in Figs. 3 and 4. The epoxide band at 915 cm^{-1} has disappeared in all spectra. The epoxide conversion is not influenced by the composition of the sample (r_2 = 1, 0.5 or 0.25). It is nearly complete in all formulations.

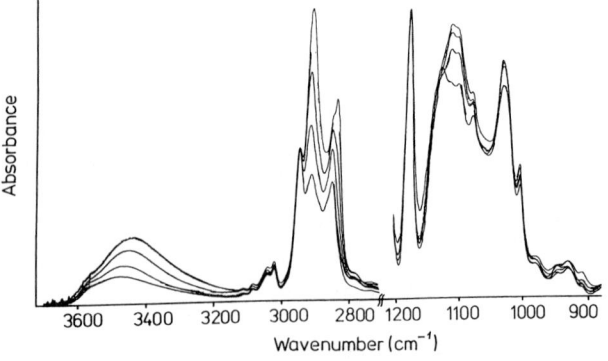

Fig. 3. FTIR transmission spectra of DGEBA modified with diols of different chain length ([OH]$_0$/[Epoxide]$_0$ = 0.5, from top to the bottom n = 12, 8, 6, and 2, respectively)

The formation of ether structures depends strongly on the diol chain length and the quantity used. With increasing diol concentration a higher degree

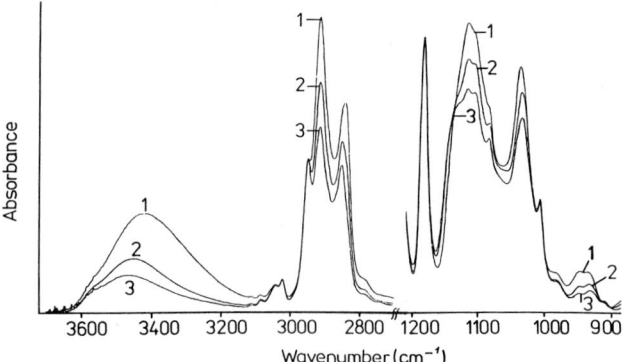

Fig. 4. FTIR transmission spectra of DGEBA modified with different amounts of 1,8-octanediol (r_2 = [OH]$_0$/[Epoxide]$_0$, 1) r_2 = 1, 2) r_2 = 0.5, 3) r_2 = 0.25)

of oligomer formation and chain extension was found. This is indicated by the behavior of the bands at 1107 and 1117 cm^{-1}. These bands are assigned to both unbranched ether and secondary hydroxyl groups. Branched ether groups as a result of crosslinking appear as a shoulder at 1132 cm^{-1}. From its intensity behavior it can be concluded that stronger branching occurs in systems with short diol chain length because of the higher mobility of DGEBA/diol oligomers during the later stages of curing. The apparent chain length dependency of the unbranched ether peak is a consequence of this different crosslinking behavior.

With increasing diol chain length and concentration the hydroxyl peak centered at 3400 cm^{-1} becomes stronger and simultaneously shifts to lower wavenumbers. The latter dependency is valid for samples with more than four C-atoms in the diol and the first one was only found for networks with a least r_2 = 0.5. The rather high hydroxyl peak intensity for samples with r_2 = 0.25 may be caused by various side reactions. Cyclization and homopolymerization of the diglycidyl ether as well as reactions with water could occur. Cyclic products were found in the monofunctional system phenyl glycidyl ether/butane-1-ol [7]. A more detailed analysis of the hydroxyl absorptions will be given in a later publication [10].

Diol acts mainly as a chain starter. The DGEBA/diol oligomers are more reactive than the diol itself. Therefore, branching already occurs at early stages of reaction. Only 15...20% of the diol are incorporated into the network by covalent bonding. The residual diol (oligomers, free chain ends, and partly free diol) plasticizes the network and leads to a higher tendency of hydrogen bond formation and a stronger hydroxyl band intensity.

The longer the diol chain the more residual diol remains in the network because of the reduced time to gelation during curing and the lower mobility of these molecules. The networks become more and more flexible. Accordingly, the glass transition temperature decreases [5] and the hydrogen bonding is more favored. Little influence is observed if the number of C-atoms is larger than eight. This observation is confirmed by ultrasonic measurements [5].

Acknowledgements

We gratefully acknowledge support of this work by the BMFT under contract 33 L 10700.

References

1. Tänzer W, Fedtke M (1986) Acta Polymerica 37:24—28
2. Tänzer W, Fiedler H, Fedtke M (1986) Acta Polymerica 37:70—72
3. Wolff D, Havránek A, Schlothauer K, Fedtke M (1991) Acta Polymerica 42:151—153
4. Schlothauer K, Tänzer W, Fedtke M, Spevacek J (1991) Acta Polymerica 42:190—191
5. Alig I, Fedtke M, Häusler KG, Tänzer W, Wartewig S (1988) Progr Colloid Polym Sci 78:54—58
6. Blaszó M (1991) Polymer 32:590—596
7. Fedtke M, Tänzer W, Pospiech D, Wintzer J (1985) Z Chem 25:107—110
8. Strehmel V, Zimmermann E, Häusler KG, Fedtke M, Progr Colloid Sci (this volume)
9. Scherzer T, Strehmel V (to be published)
10. Scherzer T, Wartewig S submitted to J Appl Polym Sci

Authors' address:

T. Scherzer
Technische Hochschule Merseburg
Fachbereich Physik
D-O-4200 Merseburg, FRG

Influence of imidazole on the structure of epoxy amine networks

V. Strehmel, E. Zimmermann, K.-G. Häusler, and M. Fedtke

Department of Chemistry, Institute of Macromolecular Chemistry, Technical University Merseburg, FRG

Abstract: Networks of diglycidylether of bisphenol A and 4,4'-diaminodiphenylmethane in the presence of imidazole up to 2,5 mole-% and in the absence of an accelerator were investigated. Differences in network density, glass transition temperature, thermal expansion coefficient above T_g, and in the soluble part of the samples were found if an equimolar ratio of epoxy groups to amino hydrogen or an epoxy excess and different accelerator concentrations were used for network syntheses. Imidazole accelerates the conversion of epoxy groups, not only with amino hydrogen, but also with the hydroxyl groups formed. This is the reason for the differences in glass transition temperature, network density, thermal expansion coefficient, and soluble part between the samples investigated.

Key words: Epoxy amine network; accelerator; glass transition temperature; network density; expansion coefficient

Introduction

Epoxy resin matrices cured with aromatic amines are used as thermosetting materials in the production of composites owing to their excellent mechanical and chemical properties. Composites are applied as light-weight structural materials in metal substitution. Many factors are responsible for their properties, among them the structure of the matrix and the resulting interface between matrix and fiber, which is dependent on the epoxy system used.

Many results about the curing process and the properties of the composites themselves and the matrix material are already available in the literature [1—5], but the influence of an accelerator on the networks formed is generally disregarded. We investigated networks of diglycidylether of bisphenol A and 4,4'-diaminodiphenylmethane in the presence of imidazole and in the absence of an accelerator.

Experimental

Epoxy networks were prepared from diglycidylether of bisphenol A and 4,4'-diaminodiphenylmethane ($r = 2 [NH_2]_0/[Epoxide]_0 = 1$ and 0.5, respectively) in the absence of an accelerator and in the presence of different amounts of imidazole (0.1—2.5 mole-% imidazole relative to epoxy groups). The components were mixed at 70°C for 5 min, degassed at the same temperature for 15 min and cured in a PTFE mold at 120°C for 5 h. Cylindrical specimens (1.4 cm diameter and 0.5 cm high) were prepared by dividing a longer one on a turning machine. These specimens were used for measuring the uniaxial compression modulus. For determining of glass transition temperature by thermomechanical analysis these specimens were sawn into four pieces.

Thermomechanical analysis was carried out on a Perkin Elmer TMA7 in a temperature range of 50—200°C with a scanning rate of 5°C/min and a force of 10 mN. The linear thermal expansion of the samples was measured.

Network densities were determined by measuring the uniaxial compression modulus at 30 K above T_g [6, 7]. The samples were heated to the corresponding temperature and were held at this temperature for 30 min. Then, a weight of 50 to 500 g was put on in 50 g steps.

To analyze the soluble part 1 g of rasped sample was extracted with 320 ml chloroform on a soxhlet-extraction for 16 h.

Results and discussion

During the curing of diglycidylether of bisphenol A (DGEBA) with 4,4'-diaminodiphenylmethane (DDM) three main reactions occur (Eqs. (1)—(3)):

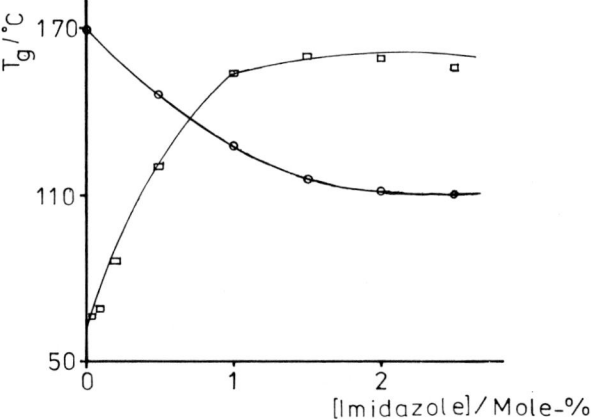

The networks synthesized in the way described show differences in glass transition temperature (Fig. 1), soluble part of the samples (Fig. 2), network density (Fig. 3) and expansion coefficients (Fig. 4).

In the absence of an accelerator the glass transition temperature is considerably higher in the stoichiometric system than in the case of an epoxy excess (Fig. 1), because of incomplete conversion in the latter case, which was shown by a soluble part of this sample of about 25% (Fig. 2) and FTIR spectroscopy results [8].

Fig. 1. Glass transition temperature (T_g) of DGEBA-DDM-networks as a function of imidazole concentration: $r = 2\,[NH_2]_0/[Epoxide]_0 = 1$ (o) and 0.5 (□), respectively

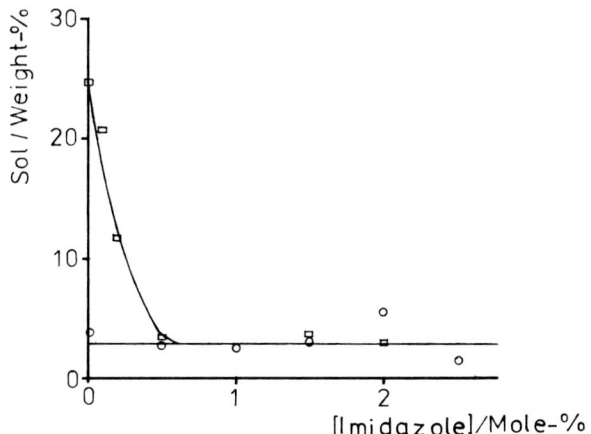

Fig. 2. Chloroform soluble part of DGEBA-DDM-networks as a function of imidazole concentration. $r = 1$ (o) and 0.5 (□), respectively

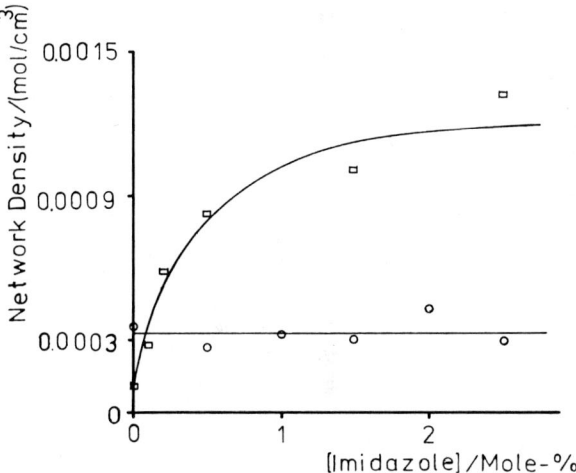

Fig. 3. Network density of DGEBA-DDM-networks as a function of imidazole concentration: $r = 1$ (o) and 0.5 (□), respectively

Adding only a small concentration of imidazole to the system cured with epoxy excess leads to a reduction of the extractable part of the sample cured (Fig. 2) and to an increase of glass transition temperature (Fig. 1) and network density (Fig. 3). This is attributed to better conversion of epoxy groups accelerated by imidazole [8].

Adding imidazole to a DGEBA-DDM-mixture with a stoichiometric ratio of functional groups leads to cured samples having a decreased glass transition temperature (Fig. 1), but network density

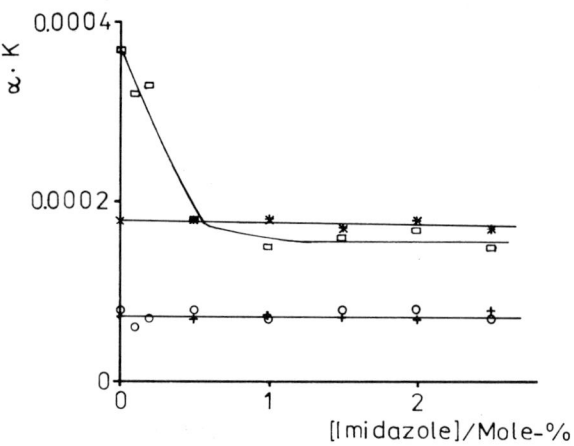

Fig. 4. Expansion coefficients of DGEBA-DDM-networks as a function of imidazole concentration: $T < T_g$: $r = 1$ (+) and 0.5 (○), respectively $T > T_g$: $r = 1$ (∗) and 0.5 (□), respectively

(Fig. 3) and extractable part of the samples (Fig. 2) remain constant. FTIR spectroscopy studies show a rise of intensity of ether bands and a decrease of band intensity of primary and secondary amino groups [8]. It may be concluded that imidazole accelerates etherification (Eq. (3)) more than the reaction of a glycidyl group with a secondary amino group (Eq. (2)). Thermal linear expansion coefficients below glass transition temperature are the same in all cases, but above this temperature the linear expansion coefficients show a significant decrease only up to an accelerator concentration of 1 mole-% using epoxy excess in the samples cured (Fig. 4), which is confirmed by increasing network density.

These results show that an accelerator influences the properties of the networks obtained.

References

1. Rozenberg B (1986) Adv Polym Sci 75:113—165
2. Le May J, Kelley F (1987) Adv Polym Sci 78:115—148
3. Dusek K (1987) Adv Polym Sci 78:1—60
4. Oleinik E (1987) Adv Polym Sci 80:49—99
5. Strehmel V, Krukowka L, Fedtke M (1990) Chem Techn 42:416—419
6. Häusler K-G, Hube H (1980) Plaste Kautschuk 27:78—81
7. Häusler K-G, Tänzer W, Kunze A (1991) Acta Polymerica 42:565—570
8. Scherzer T, Strehmel V, Tänzer W, Wartewig S (1992) Progr Colloid Polym Sci (this volume)

Authors' address:

Dr. Veronika Strehmel
Technical University Merseburg,
Department of Chemistry
Institute of Macromolecular Chemistry
Geusaer Straße
D-O-4200 Merseburg, FRG

Epoxy polymer networks: Relaxation processes and crack resistance

S. G. Kulik, P. G. Babayevsky, and V. V. Borovko

Department of Composite Technology, Division of Composite Physics and Chemistry,
Tsyolkovsky Institute of Aviation Technology, Moscow, Russia

Abstract: Time-temperature dependence of mechanical loss factor and crack resistance parameters — critical stress intensity factor K_{Ic} and critical strain energy release rate G_{Ic} for the cured epoxy/amine polymer have been studied. Mechanical loss and crack resistance master curves have been obtained using time-temperature superposition principle. Activation energies of relaxation processes and crack growth have been estimated. It was shown that the crack resistance is defined by the relaxation spectrum of the epoxy polymer.

Key words: Epoxy polymers; relaxation processes; crack resistance; master curves

Introduction

The ability of polymers to resist crack initiation and growth is determined by two basic factors: the energy of physical and chemical links breakage in the plane of the crack propagation and by the local nonelastic (plastic or viscoelastic) deformation in the vicinity of a crack tip [1—10]. The contribution of the first factor (an elastic constituent of crack resistance) is not decisive and does not depend noticeably on polymer structure and time-temperature loading conditions. Its value for different formulations of epoxy polymers is 1.9—3.2 J/m² and is similar to that for the linear glassy polymers such as PS, PMMA, PVC, PC [10, 11]. A more important contribution to polymer crack resistance gives the second, nonelastic constituent, the value of which is 2—3 decimal orders higher than the elastic one. In contrast to the latter, it strongly depends on both polymer structure and time-temperature loading conditions. Such dependence is a result of the viscoelastic character of local deformation in the vicinity of a crack tip before and during its propagtion. The local zone size and the plastic deformation intensity are related to molecular motions activated by mechanical stresses at given loading conditions and, therefore, to corresponding relaxation process [5, 8—10].

In this work experimental data of the dependence of crack resistance parameters on the temperature — real deformation rate in the vicinity of a crack tip and the master curves for a typical crosslinked glassy polymer are presented and their relationships with polymer relaxation spectrum are shown.

Materials and experimental procedures

The cured epoxy composition investigated was a mixture of ED-20 resin (Russian analog of Epon 828 resin) and hexamethylenediamine (17 phr), a convenient model of a glassy polymer network with long chains between crosslinks. Epoxy polymer sheets 1- and 5-mm thick were prepared by casting the initial composition into a mold consisting of two glass plates with antiadhesive coating, and then curing for 12 h at 298 K, 2 h at 353 K, 2 h at 393 K, and 2 h at 413 K.

Mechanical loss factors as a function of temperature in the range from 173 K to 393 K at different frequencies (7.8—1000 Hz) of tension-compression dynamic loading with constant deformation amplitude ($5.3 \cdot 10^{-4}$) were obtained using dynamic mechanical analyzer ("Metravib").

Polymer crack resistance was characterized by two parameters — critical stress intensity factor K_{Ic} and critical strain energy release rate G_{Ic}. For their estimation the double cantilever beam (DCB) specimens [10] with grooves along the center-line on both sides were used. Specimens were fractured in an Instron testing machine at a cross-head speed from $8.3 \cdot 10^{-7}$ m/s to $8.3 \cdot 10^{-4}$ m/s in the same temperature range 198—393 K. Load P — displacement y diagrams were recorded and used to determine the load P_c corresponding to the start of crack propagation. For linear $P - y$ diagram the values of P_c were taken as maximum loads. For nonlinear diagrams the values of P_c were evaluated as their intersection with 5% secant line [12].

Parameter K_{Ic} was calculated using the equation [10]:

$$K_{Ic} = \frac{P_c \cdot a}{(hw)^{1/2} H^{3/2}} \cdot \left(3.42 + 2.32 \frac{H}{a}\right) \quad (1)$$

where a and w are the crack length and width, respectively, h is the specimen thickness and H is half of the specimen height over the crack tip.

For calculation of parameter G_{Ic} the following equation was used [10]:

$$G_{Ic} = \frac{P_c^2}{2w} \cdot \frac{dC}{da}, \quad (2)$$

where C is the specimen compliance: $C = y/P$.

Earlier [13], it had been shown that DCB specimen testing at constant cross-head speed does not resulting in a constant deformation rate in the vicinity of a crack tip at different crack lengths and test temperatures. Therefore, the parameter $\dot{\varepsilon} = (\dot{K}_I/K_{Ic}) \cdot n$ [14] is used as a measure of the real deformation rate under monotonous loading condition. Here, $\dot{K}_I = dK_I/dt$ is the rate of the crack moving force under monotonous loading, and n is a strain hardening factor determined by tensile stress-strain curves for standard smooth specimens.

Results and discussion

The plots of $tg\delta$ versus temperature at different frequencies are shown in Fig. 1. There are three main loss peaks (α, β, γ) for which location, height, and width depend on frequency. Effective activation energies were obtained using the temperature dependence of the peak position on frequency; they are 380, 83, and 36 kJ/mole for α-, β- and γ-relaxation processes, respectively. These transition temperatures reduced to a frequency of 1 Hz are

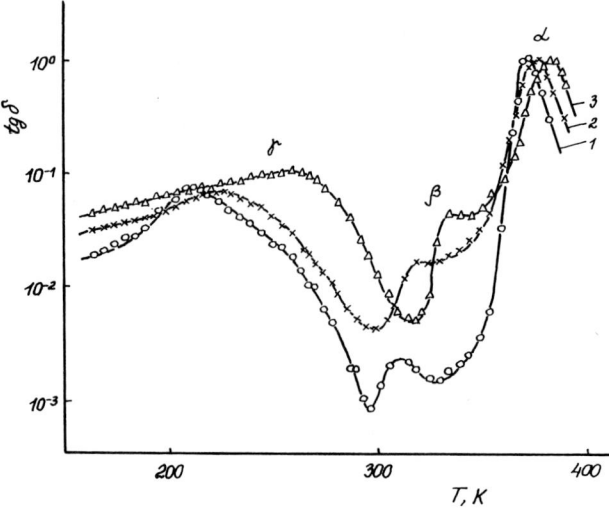

Fig. 1. Mechanical loss factor dependence on temperature at frequencies 7.8 (1), 31.3 (2) and 250 (3) Hz

365, 288, and 192 K, respectively. The ratio of α- to β-relaxation process activation energies is 4.6 and the ratio of β- to α-transition temperature is 0.79 and in agreement with known experimental data for other polymers [15—17].

It is known [17] that α- and β-processes in polymers are related to the segment mobility as individual (β-relaxation) or co-operative (α-relaxation) movements. The γ-relaxation process in cured epoxy polymers is usually related to the motion of the —CH_2—$CH(OH)$—CH_2—O— groups.

The deformation rate dependencies of the crack resistance parameters at different temperatures are shown in Figs. 2 and 3. The plots of K_{Ic}/T vs. $\ln\dot{\varepsilon}$ and G_{Ic}/T vs. $\ln\dot{\varepsilon}$ are linear in most cases. A strongly pronounced deviation from linearity is observed at temperatures close to T_α. The curves are nonlinear also at the temperatures about T_γ. In the temperature range $T_\gamma < T < T_\alpha$ crack resistance parameters increase with decreasing $\ln\dot{\varepsilon}$ and this effect is greater at higher temperatures. At temperatures $T > T_\alpha$ the parameters K_{Ic} and G_{Ic} decrease with decreasing deformation rate. Master curves for temperature — deformation rate dependence of the mechanical loss factor and the crack resistance parameters of the cured epoxy polymer were obtained using the data of Figs. 1—3 and time-temperature superposition principle [18]. Shift factor dependence on temperature was also determined. To obtain the $tg\delta$ master curve the fre-

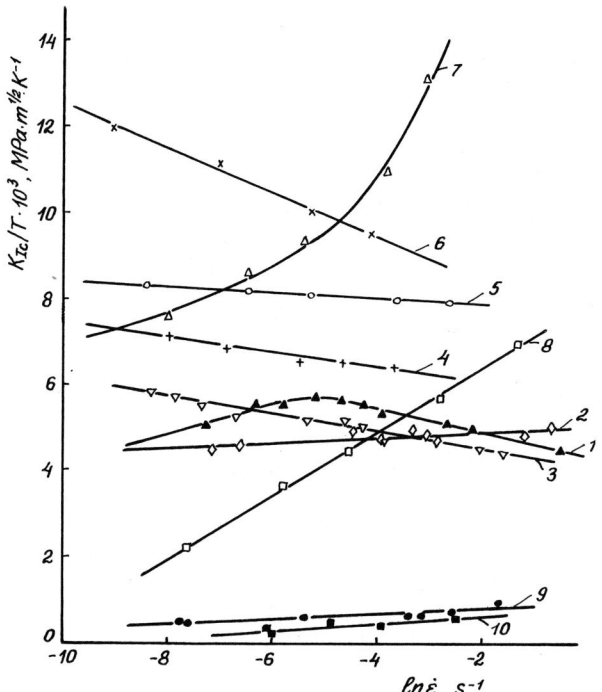

Fig. 2. Critical stress intensity factor K_{Ic} dependence on the deformation rate at temperatures 198 K (1), 213 K (2), 238 K (3), 253 K (4), 298 K (5), 318 K (6), 333 K (7), 353 K (8), 373 K (9) and 393 K (10)

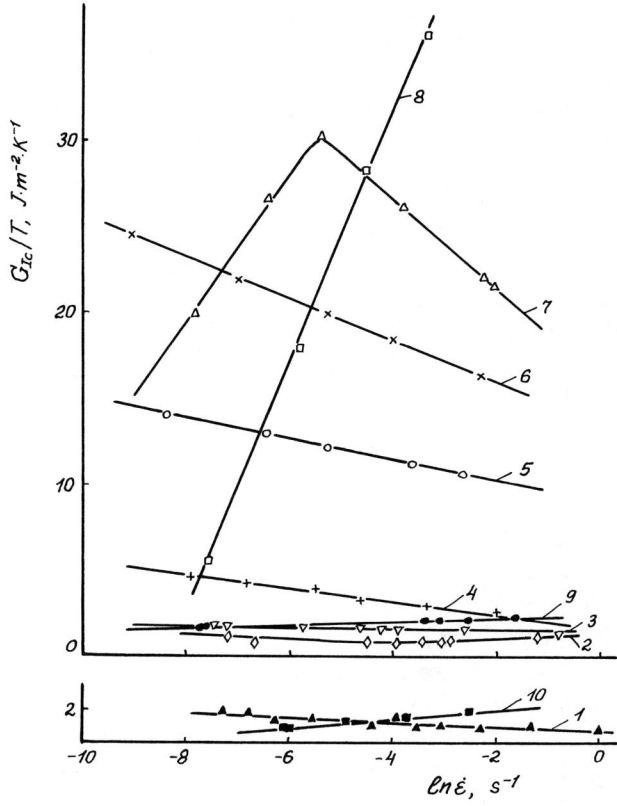

Fig. 3. Critical strain energy release rate G_{Ic} dependence on the deformation rate at the same temperatures as for Fig. 2

quencies were recalculated in deformation rates taking into consideration the constancy of dynamic deformation amplitude.

The master curves at reference temperature 298 K are shown in Fig. 4. All the curves have sharp peaks at low deformation rates (high temperatures) although its location for crack resistance parameters and tgδ does not coincide. It seems to be obvious that the peaks are connected with α-relaxation. Therefore, the crack resistance of the cured epoxy polymer at low deformation rates (high temperatures) is determined by this relaxation process. The difference of the peak location in K_{Ic} or G_{Ic} and tgδ master curves concerning with α-process (Fig. 4) is due to the high level of local deformation in the vicinity of a crack tip.

At higher deformation rates (lower temperatures) there is a wide shoulder on the master curves for crack resistance parameters corresponding to a similar shoulder on tgδ master curve in the β-relaxation region. Therefore, it can be postulated that crack growth processes and crack resistance parameters are controlled by the β-relaxation process activated by mechanical stresses in a wide range of deformation rates and temperatures.

At high deformation rates (low temperatures) there is a small peak on the K_{Ic} master curve similar to γ-relaxation peak in the tgδ master curve. It points to an influence of γ-relaxation process on crack growth processes and resistance under such loading conditions.

The data on the temperature dependence of the shift factor (Fig. 5) and the comparison of activation energies of crack growth and α-, β- and γ-relaxation processes (Fig. 6) confirm a strong relationship between crack resistance and relaxation processes in the epoxy polymer. The plot of shift factors as a function of temperature has a pronounced nonlinear character. It indicates that crack growth processes and crack resistance are controlled by a few relaxation processes. The smooth shape of the plot allows to assume that with temperature increase (deformation rate decrease) the slower relaxation processes "switch on" gradually.

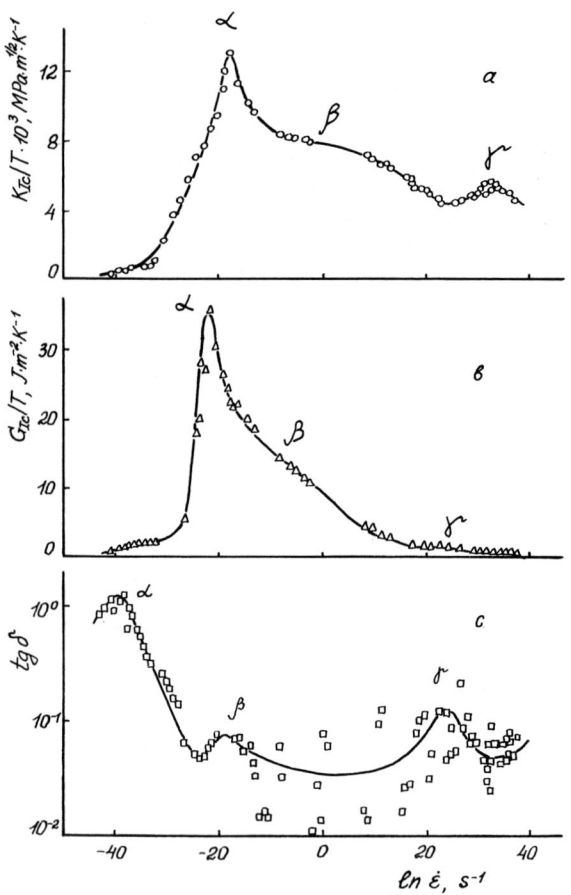

Fig. 4. Master curves for K_{Ic} (a), G_{Ic} (b) and $tg\delta$ (c) at reference temperature 298 K

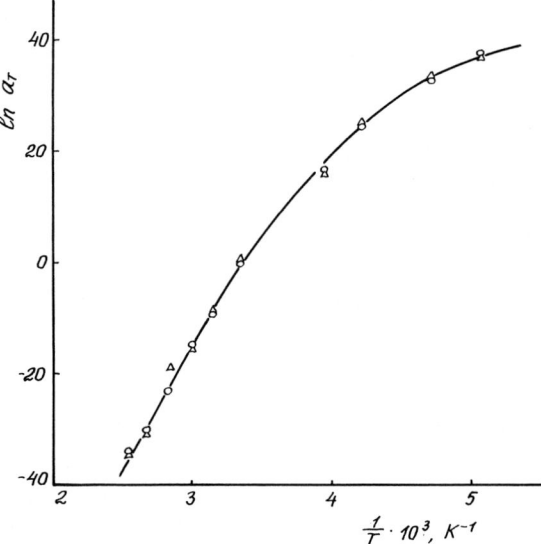

Fig. 5. Dependence of shift factors obtained from the data of Figs. 2 (○) and 3 (△) on temperature

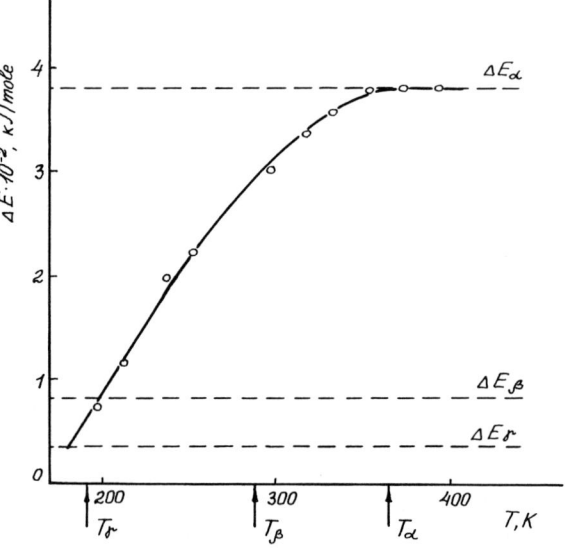

Fig. 6. Crack growth activation energy dependence on temperature

The data in Fig. 6 show that the activation energy of the processes which control crack resistance reaches the value of γ-relaxation activation energy at $T_\gamma - (20-30)$ K. When the temperature increases, the crack growth activation energy increases also and comes up to the level of the β-relaxation activation energy at temperatures much smaller than T_β. In a wide temperature range the activation energy of processes which control the crack resistance of cured polymer have values between activation energies of β- and α-relaxation. The crack growth activation energy reaches the value of the α-relaxation process at temperatures near T_α.

Conclusions

An experimental study of the crack resistance parameters and mechanical loss factors of epoxy resin cured with a hexamethylenediamine curing agent have been carried out as a function of deformation rates and temperatures. By using the time — temperature superposition principle the master curves were determined and their identity for crack resistance and mechanical loss is revealed. Activation energies of crack growth and relaxation processes were evaluated for a wide range of temperatures and deformation rates. The following conclusions were reached from this study.

There is a direct relationship between crack growth and relaxation processes in crosslinked glassy polymers such as cured epoxy resins. Their ability to resist crack propagation is not a material constant and is determined by molecular motions in glassy polymer activated by mechanical load under given time-temperature loading conditions. At low temperatures (high deformation rates) the crack growth resistance is controlled by the γ-relaxation process related to motions of short polymer chain fragments less than a statistical segment. At intermediate temperatures and deformation rates the crack growth resistance of epoxy polymers is determined by the β-relaxation process related to individual statistical segment motions. At high temperatures (low deformation rates) the most important role is played by the α-relaxation process as the co-operative motion of statistical segments.

References

1. Berry J (1964) In: Rosen B (ed) Fracture processes in polymeric solids. Interscience, New York, pp 157 to 234
2. Broutman L, McGarry F (1965) J Appl Polym Sci 9:589—626
3. Andrews E (1968) Fracture in polymers. American Elsevier, New York
4. Kausch H (1978) Polymer fracture. Springer-Verlag, Berlin
5. Kinloch A, Young R (1983) Fracture behaviour of polymers. Applied Science, London
6. William J (1984) Fracture mechanics of polymers. Ellis Horwood, Chichester
7. Atkins A, May Y (1985) Elastic and plastic fracture: metals, polymers, ceramics, composites, biological materials. Ellis Horwood, Chichester
8. Maugis D (1985) J Mater Sci 20:3041—3073
9. Evans K (1987) J Polym Sci Polym Phys B27:353—368
10. Babayevsky P, Kulik S (1991) Crack resistance of cured polymer systems. Khimia, Moscow (in Russian)
11. King N, Andrews E (1978) J Mater Sci 13:1291—1302
12. Broek D (1974) Elementary engineering fracture mechanics. Noordhoff, Leyden
13. Babayevsky P, Kulik S, Pavlenko A, Borovko V (1987) Mechanica composit mater N1:10—16 (in Russian)
14. Parton V, Morozov E (1985) Elastic-plastic fracture mechanics. Nauka, Moscow (in Russian)
15. Boyer R (1975) J Polym Sci Polym Symp 50:189—207
16. Boyer R (1975) Polymer 17:996—1008
17. Bershtein V, Egorov V (1990) Differential scanning calorimetry in polymer physics and chemistry. Khimia, Leningrad (in Russian)
18. Ferry J (1970) Viscoelastic properties of polymers. 2nd edn Wiley, New York

Authors' address:

Dr. S. G. Kulik
Olonetskaya str. 23
apt 27, Moscow 127273, Russia

Fluorescence study of interpenetrating network morphology of polymer films

Ö. Pekcan

Istanbul Technical University Department of Physics, Istanbul, Turkey

Abstract: Fluorescence experiments were carried out on polymer films composed of poly (vinyl acetate) (PVAc, 91.7 mol%) and poly (2-ethylhexyl methacrylate) (PEHMA, 8.3 mol%) (all the PEHMA present as a graft copolymer), which were prepared by solvent casting. The PEHMA phase was labeled with a low concentration (0.7 mol% PEHMA) of naphthalene groups. Exposure of the films to a pentane solution of 9-anthrylmethyl pivalate (AMP) transferred the AMP molecules exclusively to the PEHMA phase of the material. Direct energy transfer studies showed that the energy transfer rate was proportional to the AMP concentration, and that the process was characterized by an effective dimensionality $\bar{d} = 1.3$. This result can be explained by the Exponential Series Method (ESM) in terms of a crossover phenomenon in restricted geometries, if the PEHMA is present in the film in the form of interconnected, long, thin (ca. 25 Å diameter) cylinders. Supporting this idea is our finding that swelling agents for PHEMA such as hexadecane are also taken up by the film from pentane solution. Experiments on these materials give $\bar{d}$ values that increase with increasing hexadecane until $\bar{d}$ becomes equal to 3.0. These final results support the idea that PEHMA forms an interpenetrating network in glassy PVAc film.

Key words: Fluorescence; restricted geometry; interpenetrating network

Introduction

The single photon counting method for measuring fluorescence decay is routinely applied to many heterogeneous systems. Using fluorescence spectroscopy with direct energy transfer (DET) technique, the distribution of probe molecules embedded in fractal lattice [1] or in restricted geometries [2] has been characterized. Special attention has been paid to porous glass [3], silica [4] and polymer membranes [5].

Recently, we reported that the local structure of interpenetrating network-like morphology can be characterized with DET technique in polymer membranes [6], and in latex particles [7]. An attractive approach to study such systems involves the idea of energy transfer in restricted geometries [8]. The distribution of donor and acceptor molecules in porous materials having known pore geometries of size small enough to influence molecular reactions has been emphasized. The idea of the consequence of DET in restricted geometries was first suggested by Yang and El-Sayed [9] and developed theoretically by Klafter et al. [10].

In such systems, one might expects distribution of donor life-times. Ware et al. [11—15] have used the exponential series method (ESM) to recover fluorescence lifetime distributions from both simulated and experimental decay data. This method is based on recovering the coefficients of an exponential series with fixed lifetime. In fact, recovery of distributions of fluorescence life times from fluorescence decay data is closely related to inversion of Laplace transforms. In other words, the fluorescence decay law can be given by Laplace transform of a known distribution function.

Inverse Laplace transform of well-known Förster decay law [16, 17] in highly viscous systems has been used to interpret some simulated and experimented decay data [14, 15].

A more general form of the Förster decay law is known as the Klafter-Blumen equation [18, 19]. (K-B) Eq. was derived to explain donor decay in a fractal object in the presence of acceptors, however, it has been used to study effective dimensionality in restricted geometries [2, 7, 8].

In this paper, we derived the Fourier transform of the (K-B) equation. The obtained distribution function was compared with the experimentally recovered lifetime distributions which are obtained from polymer films. ESM was used to recover an underlying distribution of lifetime from experimentally obtained donor decays.

The experiments involve a material composed of poly(vinyl acetate) (PVAc 96 monomer mol%) and poly (2-ethylhexylmethacrylate) (PEHMA 4 monomer %) prepared initially as a non-aqueous dispersion (NAD) latex with the latter low T_g polymer present in the form of a PVAc-PEHMA graft copolymer [22]. These latex materials dissolve completely in solvents like chloroform and ethyl acetate, and produce films when the solvent is allowed to evaporate. Phase separation occurs and a new interpenetrating network morphology appears. The minor component forms interconnected domains of very small size which are characterized by an apparent dimension of $\bar{d} = 1.30$ [7]. In these materials, donor and acceptor molecules were both present in the rubbery phase where one of them was covalently labeled.

Theoretical considerations

Exponential Series Method (ESM)

In some cases the fluorescence decay follows a sum of exponentials law which is described by

$$\phi(t) = \sum_{i=1}^{N} F_i \exp(-t/\tau_i) , \qquad (1)$$

here N is the number of fluorescent compounds, τ_i are lifetimes with F_i amplitudes. When the physical process underlying the observed decay causes a smooth distribution of lifetimes, then Eq. (1) may be written as follows,

$$\phi(t) = \int_0^\infty F(\tau) \exp(-t/\tau) d\tau , \qquad (2)$$

where $F(\tau)$ represents the continuous distribution of fluorescence lifetimes. Usually, the experimentally observed fluorescence decay $I(t)$ is convolved with the instrument response function $L(t)$:

$$I(t) = \int_0^t L(t)\phi(t-s)ds . \qquad (3)$$

Mostly, where there are only a few discrete life times underlying the fluorescence decay, N expands from 1 to about 3 or 4. In this situation the analysis of the experimental decay is to assume a sum of exponential decay law $\phi(t)$ and convolve this with an instrumental response function $L(t)$ to obtain $I(t)$. The iterative deconvolution method then can be used to obtain the free fluorescence lifetimes and associated amplitudes. ESM method, however, extends this idea to analyze the smooth distributions by assuming a set of fixed fluorescence lifetimes τ_i and an associated set of variable amplitudes F_i [11–15]. The (F_i, τ_i) set usually contains 20 to 30 pairs and experimental data are analyzed, allowing only the F_i amplitudes to vary with the iterative deconvolution method. In ESM, the spacing between the lifetimes is choosen to be equal, then the resulting set of amplitudes $F(\tau)$ maps out the smooth distribution.

Fourier Transform of K-B Equation

In the classical problem of DET, one considers an excited donor at position r_0 surrounded by acceptor molecules which occupy some of the sites r_i of a given geometry. The survival probability of an excited donor for low acceptor concentration is then given by [8]

$$\phi(t, r_0) = \exp\left| -\frac{t}{\tau_0} - PI(t, r_0) \right| , \qquad (4)$$

with

$$I(t, r_0) = dr \sim \rho(r) | 1 - \exp(-tW(r-r_0)) | , \qquad (5)$$

where τ_0 is the excited donor lifetime, P is the probability of occupying sites by acceptors, $\rho(r)$ represents the site density function, and $W(r-r_0)$ is the rate of energy transfer from a donor to an acceptor.

Klafter and Blumen [18] (K-B) derived a general equation to analyze fluorescence decay profiles of an excited donor surrounded by acceptors in a fractal object. This relatio is obtained from Eq. (4) together with Eq. (5) by chosing the site density function as $\rho(r) = \rho_0 r^{(\bar{d}-d)}$, where d is the dimension of the embedding Euclidean space, $\bar{d}$ is the fractal dimension [18, 23] ($\bar{d} < d$), and ρ_0 is the proportionality constant. The K-B equation can be written in the functional from

$$\phi(t) = \exp[-ta - b^\beta], \quad (6)$$

where

$$a = \tau_0^{-1}, \quad P = A\Gamma\left(1 - \frac{d}{6}\right)$$

and $b = P\tau_0^{-\beta}$. (7)

Here, $\beta = \bar{d}/6$, and A is a time-independent constant which is proportional to the number of acceptors within the critical transfer radius R_0, and $\Gamma(x)$ is the gamma function.

The form of Eq. (6) is quite general. In fact, the term $\bar{d}$ can be interpreted as the "apparent dimension" characterizing the donor acceptor distribution. For energy transfer in three dimensions $\bar{d}/6 = 1/2$, and one finds the Förster equation. In two dimensions $\bar{d}/6 = 1/3$, the result obtained by Blumen [24]. In the case of restricted geometries, however, $\bar{d}$ has no intrinsic meaning and is the natural consequence of temporal or geometrical crossover. The value of $\bar{d}$ can take a value between zero and 3, but this does not mean that the system is a fractal object.

From the point of view of ESM, the fluorescence decay law represented by K-B equation can be given by the Fourier transform of the distribution function $F(\tau)$ of Eq. (2)

$$\int_{-\infty}^{+\infty} dt \exp(ik't - at - bt^\beta)$$

$$= \int_{-\infty}^{+\infty} dt \int_{-\infty}^{+\infty} dk F(k) \exp(ik't - kt), \quad (8)$$

where $k = 1/\tau$.

The solution of RHS of above equation is immediate and gives the values of $2\pi i F(k)$. In order to solve the LHS of Eq. (8), Taylor series expansion is applied to the exponent. The final form of the inverse Fourier transform of K-B equation is then obtained and given by

$$F(k) = R(\beta)^{-1/2} \exp\left[-\frac{(k-a)^\beta}{\beta b}\right]^{(\beta-1)^{-1}}. \quad (9)$$

where

$$R(\beta) = 2\pi(\beta - 1)\beta b \left[-\frac{k-a}{\beta b}\right]^{\frac{(\beta-2)}{(\beta-1)}}. \quad (10)$$

For $\bar{d} = 3$ ($\beta = 1/2$), Eq. (9) gives the well-known Förster distribution of lifetimes [14, 15]

$$F(k) = R(1/2) \exp\left[-\frac{b^2}{4(k-a)}\right], \quad (11)$$

where

$$R(1/2) = (b/2) |\pi(k-a)^3|^{-1/2}. \quad (12)$$

Equation (11) is also known as the inverse Laplace transform of Förster decay law [14, 15].

Experimental

The polymer materials have been described previously [20, 21]. They were prepared by dispersion polymerization (isooctane) of vinyl acetate in the presence of PEHMA containing 5 mol% reactive-NCO groups. Small amounts of naphthalene groups were attached to the PEHMA component by reaction with 2-(1-naphthyl)-ethanol, with the remaining NCO groups then reacted with n-octanol. In a second preparation, the particles were reacted sequentially with naphthylethanol and 2-(9--anthryl)-ethanol, followed by n-octanol, to introduce both N- and An-groups into the PEHMA component. These materials were analyzed for composition, purity, and chromophore content by ^{1}H NMR (CDCl$_3$ solutions), UV spectroscopy (ethyl acetate solutions), and gel permeation chromatography (also ethyl acetate). The N-labeled particles contained 4.1 monomer mol% (8.3 wt%) PEHMA and 4.4×10^{-6} mol g^{-1} N groups (0.69 mol% N in the PEHMA phase) and are referred to as N.7.

Films were prepared by dissolving a known weight of N.7 in chloroform, placing the solution in a 12 mm o.d. quartz tube attached to a rotary evaporator and removing the solvent under vacuum. Films formed on the inside of the tube and were left under vacuum (10^{-1} Torr) overnight. Films from N.7 were subsequently exposed to solutions of 9-anthrylmethyl pivalate in pentane for 3 h and then the pentane was removed on the rotary evaporator, followed by overnight exposure to vacuum (10^{-1} Torr). The amounts of AMP added ranged from 0.85×10^{-5} mol g^{-1} to 2.35×10^{-5} mol g^{-1}.

Fluorescence decay profiles were obtained using a home built apparatus for time-correlated, single photon counting measurements. Decay curves of Eq. (3) were analyzed using an iterative least-squares analysis programs. In this analysis, Eq. (2) was used for the ESM to find $F(\tau)$ distributions. The same experimental decay curves were also analyzed using Eq. (6) to obtain β values.

Film samples were excited at 284 nm, which corresponds to the absorption band of naphthalene, and the decay profiles were measured at 337 nm. In order to eliminate the color-shift effect of the photo multiplier tube, the delta pulse convolution method was used with a response curve from a dilute solution of 1,4-bis(5-phenyl-1,3-oxazol-2-yl)benzene (POPOP) in cyclohexane ($\tau = 1.1$ ns).

In ESM analysis amplitudes were constrained to be positive in order to prevent the results from oscillating [11] and also lifetimes whose amplitudes were very small are deleted from the $F(\tau)$ distributions. The ratio between the largest and smallest amplitudes was chosen to be 1—3% to remove a lifetime from the distribution curve. Chi-square cut off is chosen as $0.001 \chi^2 < 1.20$ were obtained from both ESM and fractal analysis. Initial guesses for the amplitudes were chosen equal, which prevents biasing the least-squares analysis. We found that, even for very high values of initial amplitudes, convergence is very fast (χ^2 jump from 10^4 to 10^0 orders in a single step).

Results and discussion

In order to test the reliability of fractal analysis procedure for the determination of $\beta = (\bar{d}/6)$ the fluorescence decay of N was measured in an N.7 film with no AMP. This decay curve showed single-exponential behavior (three decades) with a lifetime $\tau_0 = 66$ ns. This test is the first prerquisite for further analysis of such systems with the DET method [6, 7].

A typical fluorescence decay curve from a film doped with AMP is shown in Fig. 1. The curvature at early times indictes the occurrence of energy transfer. When these types of data are fitted to Eq. (6), an average value of $\bar{d} = 1.30$ was obtained. In each case, P values were found to be proportional to the AMP concentration. These data are summarized in Fig. 2. Here, the linear relation in Eq. (7) is obeyed, and the line which fits the data passes through the origin.

In order to explain the dimensionality of $\bar{d} = 1.30$, we introduced the ESM analysis to our experimental data. ESM were applied to the film samples indicated by a, b, and c in Fig. 2. Lifetime distributions recovered from ESM analysis are shown in Figs. 3a, b, and c for the corresponding samples with increasing AMP concentrations. Chi-squares obtained for these samples were found below 1.20, and the weighted residuals were presented by perfectly random distributions. As shown in Fig. 3, two distinct distributions were observed for AMP

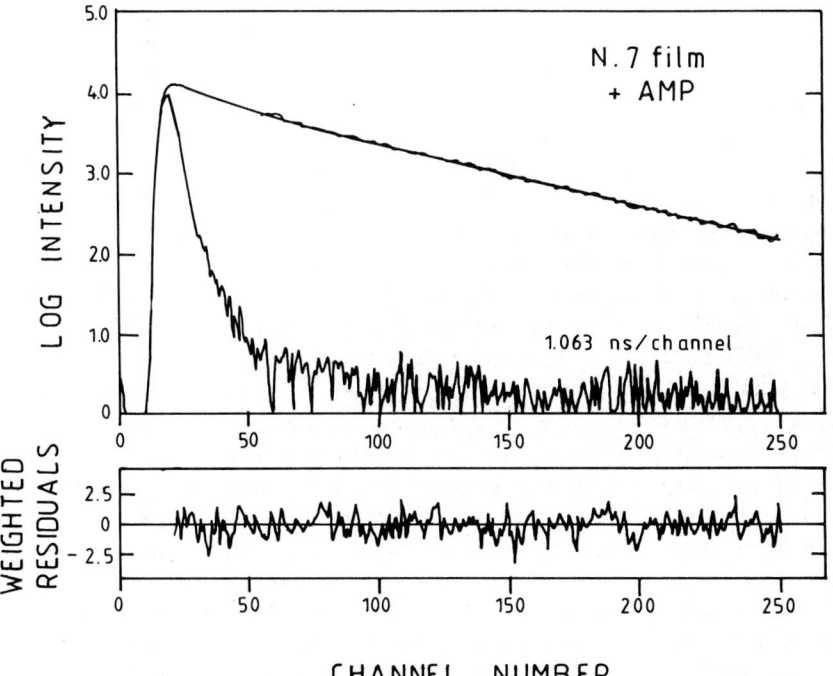

Fig. 1. Fluorescence decay curve for a film of N.7 doped with AMP. The solid line represent the best fit of the data to Eq. (2) (ESM) and Eq. (6). Both techniques resulted in good χ^2 and random-weighted residuals. The lower vurve is the response function of the SPC system. The sample is excited at 284 nm, and measured at 337 nm

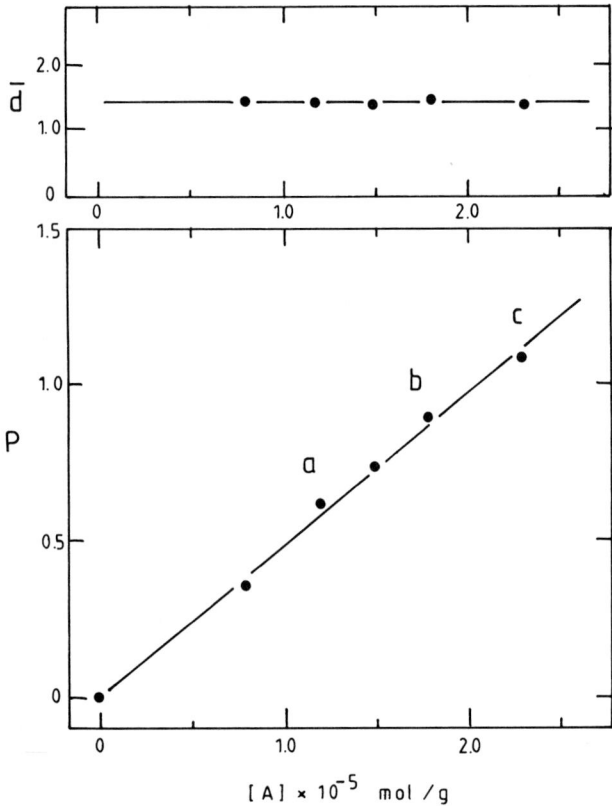

Fig. 2. Plot of d and P vs. AMP concentration for films of N.7. The parameters were obtained by fitting the fluorescence data to Eq. (6). Samples a, b, and c were also analyzed by ESM

doped N.7 film samples. A single spike at 3 ns channel was omitted from the distribution curve. Such spikes may be an artifact, which commonly arises when analyzing broad distributions [13]. As the concentration of AMP increased, amplitude of $F(\tau)$ at short times increased considerably. At the long-time region, however, $F(\tau)$ distribution curved slightly as the concentration increased.

The behavior of experimental $F(\tau)$ may be explained by the theoretical k distribution given by Eq. (9) and Eq. (11). Theoretical distributions obtained from Eqs. (11) and (9) are shown in Figs. 4a and b for $\bar{d} = 3$ and $\bar{d} = 1$, respectively $a = 0.015$ was taken in both cases. In Fig. 5a overlap of these two $F(\tau)$ distributions is presented. These distributions are quite sensitive to b and $\beta (= \bar{d}/6)$ values, and are not much affected by the a parameter. In order to keep $F(\tau)$ amplitudes in the same scale for $\bar{d} = 3$ and $\bar{d} = 1$, $b = 0.65$ and $b = 6.5$ were chosen, respectively.

When the theoretical distribution in Fig. 5a is compared with the ESM result of $F(\tau)$ distribution in Fig. 3a, it is seen that agreement is quite sartisfactory. Theoretical τ (or k) distributions were generated for different combinations of $\beta(= \bar{d}/6)$ and b parameters. These results, plotted in Figs. 5b and c are also in good agreement with the experimentally found $F(\tau)$ distributions in Figs. 3b and c, respectively. As b values increased, $F(\tau)$ distribution in $\bar{d} = 1$ region curved slightly, which kept the agreement with increase in AMP concentrations of experimental ESM distributions. At $\bar{d} = 3$ region increasing b parameter shifts $F(\tau)$ distribution to short times.

Experimental $F(\tau)$ distributions obtained from ESM analysis cannot be explained with a single dimension $\bar{d}$. This result rejects the ida that interpenetrating network morphology is a fractal in this glassy polymer membrane film. On the other hand, using the theoretical distribution, ESM results may be explained with the idea of restricted geometries. If one considers that acceptor molecules are distributed in a cylinder of radius R and the donor to be located in the axis of the cylinder, then for $r < R$ the site density function is given as $\rho(r) = 2\pi r$ and Eq. (5) becomes

$$I(t, R) = 4\pi\rho \int_0^R dr \int_0^L dz [1 - \exp[-ta(r^2 + z^2)^{-3}]] , \quad (13)$$

with

$$a = R_0^6/\tau_0 .$$

Here, R_0 is the critical transfer radius.

In the limiting case of short times $t \ll (R/R_0)^6 \tau_0$, Eq. (13) leads Eq. (4) to Förster decay in three dimensions. For the long-time region, $t \gg (R/R_0)^6 \tau_0$, Eq. (3) yields a decay corresponding to energy transfer in one dimension. Thus, the cylindrical geometry leads to a crossover between a three-dimensional and a one-dimensional behavior.

It can be argued, that the ESM results in Fig. (3) may represent geometrical crossover between three- to one-dimensional behaviour. For example, if PEHMA is present in the form of thin uniform cylinder with $R = 23$ Å, one can calculate the crossover time as $t_c = 27$ ns for the $R_0 = 26.6$ Å and $\tau_0 = 66$ ns values. This crossover time is between the three- and one-dimensional $F(\tau)$ distributions in Fig. 3 and corresponding theoretical distributions in Fig. 5. This argument supports the idea that the experimentally obtained dimension

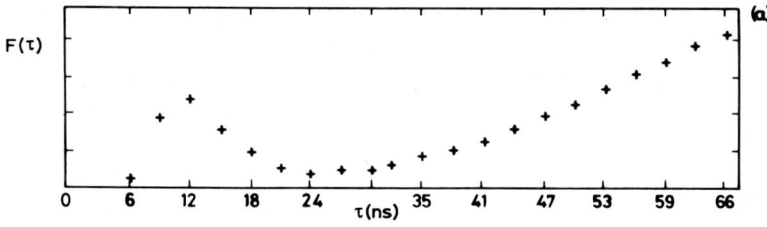
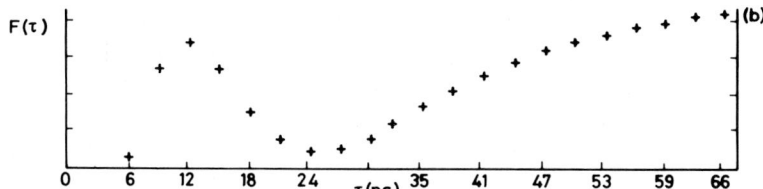
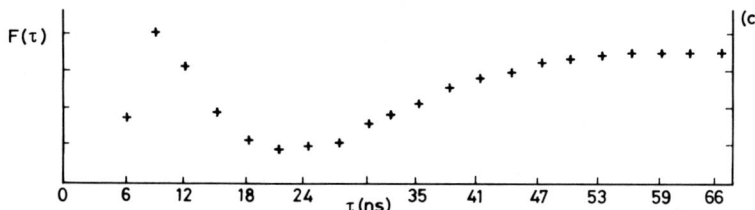

Fig. 3. Lifeime distribution recovered by ESM analysis of experimental fluorescence data for samples:
a) indicated by a in Fig. 2;
b) indicated by b in Fig. 2;
c) indicated by c in Fig. 2

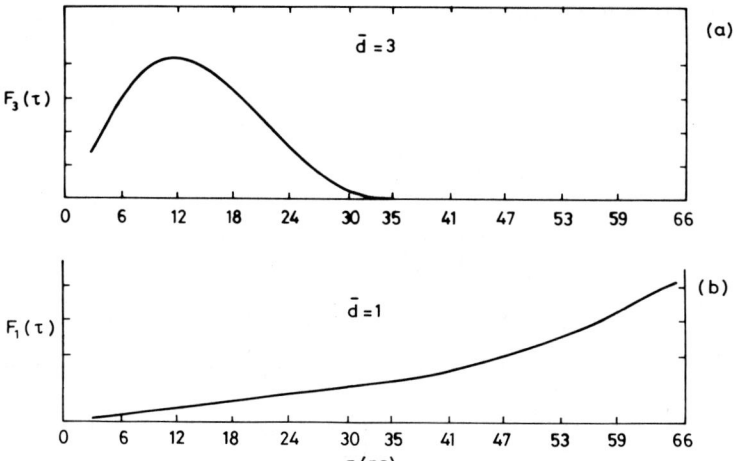

Fig. 4. Theoretical lifetime distributions obtained from Eq. (9).
a) for $\bar{d} = 3$ and $b = 0.65$;
b) for $\bar{d} = 1$ and $b = 6.5$;
In both cases $a = 0.151$ ns^{-1} was taken $(a = \tau_0^{-1})$

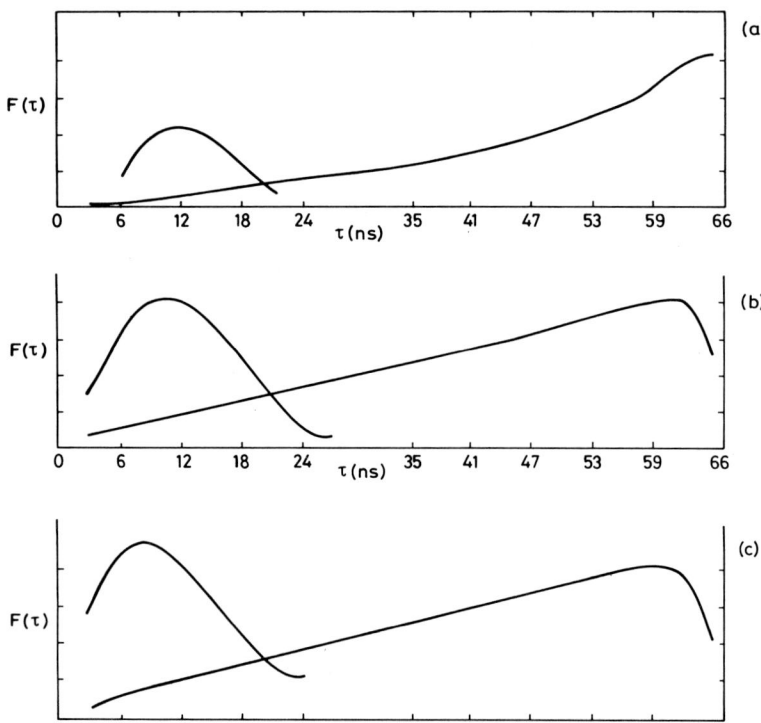

Fig. 5. Overlap of theoretical lifetime distributions obtained from Eq. (9).
a) for $\bar{d} = 3$, $b = 0.65$ and $\bar{d} = 1$, $b = 6.5$;
b) for $\bar{d} = 3$, $b = 0.70$ and $\bar{d} = 1$, $b = 8.0$;
c) for $\bar{d} = 3$, $b = 0.80$ and $\bar{d} = 1$, $b = 8.5$

$\bar{d} = 1.3$ is the apparent dimension of the local morphology of the PEHMA phase in N.7 films.

The apparent dimension can also be obtained analytically by taking the average of $\bar{d}$ over the theoretical distribution as follows:

$$d_{app} = \frac{\int_0^{\tau_0} F(\tau)\bar{d}d\tau}{\int_0^{\tau_0} F(\tau)d\tau}$$

$$d_{app} = \frac{3\int_0^{\tau_0} F_3(\tau) + \int_0^{\tau_0} F_1(\tau)d\tau}{\int_0^{\tau_0} F(\tau)d\tau}, \quad (14)$$

where $F_3(\tau)$ and $F_1(\tau)$ represent the τ distributions at $\bar{d} = 3$ and $\bar{d} = 1$ regions. Using Eq. (14), apparent dimensions were calculated from Figs. 5a, b, and c and found to be $d_{app} = 1.21$, 1.36, and 1.31, respectively. The corresponding experimental dimensions for samples a, b, and c in Figs. 3 and 4 were found as $d_{app} = 1.28$, 1.22 and 1.38.

References

1. Even U, Rodemann K, Jortner J (1984) Phy Rev Lett 52:21
2. Klafter J, Drake JM (1989) (eds) Molecular Dynamics in Restricted Geometries: New York
3. Dozier WD, Drake JM, Klafter J (1980) Phys Rev Lett 56:197
4. Levitz P, Drake JM (1987) Phys Rev Lett 58:686
5. Kopelman R, Parus S, Prasad J (1988) Phys Rev Lett 56:641
6. Pekcan Ö, Winnik MA, Croucher MD (1988) Phys Rev Lett 61:641
7. Pekcan Ö, Winnik MA, Croucher MD (1990) Chem Phys 146:283
8. Klafter J, Blumen A, Zumofen G, Drake JM (1987) J Luminescence 38:113
9. Yang CL, El-Sayed MA (1986) J Phys Chem 90:5720
10. Blumen A, Klafter J, Zumofen G (1986) Optical Spectroscopy of Glassesed. Zschokke I, Reidel Dordrecht, Holland
11. James DR, Ware WR (1986) Chem Phys Letter 126:7
12. Siemiarczuk A, Ware WR (1987) Chem Phys Letter 140:277
13. James Dr, Lin Y-S, Peterson NO, Siemiarczuk A, Wagner BD, Ware WR (1987) SPIE Fluorescence Detection 743:117
14. Siemiarczuk A, Wagner BD, Ware WR (1990) J Phys Chem 94:1661

15. Wagner BD, Ware WR (1990) J Phys Chem 94:3489
16. Förtster T (1949) Z Naturforsch 49:321
17. Birks JB (1968) J Phys B ser 2, 946
18. Klafter J, Blumen A (1984) J Chem Phys 80:875
19. Klafter J, Blumen A (1985) J Lumens 34:77
20. Pekcan Ö, Egan LS, Winnik MA, Croucher MD, Macromolecules
21. Egan LS, Winnik MA, Croucher MD (1986) J Polym Sci Chem, ed 24, 1895
22. Winnik MA (1975) Polymer Surface and Interfaces, Feast J, Muro H (eds) Wiley, New York
23. Mandelbrot B (1982) The Fractal Geometry of Nature, Freeman, San Francisco
24. Keller H, Blumen A (1984) J Biophys Soc 46:1

Author's address:

Ö. Pekcan
Istanbul Technical University
Department of Physics
Maslak 80626
Istanbul, Turkey

Kinetic regularities of polymer network thermal degradation

L. P. Smirnov and N. N. Volkova

The Institute of Chemical Physics in Chernogolovka, Russia

Abstract: The structures and kinetics of the thermal degradation of three polymer networks have been studied. In the course of the thin polyurethane network degradation, the evolution of the MWD of polymer chains has a nonmonotonous character, but the sample mass decrease is described by the equations for two parallel first-order reactions. Activation parameters are shown to depend on the polymer curing conditions. The kinetics of the dense polymer network (the polymer of dimethacrylic ester of triethylene glycol, the epoxyanhydride polymer) degradation depends on the hardness of the polymer samples: soft polymer samples are degraded following a mechanism of radical-chain depolymerization, while for hard polymer samples the degradation proceeds mainly by a decay of unstable functional groups through a molecular mechanism. The results of an NMR study of molecular mobility during PTGM degradation are explained by the microheterogeneous structure of this polymer. It has been shown that the kinetic regularities of the dense polymer network degradation strongly depend on the presence of inert fillers.

Key words: Polymer network; thermal degradation; kinetics

Introduction

The purpose of this paper is to generalize ours results on the polymer network thermal degradation in vacuum [1—10]. Three types of polymers have been studied: crosslinked polyurethanes (CPU-I and CPU-II) with low density of crosslinks; crosslinked polyepoxides (CPE) and the polymer of dimethacrylic ester of triethylene glycol (PTGM) with high density of crosslinks.

Crosslinked polyurethane

The polyurethane networks were synthesized from 2,4-toluene diisocyanate and polyester (CPU-I) or oligobutadiendiole (CPU-II) using from 0.3 to 7.0 mass % of 1,1,1-trimethylolpropane, polyoxipropylenetriols, and triethanolamine as branching agents. It has been found (see the gravimetric method [1, 2]) that the kinetics of the mass decreases ($\Delta m/m_0$) of degrading CPU-1 samples can be described in terms of two first-order differential equations

$$-\frac{d(\Delta m/m_0)}{dt} = \sum_{i=1}^{2} k_i [(\Delta m_\infty^{(i)}/m_0) - (\Delta m^{(i)}/m_0)] \ .$$

The ratio of the rate constants (k_1/k_2) amounts to 10. The total mass change ($\Delta m_\infty^{(1)}/m_0$) corresponding to the first stage amounts to 0.20—0.25. There is a number of convincing facts, which show that the two-stage kinetics is due to the different reactivity of unstable groups (probably urethane groups (UG)), depending on their location in the CPU-1 network: 1) the UG located directly near crosslinked points are strongly deformed as compared with those belonging to other parts of polymer chains (see the computer simulations of the UG configuration [3]; 2) the quantity of the low molecular weight fraction (corresponding to the original oligomer M_w) does not change during the first stage, whereas that of the high molecular weight fraction increases (see the sol-gel analysis [4]); 3) the UG decay occurs mostly near the crosslink points of the CPU-1 network (see the NMR study of the molecular weight distribution (MWD) of inter-crosslink chains [5, 6]).

The CPU-1 networks were synthesized in two ways: 1) in the presence of a catalyst at 303 K, and 2) at 313—333 K with no catalyst present. The measurements have shown (see the gravimetric method [1, 2]) that, in the first case, the degradation was characterized by the activation energy E_a approximately two times smaller than in the second case. It has been found [7] that the catalyst of the synthesis had no effect on E_a of the degradation. However, kinetic parameters of the degradation of CPU-1 samples synthesized at 303 K depend on the rate and the temperature of synthesis, higher rates and lower temperature of synthesis resulting in less stable networks.

These facts led us to the assumption that, at certain conditions of CPU-I synthesis, the use of the catalyst results in the formation of nonequilibrium network structures with stressed chemical bonds. It is known [11—12] that polar oligomer molecules form associates. The concentration of the latter can be considerable at low temperatures. Since the mobility of the associated oligomer molecules is low, the rate of synthesis (accelerated by the catalyst) may be higher than that of the structural changes, resulting in the formation of network in thermodynamically nonequilibrium conformational states.

The assumption that the reduction of E_a in CPU-1 samples synthesized at 303 K arises from the presence of structural stresses is confirmed by a number of experimental results [1, 2]: 1) The total heat produced at the first stage of the degradation of these samples, exceeds by more than two times the heat yield of the initial stage of degradation of more stable samples (see the calorimetric method). 2) As time passes (the sample has been kept at room temperature in vacuum during 2 years) E_a of the unstable samples grows, whereas the crosslink density and the thermal effect of the decomposition decrease. At the same time, in the case of stable samples synthesized at higher temperatures, these characteristics remain unchanged (see the Cluff method). 3) Factors promoting the decomposition of the associates (such as higher temperature of synthesis, addition of low molecular weight diole) increase E_a.

Analysis of the MWD of the sol and inter-crosslink chains of the network in the course of degradation has shown [4—6] that, in contrast with relatively simple kinetics of the sample mass decrease, both the sol and the inter-crosslink chains MWD change in a rather complicated, irregular manner. The simple calculations show that, at the initial stage of the degradation, the kinetics of the mass decrease of the CPU-I samples is determined by the decay of the UG followed by elimination of the oligomer fragment. The mass change can be essentially reduced if, simultaneously with the decay of the UG, formation of new crosslinks occurs: for CPU-II the mass decrease accompanying the degradation constitutes only 0.02—0.06. The reason is that, parallel to the scission of chemical bonds, there occurs a reaction of crosslinking via the double bonds of the oligomer fragments [1, 2].

Polymer of dimethacrylic ester of triethylene glycol

Analysis of the degradation kinetics of PTGM has showed that both the kinetic parameters and $\Delta m_\infty / m_0$ are strongly dependent on the initial polymerization conversion Γ_0. The thermal stability of the network undergoes strong changes at $\Gamma_0 = \Gamma_m$, where Γ_m is the polymerization conversion at which the polymer becomes monolithic. The degradation of networks with $\Gamma_0 > \Gamma_m$ is described by a first-order differential equation, while the degradation of samples with $\Gamma_0 < \Gamma_m$ is well described in terms of two stages, each described by the first-order autocatalysis equation (Fig. 1) [8].

There are several mechanisms of polyestermethacrylate degradation: decay of ester groups followed by the formation of alcohol fragments and CO_2; chain depolymerization, etc. Since not only the observed kinetics, but also the mechanical properties of PTGM undergo sharp changes in the vicinity of the Γ_m point, one may assume that the ratio of the rates of these processes is determined by the "hardness" of the polymer matrix. The latter in turn depends on the degradation temperature, degradation degree, mechanical processing and Γ_0 [8].

The PTGM structure behavior in the course of degradation was studied by analyzing the NMR proton FID's of samples with various degrees of degradation [9]. The undestructed PTGM sample has a microheterogeneous structure. Its FID can be presented by the superposition of two "Gauss-like" components corresponding to "grains" with a high

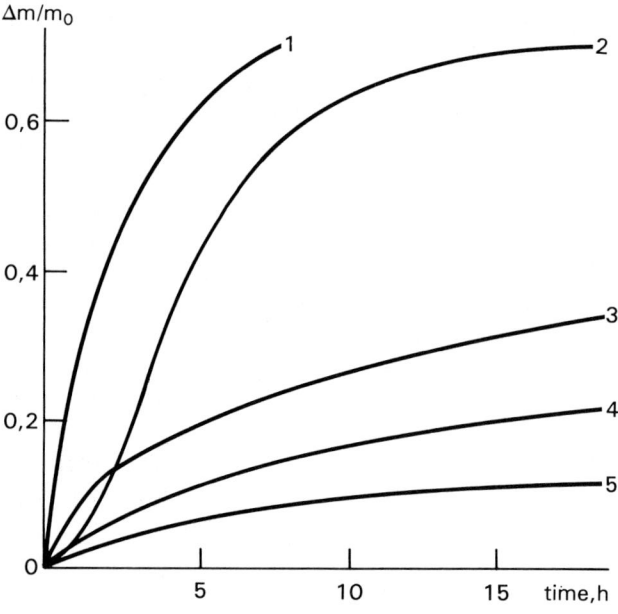

Fig. 1. The thermal degradation of PTGM. The kinetic curves of the mass decrease of samples of 1—2 mm particles at 500.5 K. 1) — $\Gamma_0 = 0.73$; 2) — $\Gamma_0 = 0.78$; 3) — $\Gamma_0 = 0.81$, 4) — $\Gamma_0 = 0.82$; 5) — $\Gamma_0 = 0.83$

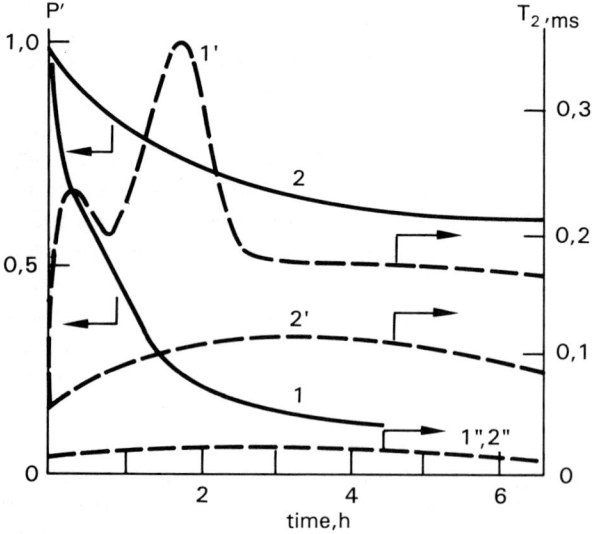

Fig. 2. The thermal degradation of PTGM at 538 K. The evolutions of the characteristic times T_2'' (1',2'), T_2' (1",2") and number protons P' corresponding to the phase with low molecular mobility. (1,2). (1,1',1") — $\Gamma_0 = 0.77$; (2,2',2") — $\Gamma_0 = 0.83$

crosslink density and an "intergrain" matrix having a higher concentration of defects. It has been found that for samples with $\Gamma_0 < \Gamma_m$ the degradation is initiated in the "intergrain" matrix. The characteristic time T_2'' of the FID corresponding to the "intergrain" matrix increases sharply at the initial stage of the degradation (Fig. 2). Presumably, this is due to the scission of weak and stressed bonds. At the second stage of the degradation the time T_2'' decreases due to partial recovery of the network by crosslinking processes. Then, T_2'' increases again. The growth of mobility results in the increase of the probability for the radicals formed at the scission of the C—C bonds to leave the "cage". Correspondingly, the depolymerization rate grows and, finally, the last stage of the degradation is characterized by a decreasing T_2'' and constant number of protons P' belonging to the phase with low molecular mobility. In the course of degradation the size of the "grains" decreases, but the structure of the "grains" remains unchanged (the time T_2' has not change).

In the case of samples with $\Gamma_0 > \Gamma_m$ the kinetic curves describing the change of P' are similar to that of the mass decrease. Since the maximum variation of P' is only about 0.3, and the time T_2' does not change in the course of the degradation, a conclusion can be made that the degradation affects primarily the "intergrain" matrix and the surface layers of the "grains" (surface layers have a higher concentration of defects). The formation of volatile substances and reduction of P' stop simultaneously. This is in accordance with the mechanism of degradation via the decay of ester groups accompanied by the yield of the alcohol fragment and the formation of a network of polymethacrylic anhydride.

Crosslinked polyepoxides

The kinetics of the degradation of CPE (synthesized from epoxy resin and anhydride) has been found to be similar to that of PTGM. The block polymer was broken into pieces of 1—2 mm. After the fragmentation the powder was separated into fractions of different sizes. It has been found that the kinetics of the mass decrease of 1—2 mm particles could be described by the autocatalysis equation (Fig. 3). With decreasing particle size the S-shaped kinetic curves eventually degenerated and for 0.3 mm particles the kinetics could be described by a first-order differential equation. Since the

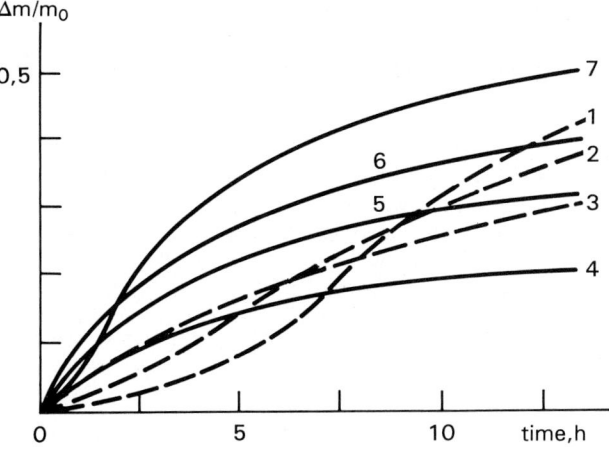

Fig. 3. The kinetic curves of the mass decrease at the thermal degradation of CPE samples at 543 K. 1) unfilled sample, 1—2 mm particles; 2) unfilled sample, 0.2 mm film; 3) unfilled sample, 0.09—0.16 mm particles; 4) composite polymer, boracic fiber, 5) composite polymer, organic fiber PABI-S; 6) composite polymer, glass-fiber; 7) composite polymer, coal-fiber

degradation of 0.2 mm film is also characterized by S-shaped kinetic curves, one can assume that the kinetics of the degradation depends on the size of particles only indirectly and is determined primarily by the structure and the particular mechanism of the CPE degradation [10].

Analysis of data available in the literature shows that there exist several degradation mechanisms: molecular mechanisms including elimination of CO_2, H_2O, low molecular weight olefin or N-dimethylamine, radical mechanism, etc. Presumably, the degradation of samples, which were not processed mechanically, at the initial stage proceeds via a molecular mechanism, while the radical mechanism acts less effectively due to the high "hardness" of the network and strong "cage effect". During the degradation the "hardness" of the matrix eventually decreases, resulting in suppression of the "cage effect" and increase of the probability for a radical to leave the "cage". This in turn results in the growth of the observed rate of the mass decrease. The fact that the radical mechanism becomes more important explains the S-shaped kinetic curves. The degeneration of the S-shaped kinetic curves and the growth of the initial degradation rate with a reduction in particle size can be explained by the "activation" of the polymer due to "loosening" of its structure with the more intense fragmentation.

The kinetics of degradation of polymer matrix of composite polymer differs significantly from that of block polymers and depends on the type of filler used (Fig. 3). The increase of the initial degradation rate of the binder can hardly be explained as due to chemical interaction between the binder and the filler. This effect is similar to the mechanical "activation" of the degradation of unfilled CPE. It can be assumed that the increase of the degradation rate of the composite is due to the fact that its polymer matrix is less dense than that of the block polymer if the latter has undergo no mechanical processing. The loosening of the polymer matrix can be explained as resulting from residual shrinkage and thermoelastic stresses. The level of these stresses obviously depends on the type of filler used and, in particular, on the adhesion between the filler and the binder and the values of expansion indices.

It has been found that there exists a correlation between the variations of E_a and the preexponential factor k_0 (the so-called compensative effect). An analysis of this correlation confirms the assumption about two possible routes of CPE degradation. Activation energies corresponding to the two mechanisms lie in the intervals 132—165 and 152—196 KJ/mol. According to modern theories [13], which consider the relation between activation energy of the chemical reaction and mechanical stresses, samples characterized by smaller E_a are more stressed. It should be noted that the decrease of E_a does not necessarily result in an increase of the initial degradation rate. From this fact it follows that there exists another factor affecting the CPE stability. Presumably, this is the "hardness" of the polymer. In the case of the radical degradation mechanism the "hardness" determines the probability for the macroradical to leave the "cage". When the degradation goes according to the molecular mechanism, the higher the molecular mobility, the more favorable are conditions for the formation of the transient complex and for the reaction. It should be noted that the considered factors determining the reaction rate effect different parameters of the Arrhenius equation. The harder the structure, the smaller the k_0. The higher the stresses, the smaller E_a. On the other hand, these factors are not independent. Samples having harder structures are, in general, characterized by higher residual shrinkage and thermoelastic stresses. A relaxation of stresses resulting from the scission of chemical bonds (thermally or mechanically activated) results in a loosening of the structure and growth of both E_a and k_0.

Conclusion

We see that for each polymer network the mass decrease accompanying the degradation is described by relatively simple kinetic equations. The particular form of the equations and corresponding activation parameters depend essentially on the conditions of the synthesis and the mechanical history of the sample.

The kinetic laws describing the degradation of thin polymer networks synthesized from polar oligomers are determined by the relation between the rates of the network formation and that of degradation of associates of oligomer molecules. At low temperatures and high rates of synthesis a crosslinked polymer with structurally stressed chemical bonds is formed, resulting in the lowering of the thermal stability. The kinetics of dense polymer network degradation depends on the relation between the rates of reactions corresponding to the radical and molecular mechanisms, and the ratio of the two rates depending on the molecular mobility.

References

1. Volkova N, Ol'khov J, Baturin S, Smirnov L (1978) Vysokomol Soedin A20:199—206
2. Volkova N, Smirnov L (1984) In: International Rubber Conference, Moscow, A25
3. Nudelman E (1992) In: ACS Abstract, San Francisco (to be published)
4. Nudelman E, Gorbushina G, Ol'khov Y, Smirnov L (1992) Vysokomol Soedin A34:N4, 69—78
5. Sandacov G, Smirnov L, Sosicov A, Summanen K, Volkova N (1992) Colloid Polym Sci (to be published)
6. Sandacov G, Tarasov V, Volkova N, Ol'khov Y, Smirnov L, Eroffeev L, Khitrin A (1989) Vysokomol Soedin A31:821—825
7. Volkova N, Ol'khov Y, Baturin S, Smirnov L (1978) Vysokomol Soedin B20:827—830
8. Volkova N, Berezin M, Korolev G, Smirnov L (1983) Vysokomol Soedin A25:871—876
9. Volkova N, Sosicov A, Berezin M, Korolev G, Yerofeev L, Smirnov L (1988) Vysokomol Soedin A30:2133—2140
10. Volkova N, Summanen E, Smirnov L, Dzhavadjan E, Ponomareva T, Rosenberg B (1990) Mekhanika Kompozitnykh Materialov: 391—397
11. Kang-Yen L, Parson J (1969) Macromolecules 2:529
12. Elias H, Solc K (1973) Journal Polymer Sci, Phys 11:137
13. Emanuel N (1979) Uspekhi Khimii 48:2113—2163

Authors' address:

Smirnov Lev P., Dr. Sci.
Institute of Chemical Physics in Chernogolovka
142432, Chernogolovka, Moscow Region, Russia

Interpenetrating polymer networks based on EVA copolymer and PMMA

U. Schulze, A. Janke, G. Pompe, E. Meyer, and M. Rätzsch*)

Institut für Polymerforschung Dresden e.V., FRG
*) PCD Linz, Austria

Abstract: Polymer blends and semi-1-IPN's based on EVA and PMMA were evaluated with regard to phase dispergation and glass transition behavior. The size of the dispersed PMMA phase in melt-mixing blends, linear blends and semi-1-IPN's was examined. It can be concluded that the prepared blends and IPN's are phase separated, but in semi-IPN's the degree of dispergation is enhanced. The dependence of the glass transition temperature on the degree of crosslinking of EVA and the EVA content in the prepared blends and IPN's is discussed.

Key words: Sequential interpenetrating polymer networks; electron micrograph; glass transition; poly(methyl methacrylate); EVA copolymer

Introduction

Interpenetrating polymer networks (IPN's) are characterized by a mixture of two or more polymer networks which have partial or total physical interlocking with each other. They are a special kind of polymer blends offering the chance to combine thermodynamically incompatible polymer systems with a high degree of dispergation of the component polymer domains. The interpenetration of the network structures impedes phase separation, but does not prohibit it. For IPN's of thermodynamically incompatible polymers it is preserved as a microphase separation. IPN's based on PMMA and elastomers, e.g., polyurethane [1—9], are characterized by improved mechanical properties, in particular, higher tensile strength, notch impact strength, and low temperature notch impact strength.

In our experiments, we used crosslinked ethylene vinylacetate copolymer (EVA) as the elastomer component in combination with PMMA as the second component.

The polymer networks are synthesized sequentially in presence of the respective other polymer.

In contrast to full-IPN's (two or more polymer networks) the semi-IPN's contain only one polymer component in crosslinked state. Linear blends and melt-mixing blends do not have crosslinked structures.

Experimental

Synthesis of polymer blends and semi-1-IPN's

For the synthesis of semi-1-IPN's EVA-copolymer was molded in sheets by means of an injection-molding machine. Afterwards, the sheets were crosslinked by electron irradiation of varied doses (50...300 kGy). The crosslinked EVA sheets were immersed in MMA containing initiator (AIBN). After the required increase in weight by MMA had been achieved, the swollen EVA sheets were stored for reaching uniform monomer diffusion. They were clamped between metal plates with teflon insert and sealing ring. The plates were hold together by spring clips. In a heating oven the monomer was polymerized at 55°C for 8 h, at 120°C for 4 h, and at 130°C for 1 h.

The linear blends were synthesized just as semi-1-IPN's, except for the fact that crosslinked EVA was replaced by linear EVA. Melt-mixing blends were prepared in a Werner & Pfleiderer ZDSK 28 extruder at 200°C.

Characterization

- Transmission electron microscopy (TEM)
  Ultrathin sections were stained in OsO_4 vapor for 24 h.
- Differential Scanning Calorimetry (DSC)
  Perkin Elmer DSC-2 calorimeter, cycle: 1st heating — cooling with 80 K/min — 2nd heating, heating rate: 20 K/min
  The 2nd heating was used for determination of T_g (PMMA) based on the midpoint method
- Dynamic mechanical analysis (DMA)
  Rheometrics spectrometer RMS 800 to measure the loss tangent
  Temperature range: —80°C to 180°C
  Frequency: 62.8 rad/s, strain: 0.05%
  Heating rate: 5 K/min

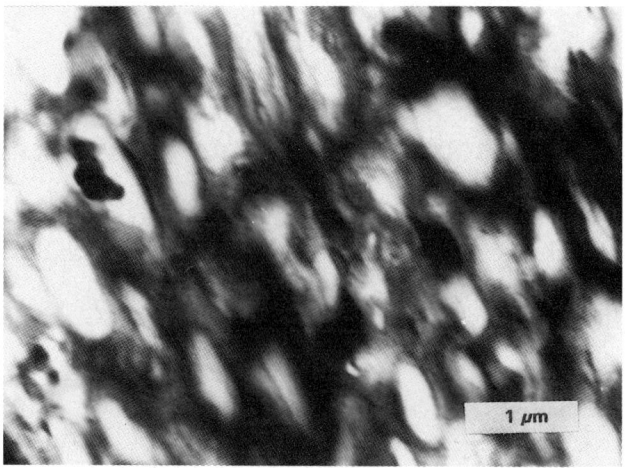

Fig. 1b. Linear blend, EVA/PMMA 34/66

Results and discussion

Figure 1a and 1b show TEM micrographs of EVA/PMMA blends; the dark phase is EVA and the light phase PMMA. Generally speaking, the two-dimensional electron micrographs do not reflect the real phase ratio in the bulk, especially if the phases have unregular shapes. Therefore, estimation of the phase ratio of the bulk is not reasonable in this way; measurement of the size of interfacial layers via comparison of the ratio of stained and unstained regions with the ratio of EVA to PMMA is not possible either. The micrograph 1a shows distinct phase boundaries in a melt-mixing blend consisting of roughly dispersed PMMA particles in an EVA matrix. The micrograph 1b shows a linear blend, the phase boundaries are not as distinct as in melt-mixing blends. This difference can be explained by the high grafting capability of MMA on EVA proved by us. This will be reported later.

Preparation of semi-1-IPN's brought about a considerable reduction in particle size compared to linear blends and melt-mixing blends. The series of Fig. 2 shows the influence of EVA crosslinking. The particle size is decreased with increasing degree of EVA crosslinking. The reason for this is the restriction of the phase separation of the PMMA macro-

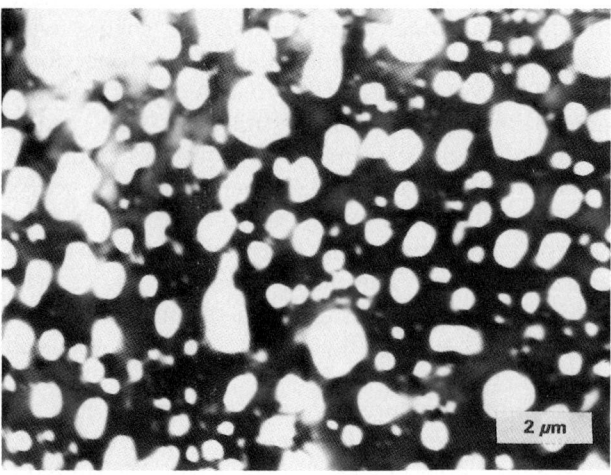

Fig. 1a. Melt-mixing blend, EVA/PMMA 40/60

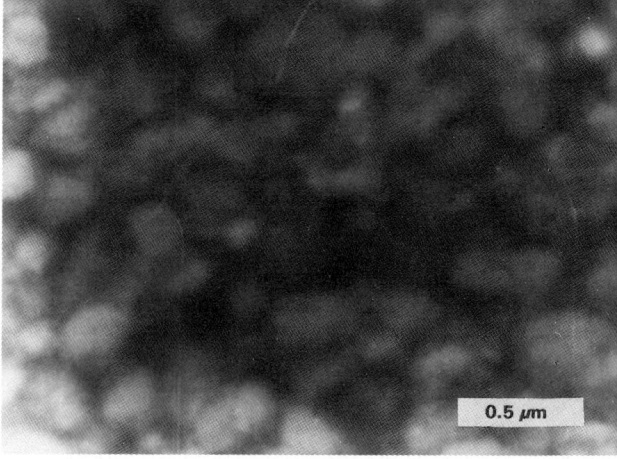

Fig. 2a. Semi-1-IPN, EVA/PMMA 33/67. EVA crosslinking with 50 kGy

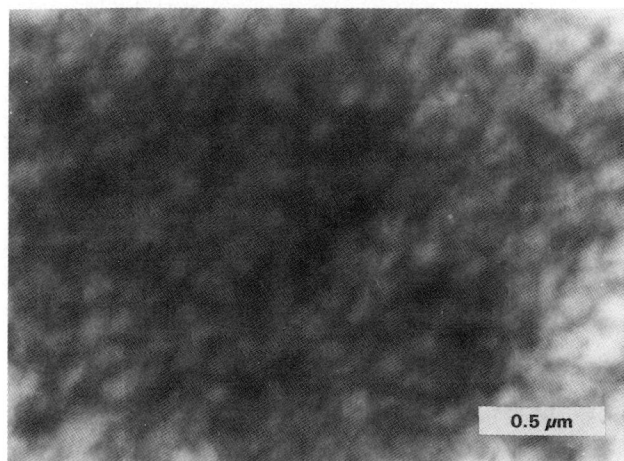

Fig. 2b. Semi-1-IPN, EVA/PMMA 33/67. EVA crosslinking with 100 kGy

The change of T_g in the range up to 30 wt.-% EVA as against pure PMMA is caused by the hindrance of the mobility of PMMA chain segments via grafting of MMA on EVA. The observed independence of the T_g for higher contents of EVA is an expression of the thermodynamic incompatibility of the polymer components. The T_g of PMMA within

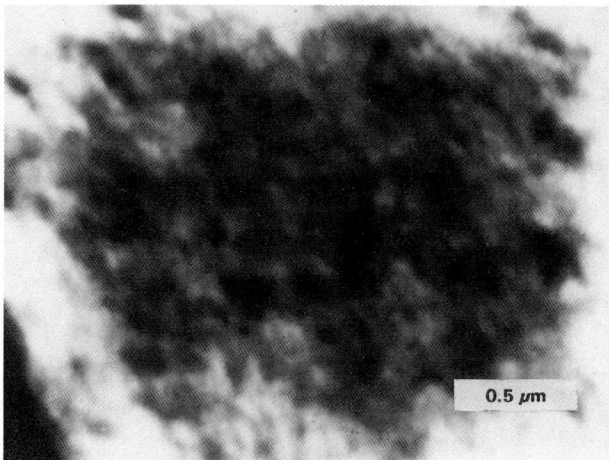

Fig. 2c. Semi-1-IPN, EVA/PMMA 42/58. EVA crosslinking with 200 kGy

molecules by the EVA primary network, particularly by entanglements. For semi-1-IPN's with the highest degree of EVA crosslinking (Fig. 2c), PMMA particles cannot be identified clearly any longer. This suggests the conclusion that the phase dispergation of the polymer components is enhanced.

DSC was applied to measure the glass transition temperatures of PMMA (Fig. 3) within the blends.

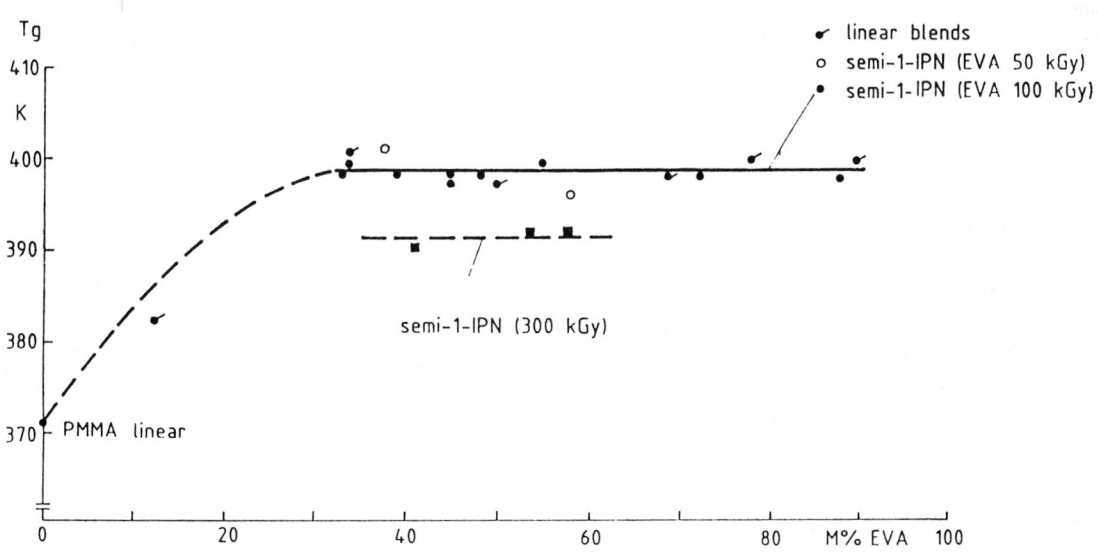

PMMA-Tg of EVA/PMMA blends and IPN's

Fig. 3. Results of DSC

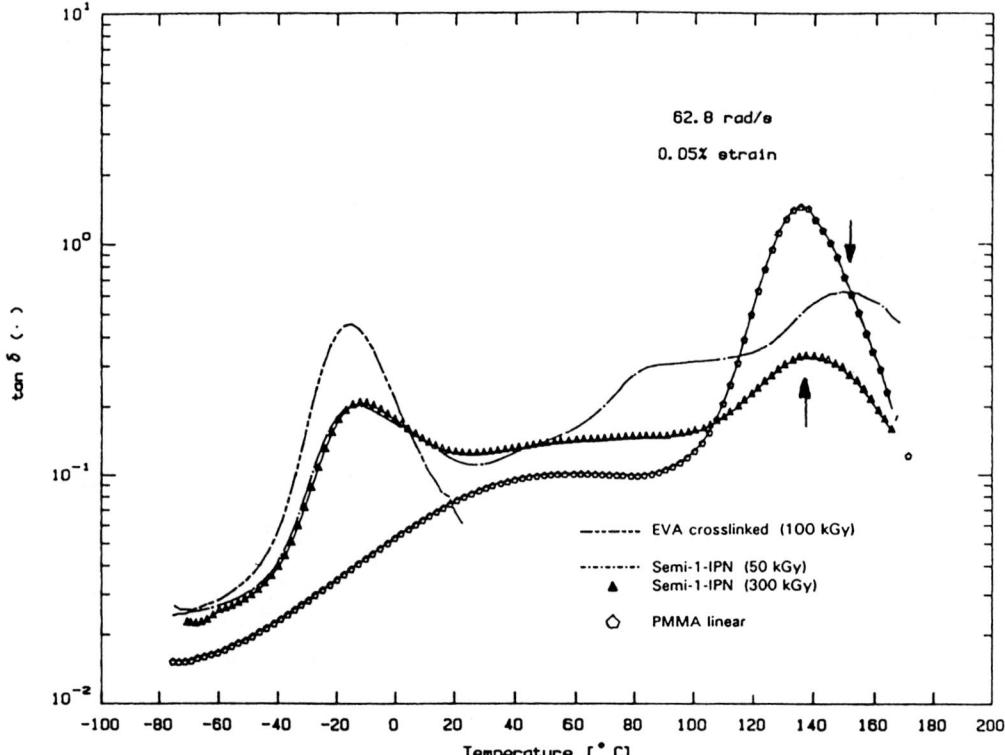

Fig. 4. Results of DMA

linear blends and within semi-1-IPN's at low degree of crosslinking is constant at T_g = 398 K if the content of EVA is at least 30 wt.-%.

Dynamic mechanical measurements confirm the results of DSC. In addition, we observed that the position of the EVA relaxation maximum is nearly constant at all samples. Figure 4 shows the loss tangent versus temperature of semi-1-IPN's based on weakly (50 kGY) and strongly (300 kGy) crosslinked EVA. The shoulder in the curve of semi-1-IPN (50 kGy) is explained by melting of EVA crystallites. The value of PMMA-T_g of semi-1-IPN (300 kGy) is lower than semi-1-IPN (50 kGy). We explain this behavior with an increased effective interface as a result of a high degree of dispergation. The degree of miscibility compelled by that causes a reduction of the PMMA-T_g (see also Fig. 3).

T_g reduction can also be caused by a lower degree of grafting in the semi-1-IPN based on strongly crosslinked EVA (300 kGy). However, we observed a high degree of grafting of PMMA onto EVA at all samples being nearly independent on the degree of crosslinking. Consequently, this factor of influence may be excluded. Furthermore, the T_g depends also on the molar mass of PMMA. Direct determination of the molar mass of the PMMA chains, which are mainly grafted, is very difficult and was not performed. However, there is not a plausible reason to assume that the molar mass of the PMMA polymerized in the presence of the strongly crosslinked EVA decrease.

Conclusion

Structures of less than 0.2 μm with diffuse phase boundaries were obtained by preparation of semi-1-IPN's. With increasing degree of crosslinking of EVA, the phase dispergation is enhanced. Consequently, the T_g of PMMA is decreased. An additional effect of grafting of MMA on EVA correlating with an increase of T_g explains the diffuse phase boundaries.

References

1. Sperling LH (1981) Interpenetrating Polymer Networks and Related Materials, Plenum Press, New York and London

ABL# Thermally and mechanically activated degradation of polyesterurethane networks. Analysis of molecular weight distribution functions

G. I. Sandakov, L. P. Smirnov, A. I. Sosikov, K. T. Summanen, and N. N. Volkova

Institute of Chemical Physics, Chernogolovka, Russia

Abstract: A novel technique for the analysis of the molecular weight distribution functions of crosslinked polymer chains is proposed. The technique is based on the relation between the high-temperature nuclear magnetic resonance free induction decay and the network structure. The method is used to study thermally and mechanically activated degradation of polyesterurethane networks.

Key words: Networks; NMR; structure; degradation

Introduction

The structure of a crosslinked polymer determines, to a considerable extent, its physical and chemical properties. One of the most detailed structural characteristics of the network is the molecular weight distribution (MWD) function which describes the distribution of lengths of molecules that form the network (here the word "molecule" is used to refer to that part of the polymer chain between the two crosslinks which fixes its ends).

The proper understanding of degradation mechanisms can hardly be achieved without knowing 1) the behavior of MWD in the course of degradation, and 2) the way the MWD affects the chemical processes. In this work, we restrict ourselves to the first part of the problem, and our aim is to follow the MWD evolution.

Another problem that we tackle indirectly in this work is that of the mechanisms which underlie stress-activated degradation of crosslinked polymers. In linear polymers the situation is, more or less, clear, and it is agreed that both thermally and mechanically activated degradation processes have common basic mechanisms, whereas, in the case of networks, this question is still under discussion. Our results show that, judging by the evolution of the molecular weight distribution functions, the two types of degradation are governed by similar mechanisms.

Calculation of the MWD

Calculation of the entire MWD function, and not only of its moments, was not possible until the technique for direct experimental determination of MWD from high temperature NMR spectra was developed [1]. Below, we consider briefly the basic ideas underlying this method. The nuclear magnetic relaxation of dipolar coupled spins in crosslinked polymers is known to reveal at high temperatures pseudo-solid-like behavior [2—6]. This feature reveals itself in a number of effects, which one would normally expect to find in solids, such as pronounced solid-echo, independence of the spectra of the temperature ("high temperature plateau"), narrowing of the spectra by magic-angle spinning. To understand the peculiar behavior of the NMR spectra in polymer networks, the concept of partial averaging of the dipolar interactions was proposed [1, 7—12]. This concept, also called concept of effective interactions, constitutes the theoretical basis of our method.

Consider a crosslinked polymer built from end-linked polymer molecules which bear dipolar coupled [13] magnetic $S = 1/2$ nuclei (from now on, called "protons"). At temperatures above the vitrification temperature the polymer molecules move chaotically and these motions make the dipolar interactions oscillate randomly. When the temperature is sufficiently high, the averaging of the dipolar interactions and narrowing of the spectral line occur [13]. At this point, a new factor, specific for polymer networks, comes into play. Namely, polymer molecules in the network, unlike molecules in a liquid, are not free, because their ends are attached to the crosslinks. The presence of crosslinks introduces anisotropy into the segmental motions. However small, this anisotropy entails drastic changes in the FID. Indeed, in the presence of anisotropy of chain motions, the Hamiltonian of dipolar interactions $\hat{D}(t)$ can be presented as a sum of a "fluctuating" time-dependent component $\hat{D}^{(f)}(t)$, which has zero time average $\langle \hat{D}^{(f)}(t) \rangle_t = 0$, and time-independent "effective" component $\hat{D}^{(e)} = \langle \hat{D}(t) \rangle_t$. The latter component can be found either by averaging the dipolar interactions over time or over conformations of the polymer chains. In this latter case, an additional constraint is imposed in that the conformations has to be compatible with the existence of the network. As the experiments show, the effect of the "fluctuating" component $\hat{D}^{(f)}(t)$, which dominates at low temperatures, rapidly decreases with increasing temperature and becomes negligibly small at temperatures ca. 100 degrees above the vitrification temperature. Thus, at sufficiently high temperatures, the NMR spectra of networks are determined essentially by time and temperature independent residual effective interactions $D^{(e)}$. It is this transition to the "spectra of effective interactions" that manifests itself as a plateau in the temperature dependence of the characteristic time of the FID.

The approach to the calculation of FID we use in this paper, is slightly different from that presented in [1, 14—16]. It was recently shown [17] that the spectra "on effective interactions" are determined by dipolar couplings between nuclei, that 1) are located on the same chain (intramolecular interactions), and 2) are not too far from each other. The last statement can be made more accurate by saying that it is enough to account for interactions between protons belonging to the same statistical segment [18] (intrasegment interactions). Consider two arbitrary protons, say i-th and j-th, located on one segment. The averaging over the chain conformations can be accomplished in two steps. First, the interactions should be averaged over rotations around the segment axis. This gives for the constant of dipolar interactions

$$b_{ij}(\bar{n}_C) = \frac{\gamma^2 \hbar \eta_{ij}}{d_{ij}^3} (1 - 3\cos^2(\theta(\bar{n}_C \bar{n}_H))), \quad (1)$$

where $\bar{n}_C$ and $\bar{n}_H$ are the unit vectors in the directions of the segment axis $\bar{C}$ and the constant magnetic field $\bar{H}$; $\theta(v_1, v_2)$ is the angle between vectors v_1, v_2; $O < \eta_{ij} < 1$ is a factor, which accounts for the partial averaging of the dipolar constant by uniaxial rotation, the value of η depends on the geometry of the molecule; d_{ij} is the distance between i-th and j-th protons; γ and $\hbar$ are the gyromagnetic ratio and Planck's constant, respectively. The effect of the chain on the orientations of the segment, which is, actually, a part of the chain, can be described in terms of effective "average" force f applied to the ends of the segment. The distribution of orientations of the segment of length a is, in this case, given by

$$P(\bar{n}_C) \propto \exp\left(- |f| a \frac{(\bar{n}_C \bar{n}_R)}{kT}\right), \quad (2)$$

where the force f is given by

$$f = kT \frac{\partial (\ln(W(\bar{R})))}{\partial \bar{R}}. \quad (3)$$

Averaging over the distribution of segment orientations with the distribution function (2) can be easily performed to give [17]

$$b_{ij}^e = \int d\bar{n}_C P(\bar{n}_C) b_{ij}(\bar{n}_C)$$
$$= \frac{\gamma^2 \hbar \eta_{ij}}{d_{ij}^3} (1 - 3\cos^2(\theta(\bar{n}_R, \bar{n}_H))) g_2(\lambda), \quad (4)$$

where $\lambda = a |f|/2kT$, and $g_2(\lambda) = 1 - 3\lambda^{-1}(cth(\lambda) - \lambda^{-1})$. For a Gaussian chain $W(\bar{R}) \propto \exp(-3R^2/2Na^2)$. Substituting this distribution function into (3) and using the fact that at $\lambda \ll 1$ $g_2(\lambda) \approx \lambda^2/15$, we get

$$b_{ij}^e = \frac{3\gamma^2 \hbar}{20} \frac{\eta_{ij}}{d_{ij}^3} (1 - 3\cos^2(\theta(\bar{n}_R, \bar{n}_H))) \frac{R^2}{N^2 a^2}. \quad (5)$$

Using the Anderson-Weiss model [13], the FID of protons of the considered segment can now be approximated by a product of cosines which, in turn, can be replaced by a Gaussian function:

$$G(t, N, \theta) = \exp(-\omega_{loc}^2 (1 - 3\cos^2(\theta(\bar{n}_R, \bar{n}_H)))^2 \cdot t^2 R^4 N^{-4} a^{-4}), \quad (6)$$

where the local frequency ω_{loc} is given by

$$\omega_{loc}^2 = \frac{81 \gamma^4 \hbar^2}{3200} \sum_j \eta_{ij}^2 d_{ij}^{-6}. \quad (7)$$

Assuming now that for an undisturbed macromolecule in the network the end-to-end distance is related to the molecule length as $R^2 \propto Na^2$, we find that

$$G(t, N, \theta) = \exp(-\omega_{loc}^2 (1 - 3\cos^2(\theta(\bar{n}_R, \bar{n}_H)))^2 \cdot t^2 N^{-2}). \quad (8)$$

The important result is that the FID of a macromolecule with fixed ends depends essentially on the length of the macromolecule. It is at this point that the distribution of chain lengths becomes related to the NMR spectra.

As follows from Eq. (8), the amplitude of the effective interactions does not depend on the length of statistical segment. This means that we need not use the assumption of absolutely rigid statistical segment, and Eq. (8) can be applied to the description of flexible molecules. In the latter case, the "local frequency" ω_{loc} is no longer given by Eq. (7).

The FID of the whole polymer body is found by averaging (8) over orientations of the end-to-end vector directions and distribution of chain length

$$G(t) = \int_0^\infty dN P_W(N) G_0(t, N), \quad (9)$$

where

$$G_0(t, N) = \int_0^1 d(\cos\theta) G(t, N, \theta), \quad (10)$$

and $P_W(N)$ is the molecular weight distribution function of the polymer network. Equations (9, 10) constitute the basis of our technique. The MWD function was calculated numerically by varying $P_W(W)$ in Eq. (9), and a simple iterative procedure was used to achieve best agreement between the theoretical FID given by Eqs. (9, 10) and the experimentally measured transverse relaxation.

An important assumption that we used implicitly in deriving Eqs. (9, 10) is that the FID is dominated by intramolecular interactions, whereas intermolecular interactions can be ignored. This assumption can be supported by the following simple estimations. In the presence of the network, each proton can be imagined as "walking" near some "equilibrium" point. After averaging over the chain conformations, the effective interactions look as if the protons were located at these points. The amplitude of effective interactions, however, differs from that of original dipolar interactions by a factor $0 < \xi < 1$, which reflects the partial averaging. As the distance between the equilibrium points is decreasing, ξ tends to zero, whereas at distances greater than $R_0 = aN^{1/2}$, ξ is of the order of unity. As a result, the intermolecular effective interactions are essential only for protons whose equilibrium points are at distances larger than R_0. This gives, for the amplitude of the intermolecular interactions, $(\omega_{loc}^{inter})^2 \propto N^{-3}$. Comparing this result to Eq. (8), we see that for long chains with $N \gg 1$ the intermolecular interactions are N times weaker than the intramolecular ones.

To end this section, we discuss, briefly, the main limitations of the method.

1) The lower limit for the chain length that can be detected with our method is set by the intermolecular interactions. Unfortunately, we have no accurate estimations on this score. Experiments show, however, that for chains with lengths of about 10 statistical segments, the intermolecular contribution does not essentially affect the FID.

2) The upper limit is set by entanglements, which, from the NMR viewpoint, look exactly like chemical crosslinks. This limit was experimentally found to be about 300 statistical segments.

Experimental. Samples, measurements, and presentation of results

The method was applied to study degradation of polyesterurethane networks synthesized from poly (diethylene adipate diol), 2,2-toluylene diisocyanate and 1,1,1-trimethylol propane, in the presence of a catalyst, hexabutildistannic oxide. The reaction was carried out at 60 °C until full conversion of NCO-groups. The initial molar ratio [NCO-]/[-OH] was equal to unity. The amount of triol was 2 wt.% in the case of thermally degraded samples, and 1 wt.% in the case of mechanically degraded samples. The initial copolymer was characterized by a narrow distribution of molecular weights, with the average molecular weight $M = 900$. The thickness of the samples was around 1 mm. Earlier, it was shown that the size of the samples did not affect the degradation kinetics. Prior to degradation, the samples were vacuum dried in the following regime: 2 h at 80 °C, 16 h at 25 °C, and 20 min at 140 °C.

Free induction decays were measured by using a multipulse NMR spectrometer RI-2303 with Larmor frequency (for protons) of 56 MHz. To eliminate the effects of the magnetic field inhomogeneity, the Hahn echo technique was used. The measurements were carried out at temperatures corresponding to the high-temperature plateau (~140°C for our samples).

The MWD functions covered in some cases up to two orders of magnitude, and it was found convenient to plot them as a function of $\ln(N)$. To make the area under the peak in the semilogarithmic coordinates to give the total weight of chains, corresponding to this peak, the distributions were multiplied by N.

The frequency ω_{loc} in Eqs. (9, 10) was not calculated. As a result, the positions of MWD functions in the semilogarithmic coordinates are given with an accuracy up to a shift.

In the interpretation of the results presented below, of importance is the question of homogeneity of our samples. Analysis carried out with the use of NMR and ESR showed that the samples studied, indeed, revealed microphase separation at temperatures below 70°C. This manifested itself in the presence of "hard" and "soft" domains, with low and high molecular mobility, respectively, and the "hard" regions resulted in fast decaying component in the FID. At higher temperatures, however, NMR showed no fast component in the FID, and the FID was independent of the temperature. These observations enabled us to assume that the hard regions disrupt upon increasing the temperature, so that at temperatures corresponding to the high temperature plateau (140°C), at which we measured the FIDs, the sample were homogeneous.

Analysis of thermal degradation

The experimental cycle included two stages. During the first stage, the sample was placed into a thermostat, where it was heated up to the degradation temperature (225°C for the experiments presented below), and kept at this temperature during a certain period of time. At the second stage, the sample was cooled to 140°C and the MWD function was measured; then the cycle was repeated. The set of MWD's corresponding to the increasing total degradation time is presented in Fig. 1. The initial (undegraded) network has a unimodal MWD with lengths of molecules distributed in a narrow interval near the average length N_0. The distribution corresponding to 15 min of degradation (the shortest degradation time) differs drastically from that of the undegraded network. As one can see from Fig. 1, it is much broader and has apparent

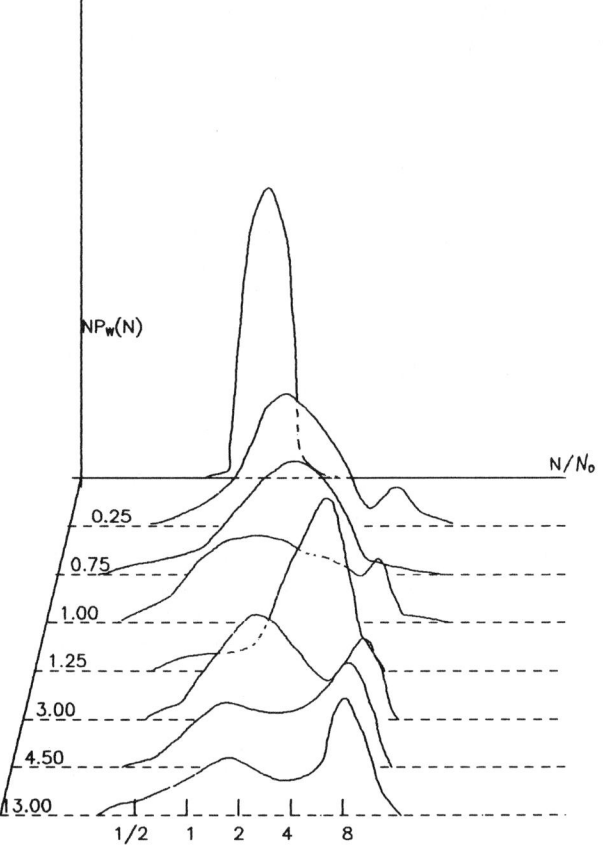

Fig. 1. Evolution of the molecular weight distribution function of a crosslinked polyesterurethane network in the course of thermal degradation. Degradation times (in hours) are shown to the left of the curves

bimodal structure. The two visible peaks correspond to molecules with lengths roughly equal to $2N_0$ and $8N_0$; molecules with other lengths are also present, though. The enormous difference between MWD of the initial network and that corresponding to $t_d = 15$ min shows that most of the changes to the network structure occur during the first, relatively short, period of degradation. It is interesting to note that not only longer molecules appear, as one would expect to occur in the result of the destruction of crosslinks, but also a certain amount of shorter polymer chains is obviously present, as indicated by Fig. 1. The appearance of shorter molecules shows that, simultaneously with the reactions of chain scission and decay of crosslinks, there exist "reverse" processes of crosslinking.

At longer degradation times MWD changes in a more regular way. This is most evident from the behavior of the distribution functions at degradation times exceeding 1 h. In this case, all molecular weight distribution functions have bimodal structure, with the two peaks corresponding to short ($N \approx 2N_0$) and long ($N \approx 8N_0$) molecules. The degradation does not affect the positions of these peaks, but it does change their areas. In general, these results show that thermal degradation has a nonmonotonic, "oscillating" character.

Stress-activated degradation

To study the variations of the network structure in the process of mechanically activated degradation, the following experiment was carried out: A sample of crosslinked polymer (prepared as was described above) was cut in the form of a ring, and the ring was stretched (the stretching degree was about 4) by slipping the ring over a glass tube. In this state, the polymer stayed in vacuum at room temperature. At regular intervals the sample was taken off the tube and relaxed 24 h at room temperature, and 2 h at 80°C, to let the network to come to an equilibrium state. Then the free induction decay was measured at the temperature of the "high temperature plateau", and the sample was returned to the stressed state. Repeating this cycle gave a number of MWD's, corresponding to increasing degradation times. To exclude degradation mechanisms other than the stress assisted one, a control non-stressed sample, which was kept at room temperature, was studied. Analysis showed that the MWD function of the control sample did not change. Figure 2 shows MWD functions measured after degradation under the stress during 0, 114, 174, and 254 days. For degradation times of 10 days the MWD functions coincides with that corresponding to the initial sample. From these data, one can see that, under the conditions of mechanically activated degradation, the network structure changes in the same manner as it does in the case of thermal degradation. The MWD function is transformed (see Fig. 2) from a slightly biomodal one to apparently bimodal, and further to MWD with three visible peaks. Along with the formation of long polymer molecules (peaks to the right from the central one), the number of molecules with lengths smaller than that of the undegraded polymer also increases. One of possible mechanisms of formation of shorter chains could be, probably, trans-

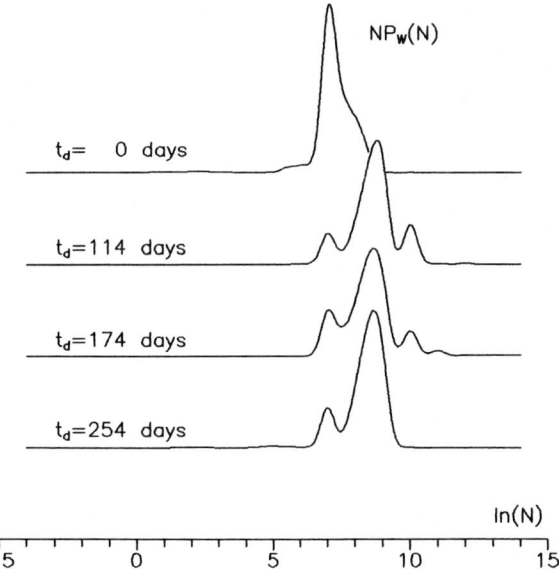

Fig. 2. Evolution of the molecular weight distribution function of a crosslinked polyesterurethane network in the course of mechanically activated degradation. Degradation times are indicated to the left of the curves

urethanization reaction. Just as the thermal activation of the degradation, the mechanical stress results in nonmonotonic degradation, accompanied by "oscillations" of the MWD. These facts show the similarity of processes underlying thermally and mechanically activated degradation.

References

1. Kulagina TP, Provotorov BN, Martchenkov VV (1989) Polym Sci USSR 31:420
2. Folland R, Steven JH, Charlesby AJ (1978) J Polym Sci A2:1041
3. Powles JG, Hartland A (1960) Nature 186:26
4. McCall DW, Douglass DC, Anderson EW (1962) J Polym Sci 59:301
5. Slichter WP, Davis DD (1964) J Appl Phys 35:3103
6. Fedotov VD, Schneider H (1989) NMR. Basic principles and progress 21:1
7. Gotlib YI, Lifshits MI, Ievlev VA (1976) Vysokomol Soed (in Russian) A18:2299
8. Cohen-Addad JP, Dupeyre R (1983) Polymer 24:400
9. Cohen-Addad JP, Domard M, Herz J (1982) J Chem Phys 76:2744
10. Cohen-Addad JP, Domard M, Boileau S (1981) J Chem Phys 75:4107
11. Cohen-Addad JP (1976) J Chem Phys 64:3438
12. Lundin AA, Khazanovich TN (1989) Polym Sci USSR 31:363

13. Abragam A (1961) The Principles of Nuclear Magnetism, Clarendon, Oxford
14. Volkova NN, Erofeev LN, Sandakoff GI, Smirnov LP, Summanen KT (1990) Modern NMR and ESR Techniques in the chemistry of solids (in Russian), Chernogolovka, USSR
15. Kulagina TP, Summanen KT (1989) In: Magnetic Resonance in Polymers, Proceedings, 9-th Specialized Colloque Ampere, Prague
16. Khazanovich TN, Summanen KT, to be published
17. Flory PJ (1969) Statistical Mechanics of Chain Molecules, Interscience Publishers, New York

Authors' address:

Dr. K. T. Summanen
Institute of Chemical Physics
Moscow region
Noginsk district
Chernogolovka, 142432 Russia

The solidification of bulk and solution cast segmented polyurethanes

H. Janik and J. Foks

Technical University of Gdansk, Institute for Organic and Food Chemistry and Technology, Gdansk, Poland

Abstract: The bulk PUs obtained with the use of diol and triol have been studied by electron microscopy. The fracture surface morphology of these samples has been compared. It has been found that the degree of crosslinking can be responsible for different mechanisms of fracture and, thus, various fracture curves are observed. The solvent cast PUs (obtained with the use of diol) have been studied by polarized light and electron microscopy. The birefringent and non-birefringent elements have been found under polarized light. Both appear to have a spherulite-like feature when observed under parallel polarizers and care must be taken in their interpretation.

Key words: Morphology of PUs; electron microscopy of PUs; polarized microscopy of PUs; fracture curves

Introduction

Segmented polyurethanes (PU) are multiblock copolymers with particular properties directly related to their microstructure. They are built of soft and hard segments which phase separate easily. The soft segment is often polyester or polyether chain, while the hard segment is usually the reaction product between diisocyanate and diol or triol.

It is well known that micro- and macro-phase separation takes place during obtaining of segmented polyurethanes (PUs) [1—4]. Extensive investigations on microphase separation were initiated by Cooper and Tobolsky in 1966. They found that the phase separation into domains of hard and soft segment of the order of 10 nm takes place in segmented PUs [1].

Morphological forms like globules or spherulites of a size considerably exceeding that of domains were also observed in segmented PUs by some authors [2—5].

The type of the technological procedure strongly influences the polymer structure [5—6]. Microscopy is one of the methods allowing direct observation of the structure formation. In the paper the results of some microscopic observations of the structure for different PUs are discussed.

We have chosen and describe only some microscopic results which illustrate the complexity of solidification of PUs the best. Thus, this paper is not an overview of morphological studies of PUs by microscopy. More information about different microscopic studies of segmented PUs can be found in our other papers [4, 5, 7, 8].

Experimental

The bulk PUs containing diol (butanediol-BDO) [7] and triol (L-allyl-glycerin ether) [9] as chain extender have been studied. The homopolymers of the soft segment (derived from polyethylene adipate glycol and MDI) and the hard segment (derived from MDI and BDO) have been studied as well. The samples were prepared according to the same procedure (cast technique, two-step prepolymer synthesis). The influence of the parameters of the synthesis on the morphology of PUs was described earlier by us in [5, 10]. Here, the emphasis is put on the similarity found in the surface fracture appearance of different PUs. Despite this the results of solution cast films observation are discussed. These two techniques have been found to be those which are very useful to reveal the complexity of segmented PUs.

The investigation of solution cast PUs (BDO as chain extender) were carried out on films cast on glasses from 5% solution in dimethylformamide

(DMF) at the temperatures of 70°, 100°, and 145°C. The evaporation of the solvent was carried out in a covered petri dish that contained pure solvent in the bottom. The morphology was investigated by polarizing and electron microscopy with the accomplishment of the earlier described technique [7, 8]. Swelling [11] and allophanate bonds [12, 13] measurements were carried out for some samples.

Results

1. The fracture surface appearance

Polyesterurethanes obtained with the use of diol (PU-D): The cast PU-D are the materials most frequently studied by us [5, 7, 10, 12, 14].

The samples without spherulites or the interspherulitic volume in the crystalline PUs are of different nature when observed by cryogenically fractured technique. Generally, three types of fracture curves have been found under electron microscopy (Fig. 1) in all investigated samples. We have found some rules which are common for different PUs. Those samples which were soluble in DMF possessed patch-like fracture curves (Fig. 1a). Unsoluble PU samples have spherulite-like (Fig. 1c) or patch-radial (Fig. 1b) fracture curves. The samples with a patch-radial feature on the fracture surface have always been swollen to a higher degree than the samples characterized with spherulite-like fracture curves. Some explanation of the nature of fracture curves in PUs can be obtained from the measurements of allophanate groups carried out for these samples. Those which possessed less allophanate bonds characterized with patch-radial fracture curves and those which had more allophanate groups had spherulite-like fracture curves.

For example, we have compared the amount of allophanate groups for two samples which were obtained from the same chemicals but differed only by thermal treatment after the synthesis (PU-D). The amount of allophanate groups was two times less for the sample with path-radial structure than for that with spherulite-like one.

Polyesterurethanes obtained with the use of triol (PU-T): Thin sections of crosslinked PUs observed under polarized light are nonbirefringent. The fracture surface appearance of cryogenically fractured samples are of spherulite-like type regardless of the ratio and the components used in the synthesis. The only difference found between the samples was the size of polygonal elements. Thus, in the case of crosslinked PUs spherulite-like type fracture curves are only observed.

2. Solution cast thin films (PU-S)-birefringent and non birefringent elements

A film obtained from the soft segment (e.g., at 70°C — Fig. 2b) shows only the characteristic feature when polarizers are parallel. The entire volume of the film is grainy and in many places it

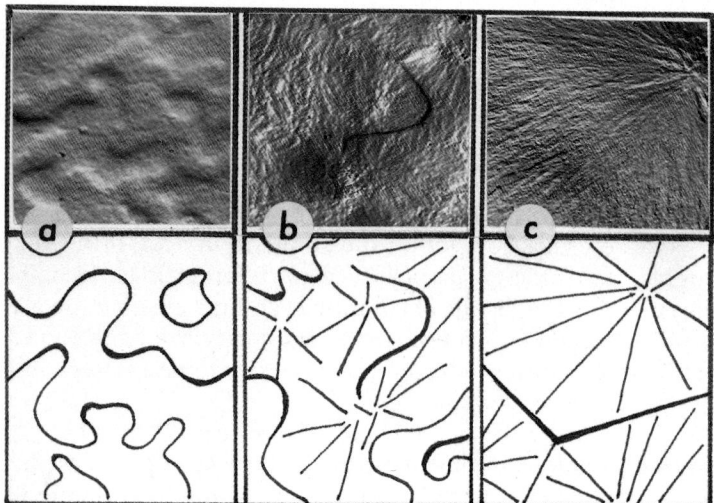

Fig. 1. Three types of fracture curves found for segmented polyurethanes (bottom: schematic, top: an example of an original picture). These pictures are suitable for noncrystalline PUs. When the hard-segment crystallization takes place the dispersed phase appears [19].
a) patch-like; b) patch-radial; c) spherulite-like

has spherulite-like appearance. The film is distinctly spherulite-like when the solvent is evaporated very slowly (Fig. 2c). The grains disappear at 110—120 °C during heating of the sample observed under a microscope with parallel polarizers. A film obtained from the hard segment at 70—145 °C is birefringent in the entire volume with the typical feature of a Maltese cross present in the spherulite (Fig. 2a). Melting temperature, determined as disappearance of birefringence, is 225 °C.

A film obtained from segmented polyurethanes at 70 °C is, to some extent, similar to that of the soft-segment homopolymer. It does not display any birefringence and the very distinct spherulite-like grains are observed under the microscope when the polarizers are parallel (Fig. 2d). The temperature of their disappearance is 190 °C. A film obtained from segmented PU at 100 °C has the faint feature of spherulite-like grains when observed under a microscope with parallel polarizers. In polarized light the film has only separate spherulites with the distinct feature of a Maltese cross (Fig. 3b). In quasi-polarized light (Fig. 3a) and under the electron microscope (Fig. 3c) it has a visible superposition of birefringent spherulites and nonbirefringent spherulite-like grains; the latter disappear at about 170 °C. The melting temperature of spherulites is in the range of 185—195 °C.

A film obtained from segmented PU at 145 °C (Fig. 4), when observed between crossed polarizers, displays two kinds of spherulites (at medium rate of solvent evaporation); one melting at 212 °C and the other at about 160 °C (Fig. 4a).

The latter has the typical Maltese cross, but needs a very strong light source to be visible. The spherulites melting at 212 °C are heavily birefringent and they are course and open spherulites without the typical feature of a Maltese cross. The polymer is not volume filled by spherulites. They are separately distributed in the nonbirefringent matrix (disappearance temperature of the matrix is 153 °C). The film observed under a microscope with parallel polarizers is grainy (Fig. 4b). The grains are of different size. Those which are needle-like correspond to the spherulites melting at 212 °C. When the solvent is evaporated at a very high rate (Fig. 4c, d) the only distinctly visible elements found in the polymer formed at 145 °C are birefringent spherulites. The matrix in which the spherulites are ambedded is featureless, both in parallel and crossed polarizers. The solvent-cast films observed under electron microscope are of different types depending on the casting temperature. The spherulites exhibiting the lowest melting temperature are featureless globules while those with higher melting temperature are characterized by

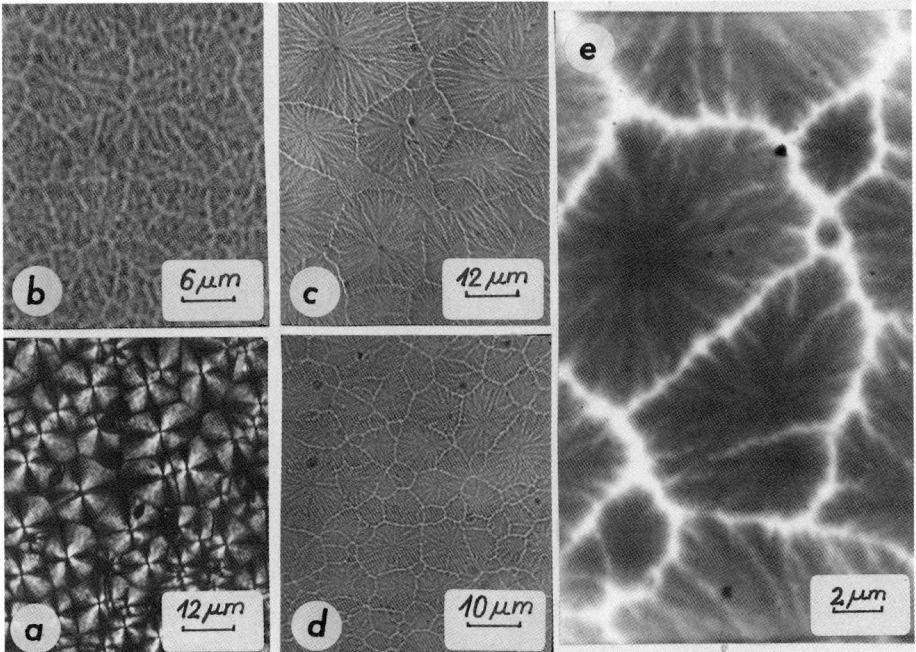

Fig. 2. A thin film of hard (a) and soft segment (b, c) and segmented polyurethane (d, e) cast at 70 °C from solution. a) crossed polarizers; b, c, d) parallel polarizers; e) electron micrograph; for b, c and d there are no birefringent elements visible; a, b, d, e) medium rate of solvent evaporation; c) very low rate of solvent evaporation

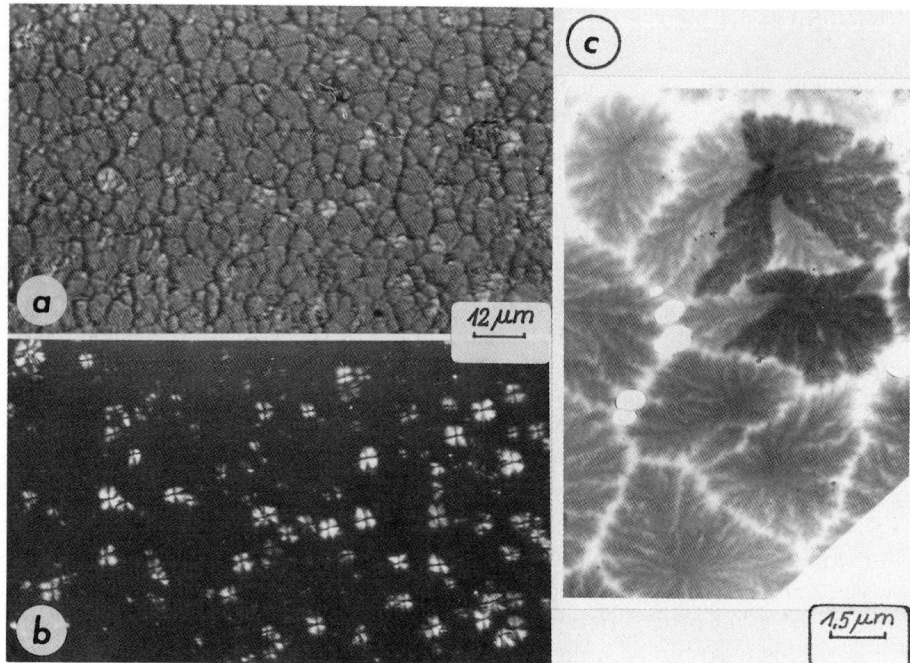

Fig. 3. A thin film of segmented polyurethanes obtained at 100°C.
a) quasi-crossed polarizers;
b) crossed polarizers;
c) electron micrograph

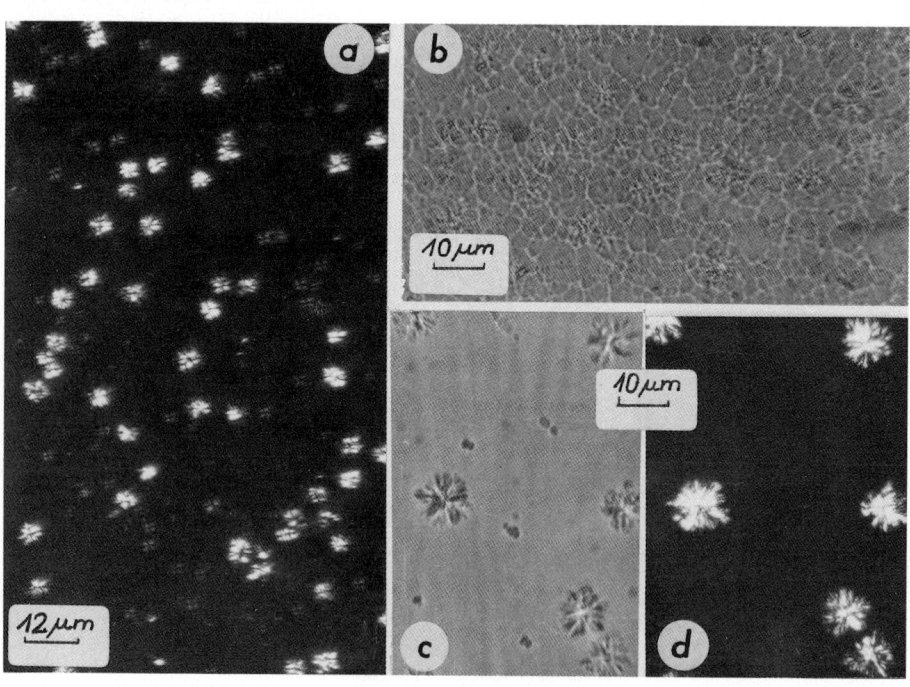

Fig. 4. A thin film of segmented polyurethanes obtained at 145°C.
a, d) crossed polarizers;
b, c) parallel polarizers;
a, b) medium rate of solvent evaporation; c, d) high rate of solvent evaporation

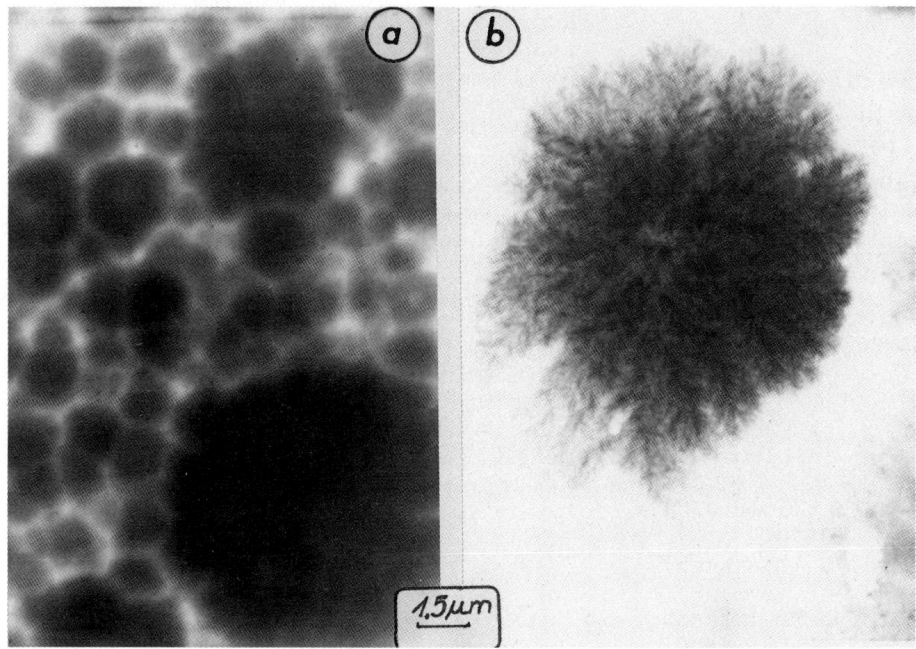

Fig. 5. Electron micrographs of segmented polyurethanes cast from 1% solution in DMF.
a) The appearance of spherulites with lower melting temperature (featureless globules under electron microscope); b) The appearance of spherulites with higher melting temperature (fine fibrils well visible)

fine fibrils inside spherulites (Fig. 5). The stability of the spherulites observed under EM is very low in PUs and the birefringent places after exposition under EM are no longer birefringent.

Discussion

The solidification of segmented PUs can proceed with or without crystallization. The melting temperature of observed spherulites ranging from 160—212 °C points to the presence of the crystalline phase of the hard segment. The crystallization of the hard segment derived from MDI and BDO is described in the literature and will not be a subject of the broad interest in the paper. However, it should be noted that according to our results concerning spherulites found in solvent-cast films and bulk samples the thus far observed globules in PUs [15, 16] are of crystalline nature as well. Globules have relatively lower melting temperature and are not stable in the electron beam and that is why they are featureless.

The more interesting problem, which has not been sufficiently discussed in the literature, appears in nonbirefringent samples or nonbirefringent part of the samples when observed under optical microscope. In the case of solvent cast films the spherulite-like elements are of interest. Koberstein [17] observed a similar morphology in PUs as-synthesized and found it non-crystalline. Briber and Thomas [18] consider this form as crystalline. The temperature observed by us at which the spherulite-like forms disappear in the soft-segment homopolymer is much higher (110—125 °C) than the melting temperature of the crystalline phase of the polymer (Tm = 50 °C).

Therefore, crystallinity cannot be responsible for the occurrence of that feature in the homopolymer and probably also not in the copolymer. Disappearance temperatures of those elements in the copolymer are in the range of 153—190 °C and decrease with increase of the casting temperature. The latter fact, DSC studies, and the observations made for soft-segment spherulite-like morphology suggest that these forms are amorphous.

The spherulite-like feature (of completely different origin than that observed in solution cast films) is also present on the surface of some cryogenically fractured PUs. Comparison of the fracture curves observed for PU-T with PU-D suggest that the spherulite-like feature found in PU-D may be due to some extent of crosslinking. Different amounts of crosslinking can be responsible for different mechanisms of fracture mode and, thus, the diverse appearance of fracture surface.

Conclusions

The solidification of segmented PU is a very complicated process, especially in the case of using butanediol as chain extender. Then, the crystallization of hard segment can take place. In the case of the triol used by us, the crystallization does not take place. Microscopy can be helpful in the control of the polymer solidification, both for solvent-cast and bulk PUs.

References

1. Cooper SL, Tobolsky AV (1966) J Appl Polym Sci 10:1837—1842
2. Chang AL, Briber RM, Thomas EL, Zdrahala RL, Critchfield FE (1982) Polymer 23:1060—1068
3. Camargo RE, Macosco CW, Tirrell M, Wellinghoff ST (1982) Polym Eng Sci 22:719—728
4. Foks J, Janik H, Potocki A (1977) 15th Czechoslovak Conference on Electron Microscopy, Praque, A:491
5. Foks J, Janik H, Russo R, Winiecki S (1989) Eur Polym J 25:31—37
6. Cawse JL, Stanford JL (1987) Polymer 28:356—366
7. Foks J, Janik H (1989) Polym Eng Sci 29:113—119
8. Janik H, Foks J (1992) Proc Inter Polyurethane Industry Exhibition and Conference UTECH 92:170-172
9. Rutkowska M, Tekely P (1982) Polym Bulletin 7:481—487
10. Foks J, Janik H, Russo R (1990) Eur Pol J 26:309—314
11. Rutkowska M, Kwiatkowski A (1975) J Polym Sci 53:141—149
12. Foks J, Janik H, Palka G (1979) Proc 9th EPS Conference 3C:103
13. Foks J, Janik H, Chmielecka R (1984) Proc International Rubber Conference, Moscow 84A:17.1—17.6
14. Foks J, Janik H, Pohl M (1991) Eur Polym J 27:729—733
15. Fridman JD, Thomas EL, Lee LJ, Macosko CM (1980) Polymer 21:393—401
16. Russo R, Thomas EL (1983) Macromol Sci-Phys B22:553—575
17. Koberstein JT, Stein RS (1984) Polymer 25:171—177
18. Briber RM, Thomas EL (1984) Polymer 25:171—177
19. Janik H, Foks J (1990) Proc Rolduc Meeting, Kerkrade

Authors' address:

Dr. H. Janik
Technical University of Gdansk
Institute for Organic Food Chemistry
and Technology
80-952 Gdansk, Poland

Author Index

Adam M 37
Anwand D 70
Apekis L 144
Axelos MAV 66

Babayevsky PG 57, 209
Bastide J 115
Baumann K 97
Bazhan L 186
Beckert W 47
Birshtein TM 177
Borisov OV 177
Borovko VV 209
Boué F 115
Brereton MG 90
Buzier M 115

Chapellier B 111
Charlet G 61
Christodoulides C 144
Coté I 61

Deloche B 111
Doublier JL 61
Duering ER 13
Durand D 37

Eicke H-F 165
Emery JR 37

Fedtke M 78, 206
Foks J 241
Fytas G 232

Garnier C 66
Goering H 144
Grela S 182
Grest GS 13
Gronski W 97

Halperin A 156
Häusler K-G 78, 206
Heinrich G 16, 47
Hofmeier U 165

Irzhak VI 174, 194
Ivashkovskaya TK 174

Janik H 241
Janke A 227

Kilian H-G 27
Kimmich R 104
Klepko V 88
Klüppel M 137
Kontou E 144
Köpf M 104
Kraus V 27
Kremer K 13
Kulik SG 209

Lairez D 37
Lartigue C 115
Llamas G 61

Mel'nichenko Yu 88
Meyer E 227
Meyer GC 232

Niaounakis M 144

Oeser R 104, 111, 131

Pekcan Ö 214
Pissis P 144
Pompe G 227

Quellet C 165

Ramik T 104
Raspaud E 37
Rätzsch M 227
Raukhvarger AB 174

Rizos AK 232
Rogovina L 151
Rozenberg BA 194

Sandakov GI 235
Scherzer T 202
Schlosser E 144
Schönhals A 144
Schulz M 52
Schulze U 227
Slonimsky G 151
Smirnov LP 222, 235
Solovjev ME 174
Sommer JU 43
Sosikov AI 235
Spathis G 144
Strehmel B 70, 83
Strehmel V 83, 202, 206
Summanen KT 235

Tamulis A 186
Tänzer W 202
Thinault JF 66
Timpe H-J 70

v. Soden W 27
Vasiliev V 151
Vilgis TA 1
Volkova NN 222, 235

Walasek J 182
Wang H 232
Wartewig S 83, 202
Weber H-W 104
Wetzel H 78

Younes M 83

Zhulina EB 156, 177
Zielinski F 115
Zimmermann E 206
Zölzer U 165

Subject Index

1,1-transition 57
^{2}H NMR 97
^{23}Na NMR 66

accelerator 202, 206
amine 202
— curing 83
amylose 61
analysis torsional braid 57
anisotropy of segmental motion 194
—, optical 182
approximation, group 52
—, mode-coupling 43

bimodal network 111
bisphenol-A-diglycidyl ether 83
bridging, SCF 156
Brillouin spectroscopy 232
butterfly pattern 131

carbazole-containing oligomers 186
carbon black 97
— — field networks 16
chain extensibility, finite 47
—, grafted 156
—, trapped 115
—, worm-like 47
charge separation, photo-induced 186
chemical networks 151
— —, physical and 144
chromatography, inverse gas 78
cis-1,4-polybutadiene networks 97
cluster 131
coefficient, expansion 206
collapse 156
compounds, styryl 83
connectivity of network 194
constraints of overall orientation 47
—, topological 47, 137
contourlength fluctuation 104
conversion, fractional 57
crack resistance 209
crosslinking, photoinduced 70
crosslinks 90
—, network 194
cure, isothermal 57
curing process 78
—, amine 83

curves, fracture 241
—, master 209

decomposition 115
defects, network 137
—, structural 177
deformations, large 27
—, nonaffine 131
degradation 235
—, thermal 222
density, network 206
deuterium NMR 111
dielectric 144
diol 202
dynamic light scattering 165
dynamical properties 37
dynamics 115
—, network 16

elasticity 1, 13
— modulus 151
—, rubber 90, 137
electric field 182
electron micrograph 227
— microscopy of PUs 241
entanglements, trapped 137
epoxy 202
— amine network 206
— and unsaturated ester systems 57
— polymers 209
— resins 78
equation, Kerr 182
EVA copolymer 227
expansion coefficient 206
experiment, SANS real-time 131
extensibility, finite 137
—, finite chain 47

field-cycling 104
field networks, carbon black 16
—, electric 182
finite chain extensibility 47
— extensibility 137
fluctuation, contourlength 104
fluorescence 214
— probes 70, 83
fractals 1
fractional conversion 57
fracture curves 241
FTIR spectroscopy 202

gas chromatography, inverse 78
gel 88, 165
— fraction 57
— point 78
gelatin 88
gelation 37, 52, 57, 61, 131
geometry, restricted 214
glass transition 78, 144, 227
— — temperature 194, 206
glass-rubber transition 16
grafted chains 156
— polymer layers 177
group approximation 52

heterogeneities 1, 115
history, thermal 61

interactions 182
interpenetrated network 131
interpenetrating network 214
— polymer networks 232
— polymer networks, sequential 227
inverse gas chromatography 78
IPN 1
—, semi 1
irreversible thermodynamics 27
isothermal cure 57

Kerr equation 182
kinetics 222

large-scale motion in networks 194
law prefactor, power 37
light scattering, dynamic 165
local order 182
long-time relaxation 43

master curves 209
matter, soft condensed 165
mechanical properties 174
— relaxation 144
melts 13
—, polymer 104
microemulsions 165
microscopy of PUs, electron 241
— — —, polarized 241
mobility, rotational 70
mode-coupling approximation 43
model networks, PDMS 131
modulus, elasticity 151
molecular photodiodes 186

Subject Index

morphology of PUs 241
motion, anisotropy of segmental 194

network 1, 47, 104, 165, 235
— crosslinks 194
— defects 137
— density 206
— dynamics 16
—, bimodal 111
—, carbon black field 16
—, connectivity of 194
—, epoxy amine 206
—, interpenetrated 131
—, interpenetrating 214
—, interpenetrating polymer 232
—, PDMS 111
—, PDMS model 131
—, physical 174
—, physical and chemical 144
—, polyelectrolyte 177
—, polymer 43, 115, 222
—, random 13
—, semiinterpenetrating polymer 70
—, van der Waals 27
NMR 90
— relaxation 104
— structure 235
—, deuterium 111
nonaffine deformation 131

oligomers, carbazole-containing 186
optical anisotropy 182
order, orientational 111
orientation 97
—, constraints of overall 47
orientational order 111
overall orientation, constraints of 47

pattern butterfly 131
PDMS model networks 131
— network 111
pectin-calcium system 66
percolation 13
— theory 52
phase transition 156, 177
photo-induced charge separation 186
— crosslinking 70
photodiodes, molecular 186
photon correlation spectroscopy 232

physical and chemical network 144
— networks 174
point, gel 78
polarized microscopy of PUs 241
poly(dimethylsiloxane) 151
poly(methyl methacrylate) 227
polyelectrolyte networks 177
polymer 13
— melts 104
— network 43, 115, 202, 222
— —, interpenetrating 232
— —, semiinterpenetrating 70
— solution 182
—, epoxy 209
polyurethane 37, 144
potentiometry 66
power law prefactor 37
prefactor, power law 37
probes, fluorescence 70, 83
process, curing 78
—, relaxation 209
properties, dynamical 37
—, mechanical 174
PUs, electron microscopy of 241
—, morphology of 241
—, polarized microscopy of 241

quality of the solvent, thermodynamical 151

random networks 13
real-time experiment, SANS 131
relaxation 27
— processes 209
—, long-time 43
—, mechanical 144
—, NMR 104
renormalization 52
resins, epoxy 78
resistance, crack 209
—, wet skid 16
restricted geometry 214
rheology 61, 66, 165
rotational mobility 70
rubber elasticity 90, 137

SANS real-time experiment 131
scaling 156
scattering small-angle neutron 115, 131
SCF bridging 156
segmental motion, anisotropy of 194
self assembly 186
semi IPN 1

semiinterpenetrating polymer networks 70
separation, photo-induced charge 186
sequential interpenetrating polymer networks 227
skid resistance, wet 16
small-angle neutron scattering 115, 131
soft condensed matter 165
sol-gel transition 66, 88
solution, polymer 182
solvent, thermodynamical quality of the 151
spectroscopy, Brillouin 232
—, photon correlation 232
stretching 115
structural defects 177
structure, NMR 235
styryl compounds 83
supermolecules 186
supramolecules 186
swelling 88, 115, 151, 174
system, pectin-calcium 66
—, epoxy and unsaturated ester 57

temperature, glass transition 194, 206
thermal degradation 222
— history 61
thermodynamical quality of the solvent 151
thermodynamics, irreversible 27
topological constraints 47, 137
torsional braid analysis 57
transition temperature, glass 206
—, glass 78, 144
—, glass-rubber 16
—, phase 156, 177
—, sol-gel 66, 88
trapped chains 115
— entanglements 137
trimethylolphenol 57

unsaturated ester systems, epoxy and 57

van der Waals networks 27
viscoeleasticity 16
viscosity 37
vitrification 57

wet skid resistance 16
worm-like chains 47